SYSTEMATICS OF CYST NEMATODES

(Nematoda: Heteroderinae)

SYSTEMATICS OF CYST NEMATODES (Nematoda: Heteroderinae)

Sergei A. Subbotin, Manuel Mundo-Ocampo and James G. Baldwin

David J. Hunt and Roland N. Perry (Series Editors)

NEMATOLOGY MONOGRAPHS AND PERSPECTIVES
VOLUME 8A

BRILL
LEIDEN-BOSTON
2010

This book is printed on acid-free paper.

Library of Congress Cataloging-in-Publication Data

Library of Congress Cataloging-in-Publication Data are available from the publisher

ISBN: 978 90 04 16225 9

Nematology Monographs & Perspectives, 2010, Vol. 8A, v-vii

Contents

Foreword

Widespread interest in cyst-forming nematodes has been largely driven by their worldwide menace to key agricultural crops. For more than 100 years there has been a deep and practical commitment to understanding cyst nematodes and their diversity. Much of this work has focused on detailed morphology and approaches for diagnosis and, to this end, there has been continuous development and adaptation of new research tools to these studies. There has been worldwide participation in the discovery and description of cyst nematodes and this has generated an extensive body of literature, particularly in systematics. However, not all of this literature is universally accessible. The primary aim of this book is to bring together this body of knowledge to generate an updated consolidation of current understanding of morphology, taxonomy, phylogeny and biology. An additional goal is to summarise basic methods in morphology, biochemistry and molecular applications in diagnostics. Current descriptions of all cyst nematode genera and species, including morphological and molecular criteria, are given. The book consists of two volumes. The first volume contains chapters on morphology, biology, and evolution of cyst nematodes. Guides to methods for extracting specimens and classical techniques for light microscopy, as well as state-of-the-art approaches, specific to cyst nematodes, for applying electron microscopy, biochemistry and molecular diagnostics are included, as are identification keys, descriptions and illustrations of 30 species as well as diagnoses of the six genera: *Globodera*, *Punctodera*, *Cactodera*, *Dolichodera*, *Betulodera* and *Paradolichodera*. The second volume is devoted to the genus *Heterodera* and summarises classical approaches together with molecular data for identification using PCR-ITS-RFLP. Comprehensive identification keys, descriptions and illustrations of 80 species are provided. Building on a legacy of books on cyst nematodes, including those of Franklin (1950), Sharma and Swarup (1984), Lamberti and Taylor (1986), Shahina and Maqbool (1995) and Sharma (1998), we hope that these new volumes will promote accessibility of species descriptions, together with classical and new tools, in the endeavour of the scientific community to work together toward further understanding of cyst nematodes and the management of these pests for worldwide benefits.

Sergei A. Subbotin, Manuel Mundo-Ocampo and James G. Baldwin
Sacramento, Riverside, USA
January 2010

Acknowledgements

The authors sincerely appreciate the support of their families, institutions and colleagues during the development of this project. We are especially grateful to colleagues and friends who read the drafts and provided helpful comments, corrections and suggestions. These include P. Castillo (IAS, CSIC, Spain), J. Chitambar (CDFA, USA), V.N. Chizhov (IEE, Russia), N. Greco (CNR, IPP, Italy), Z. Handoo (USDA, USA), R. Inserra (FDACS, USA), D. Sturhan (Germany), Z. Tanha Maafi (PPDRI, Iran), N. Viane (ILVO, Belgium) and L. Waeyenberge (ILVO, Belgium). The authors are particularly appreciative for valuable new information, some unpublished, that was generously provided by Cid Del Prado (Mexico), E. Bernard (UT, USA), N. De Sutter (ILVO, Belgium), J. Franco (Bolivia), M. Madani (CFIA, Canada), T. Powers (UNL, USA), A. Swart (ARC-PPRI, South Africa) and N. Vovlas (CNR, IPP, Italy). We gratefully acknowledge the journals and organisations for allowing reproduction of illustrations, and a number of colleagues who provided photos and figures, some of whom invested their own time to prepare originals. Those contributing images include G. Capistrano (Germany), V.N. Chizhov, I. Cid Del Prado, A.-M. Deeren (ILVO, Belgium), J.D. Eisenback (VPI & SU, USA), Z. Handoo, R. Lawson (FERA, UK), A.Yu. Ryss (ZIN, Russia), D. Sturhan, Z. Tanha Maafi, G. Tylka (ISU, USA), J. Zheng (China) and U. Zunke (University of Hamburg, Germany). Additional colleagues kindly assisted with preparation of photographs, including C. Anderson (UCD, USA), C. Blomquist (CDFA, USA), H. Ferris (UCD, USA), S. Gaimari (CDFA, USA) and T. Niblack (UI, USA). Some original monographic aspects of the project were supported by National Science Foundation grant PEET DEB 0731516. The authors appreciate the vision, guidance, assistance and patience of the Series Editors, David Hunt and Roland Perry, for their helpful insight and skill, without which this work would not have been possible. Finally, we are grateful for the generations of dedicated researchers worldwide that have invested so much of their lives in building a body of knowledge toward understanding and managing this biologically amazing and challenging group, the cyst nematodes.

Sergei A. Subbotin[1,2,3], Manuel Mundo-Ocampo[3]
and James G. Baldwin[3]
[1] *California Department of Food and Agriculture, Sacramento, CA, USA*
[2] *Centre of Parasitology of A.N. Severtsov Institute of Ecology and Evolution of the Russian Academy of Sciences, Leninskii Prospect 33, Moscow, 117071, Russia*
E-mail: subbotin@ucr.edu
[3] *University of California, Riverside, CA, USA*
E-mails: manuel.mundo@ucr.edu and james.baldwin@ucr.edu

Nematology Monographs & Perspectives, 2010, Vol. 8A, 1-5

Introduction

The cyst-forming nematodes, herein referred to simply as cyst nematodes, are taxa of great worldwide economic significance defined by their capacity to retain eggs inside the female body, which is transformed into a cyst at the completion of the female life cycle. By definition, "a heteroderid cyst is a persistent tanned sac which contains and protects embryonated eggs and is derived from some or all components of the mature female body wall" (Luc *et al.*, 1986) (Fig. 1). The embryonated eggs inside the cysts contain a dormant second-stage juvenile (J2) (Fig. 2). The unique diversity of these nematodes in morphology, geographic distribution, host range biology, and DNA is reflected by a taxonomic system that includes seven genera: *Heterodera* (80 species), *Globodera* (10 species), *Cactodera* (13 species), *Dolichodera* (1 species), *Paradolichodera* (1 species), *Betulodera* (1 species), and *Punctodera* (4 species), a total of 110 valid species. Many cyst nematodes are important pests of agricultural crops (Table 1).

Cyst nematodes are monophyletic, *i.e.*, they evolved only once, within the subfamily Heteroderinae, a subdivision of the family Heteroderidae

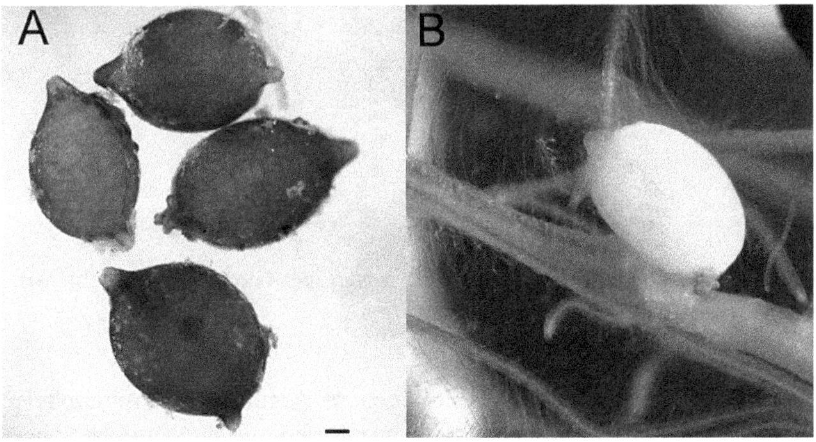

Fig. 1. *A: Cysts of sugar beet cyst nematode (original); B: White female of cyst nematode on root. (Courtesy of G. Capistrano, Germany.) (Scale bar = 100 μm.)*

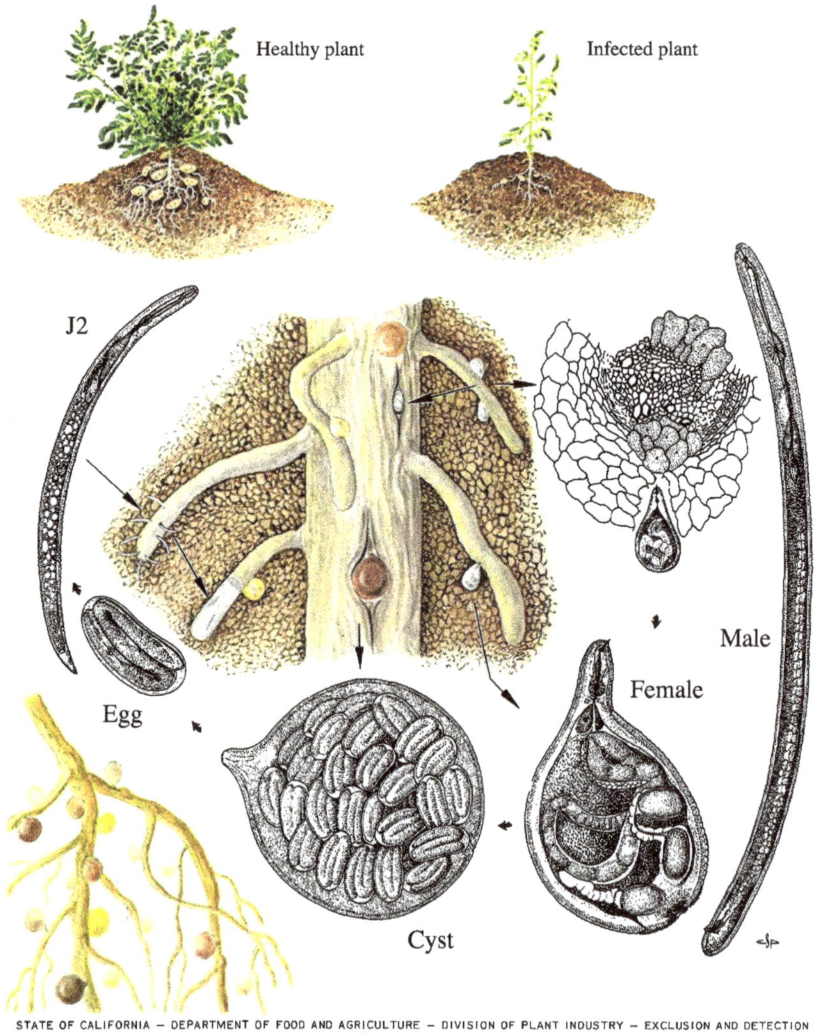

Fig. 2. *Life history of the potato cyst nematode,* Globodera rostochiensis. *After C. Papp, CDFA, USA with modification.*

that also includes non-cyst or cystoid genera. Arguments for monophyly of cyst nematodes were first based on morphology and host response (Baldwin & Schouest, 1990; Baldwin, 1992) and were later supported by molecular analyses (Subbotin *et al.*, 2001a, 2006; Ferris *et al.*, 2004; Bert *et al.*, 2008). In addition to the formation of cysts, the

Table 1. *Some agricultural crops, trees and ornamentals as host for cyst nematodes.*

Plant	Cyst nematode species
Almond	*Heterodera amygdali*
Apple	*Globodera mali*
Bamboo	*H. bamboosi*, *H. koreana*
Barley	*H. avenae*, *H. australis*, *H. bifenestra*, *H. filipjevi*, *H. hordecalis*, *H. latipons*, *H. mothi*, *H. sorghi*, *H. zeae*, *Cactodera rosae*, *C. galinsogae*
Cabbage	*H. betae*, *H. cruciferae*, *H. schachtii*
Cactus	*C. cacti*
Carrot	*H. carotae*
Chickpea	*H. cajani*, *H. ciceri*, *H. swarupi*
Clover	*H. daverti*, *H. goettingiana*, *H. galeopsidis*, *H. lespedezae*, *H. rosii*, *H. trifolii*
Clover pink	*H. daverti*, *H. lespedezae*, *C. evansi*
Eggplant	*G. rostochiensis*, *G. pallida*
Fig	*H. fici*
Hop	*H. humuli*
Lucerne	*H. medicaginis*
Maize	*H. avenae*, *H. delvii*, *H. filipjevi*, *H. oryzae*, *H. sacchari*, *H. sorghi*, *H. zeae*, *P. chalcoensis*
Oat	*H. avenae*, *H. australis*, *H. bifenestra*, *H. filipjevi*, *H. latipons*, *H. phragmitidis*, *H. zeae*
Olive	*H. mediterranea*
Pea	*H. ciceri*, *H. goettingiana*, *H. trifolii*
Pigeon pea	*H. cajani*, *H. swarupi*
Potato	*G. leptonepia*, *G. rostochiensis*, *G. pallida*
Rice	*H. graminophila*, *H. elachista*, *H. leuceilyma*, *H. oryzicola*, *H. oryzae*, *H. sorghi*, *H. skohensis*, *H. zeae*
Sorghum	*H. avenae*, *H. delvii*, *H. gambiensis*, *H. sorghi*
Soybean	*H. glycines*
Sugar beet	*H. betae*, *H. schachtii*
Sugarcane	*H. goldeni*, *H. leuceilyma*, *H. sacchari*, *H. zeae*
Tobacco	*G. tabacum*
Tomato	*G. mexicana* **sp. n.**, *G. rostochiensis*, *G. pallida*
Wheat	*H. avenae*, *H. australis*, *H. bifenestra*, *H. filipjevi*, *H. hordecalis*, *H. latipons*, *H. mothi*, *H. pakistanensis*, *H. phragmitidis*, *H. zeae*, *P. punctata*

most significant of the distinctive characters of the subfamily are those associated with sedentary feeding. These include marked sexual dimorphism. The sedentary feeding initiated in the infective J2 (Fig. 2) is accompanied by swelling of the nematode body during the third- (J3) and fourth-stage juvenile (J4) development and, in the final moult, the adult female remains swollen and sedentary. By contrast, the adult male reverts to a vermiform state and emerges from the persistent cuticles of earlier sedentary, swollen juvenile stages as active and migratory individuals. Compared with other sedentary nematode species, including *Meloidogyne* Göldi, 1887, *Rotylenchulus* Linford & Oliveira, 1940, *Tylenchulus* Cobb, 1913 and *Nacobbus* Thorne & Allen, 1944, the cyst nematodes are distinctive in that the female cuticle is thickened and, upon the death of the female, components of the mature body, especially the cuticle, acquire pigmentation (*via* polyphenol oxidase tanning) and persist as a container for the eggs – by definition this is the cyst (Luc *et al.*, 1986). Although the cuticle surface is annulated in the J2, in the adult female this annulation is usually limited to the head region where generally the cuticle pattern is greatly modified. It is especially transformed in the perineum, which lacks the fingerprint-like annulation-derived whorls characteristic, for example, of *Meloidogyne*, a genus where swollen sedentary females retain annulation (Raski, 1950). Cyst nematodes also uniquely induce in their hosts a feeding site that is a true syncytium resulting from fusion of cells and characterised by wall ingrowths (Jones & Dropkin, 1975; Jones, 1981; Wyss *et al.*, 1984); a range of other feeding sites are induced by cystoid Heteroderinae (Mundo-Ocampo & Baldwin, 1984).

The cyst nematodes are further characterised by a female reproductive system that is long, coiled and didelphic with the vulva located at the terminus, whilst males have a single outstretched testes and lack caudal alae. Males are often common, consistent with many species of cyst nematodes being amphimictic. The J2 stage, prior to hatching, has the remarkable capacity of dormancy within eggs enclosed by the cyst, a dormancy that may be sustained for many years (Jones *et al.*, 1998; Turner & Rowe, 2006). After hatch, the J2 emerge from the cyst either through openings that develop in the cyst, including where the head region deteriorates, or through one or more terminal openings, termed fenestrae, that develop where the cyst wall remains particularly thin and fibrous and eventually is likely to rupture. Cyst nematodes are also unusual with respect to sensory systems. For example, each amphidial opening in male cyst nematodes is a pore with no ampulla, rather than as in *Meloidogyne* where the aperture is a slit with a large ampulla (forming amphidial cheeks *sensu* Chitwood, 1949a). It is on the basis of such distinctive features of morphology and biology that the cyst nematodes of Heteroderinae, while diverse, nevertheless have historically been recognised as a cohesive group distinct from all other Tylenchida.

Nematology Monographs & Perspectives, 2010, Vol. 8A, 7-41

Chapter 1

Morphology

Introduction

The morphology of cyst nematodes is rendered uniquely complex by their sexual dimorphism, parasitic habits, ontogeny and dormant stages (Figs 1, 2). Aspects of the body wall, digestive, reproductive, and nervous system change with each stage, reflecting differing adaptations for phases of the life history, including developing *vs* mature, migratory *vs* sedentary, feeding *vs* non-feeding, or active *vs* dormant. Nevertheless, some features are consistent for juvenile and adult stages. The cyst nematodes, characteristic of Tylenchida, feed by a needle-like stomatostylet (Fig. 3A, B) that extends anteriorly from the pharynx, an organ that includes three pharyngeal glands that posteriorly overlap the ventral side of the intestine (Figs 4A, B; 5A, B). Apparently not varying among life stages, the excretory system is not distinctive in cyst nematodes and, like that of many other Tylenchida, the pore is near the level of the median pharyngeal valve or isthmus (Figs 4A, B; 5A-C). The specific morphology of the egg, juvenile stages, male, female and cyst, including features that vary among genera and species, is discussed below.

Egg and embryo

Apart from their association with the female and cyst (Fig. 6A), the eggs of cyst nematodes are not readily distinguishable from those of other Tylenchida (Fig. 6B). In contrast to most other Secernentea, but consistent within Tylenchida, the four cells resulting from the first cleavages in the egg align in tandem (Shahina & Maqbool, 1989; Dolinski *et al.*, 2001). Although egg dimensions are often included in species descriptions, suggesting that eggs vary in size and shape among species, these differences are not often used in diagnostics (Turner & Rowe, 2006) considering that eggs are typically found in conjunction with other, more informative, stages. Furthermore, size and shape of eggs

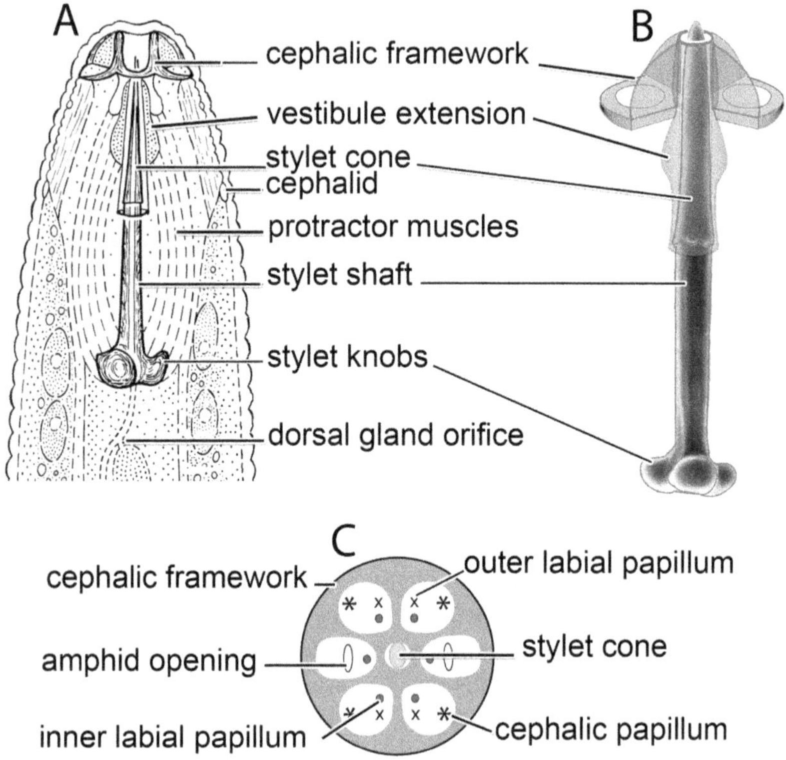

Fig. 3. *Diagrammatic interpretation of stylet region of second-stage juvenile of a cyst nematode. A: Left lateral view of head (after Hirschmann, 1956); B: 3D dorsal view of stylet in relation to labial framework and vestibule extension; C: Transverse view through labial framework in relation to stylet cone and sensilla.*

are affected by level of development and most likely other unspecified variables (Wouts & Baldwin, 1998). In some populations strikingly large 'giant eggs' may be found (Mulvey, 1959) and these might reflect a developmental aberration that is further expressed as giant and dwarf J2 (Mulvey, 1960). More promising in diagnostics is that SEM has revealed distinctive textures on the surface of eggs that may be particular to certain species, for example, those of *Cactodera* (Baldwin & Bell, 1985; Turner & Rowe, 2006) (Fig. 6C). In addition to the embryo, the first-stage juvenile (J1) stage is retained in the egg and at maturity fills the space with three to five folds, but there is little differentiation of the head

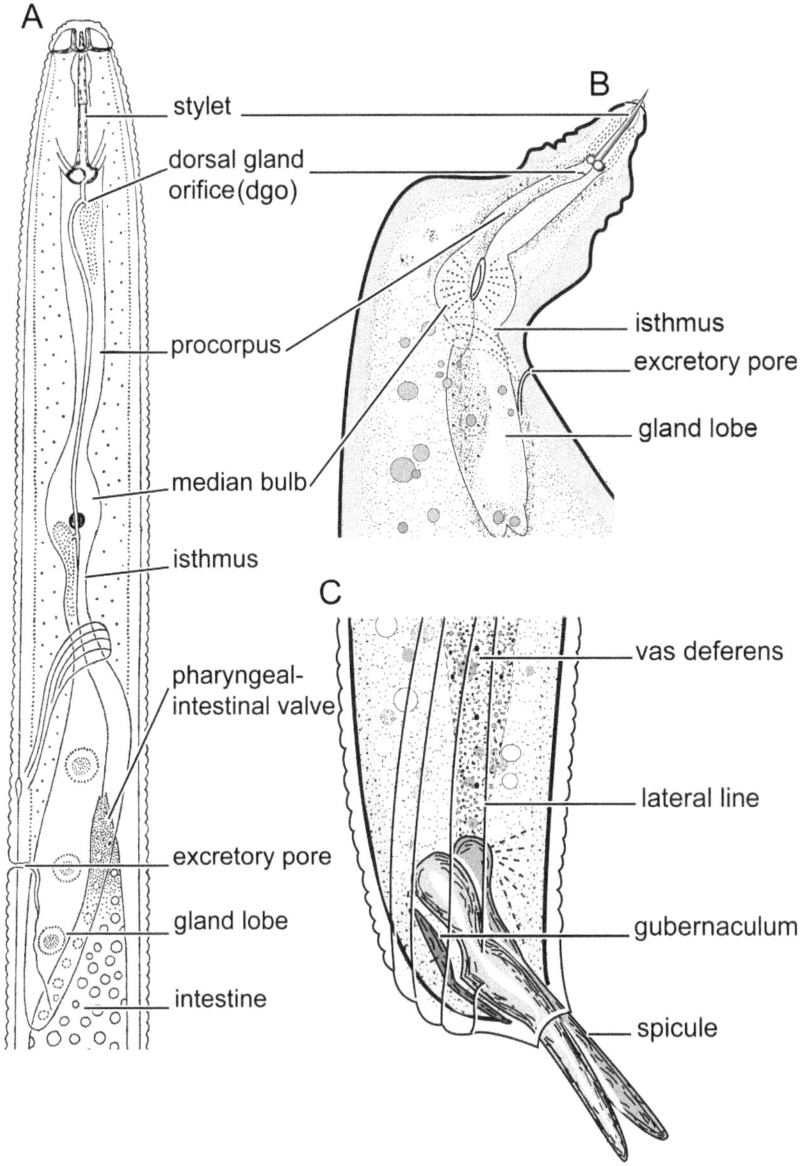

Fig. 4. *Diagrammatic interpretation of adult cyst nematodes. A: Left lateral view of anterior end of male including pharyngeal region (after Baldwin* et al., *1977); B: Right lateral view of anterior end of female including pharyngeal region; C: Right lateral view of posterior end of male.*

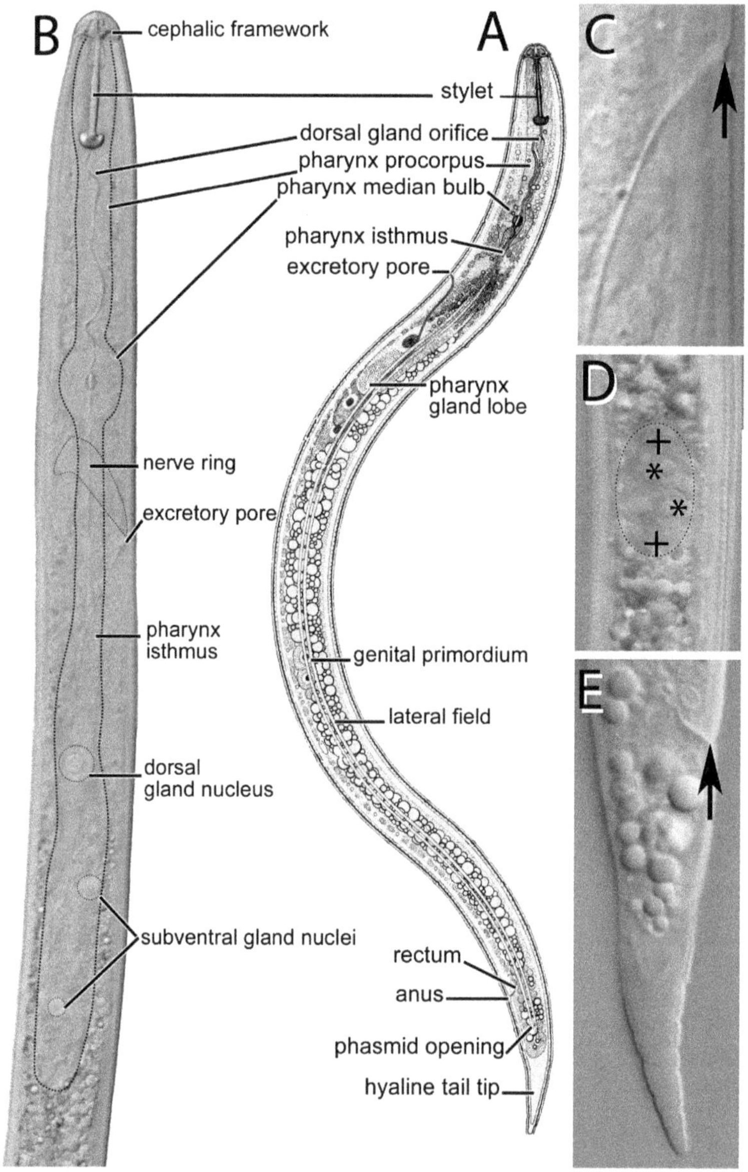

Fig. 5. *Second-stage juveniles of cyst nematodes (*Heterodera schachtii*). A: Left lateral view (after Cobb from Thorne, 1961); B: Right lateral view of anterior end including pharyngeal region; C: Excretory pore (arrow) and duct; D: Genital primordium (+ = epithelial cells; $*$ = germ cells); E: Tail region including anus (arrow) and hyaline region.*

region since the stylet does not appear until the first moult. In addition, cuticular annulation does not appear until the J2 (Raski, 1950).

JUVENILE STAGES J2-J4

Heteroderinae are often recovered as J2 directly in soil or they may be found on plants as cysts full of eggs from which the J2 are ready to emerge, or, in some species, the female, still feeding, may lay eggs into an egg sac from which J2 readily hatch (Fig. 6A) (Wouts & Baldwin, 1998; Turner & Rowe, 2006). Anteriorly, the J2 are rounded and posteriorly the tail tapers to a point (Fig. 5). The overall J2 body length (300-700 μm), as well as length and shape of the tail (30-90 μm) and distinctive hyaline region (20-40 μm), may be useful in species diagnosis (Wouts & Weischer, 1977; Wouts & Baldwin, 1998). Additional useful features may include stylet length, position and shape of the stylet knobs and number of lateral lines in each of the two lateral fields (Figs 3A, B; 5A, E; 7). The lateral lines (incisures) are also informative characters since J2 of most species have four lines, and J2 of species from the *Sacchari* and *Cyperi* groups, *H. bifenestra*, *Betulodera* and *Dolichodera* have three lines. Yet, for diagnostic purposes, it is important to consider these J2 aspects in combination with features of other stages (Wouts & Baldwin, 1998; Turner & Rowe, 2006), and consideration must be given to variation among populations of a given species (Hirschmann & Triantaphyllou, 1979). In addition, some aspects of J2 morphology, including length of body, tail, stylet or tail hyaline region, may be influenced by environmental conditions such as the host plant, temperature and nematode population densities. Furthermore, as for other stages, artefacts affecting morphology can be introduced with preservation and preparation techniques (Stone, 1971; Evans & Franco, 1977; Robinson *et al.*, 1996).

The body wall of J2, like that of the male, includes cuticle underlain and secreted by epidermal (= hypodermal) tissue. Internally there is a layer of platymyarian somatic muscles punctuated by two lateral and dorsal ventral chords (Fig. 8). The cuticle, as viewed with TEM, is the basic pattern for Secernentea, externally with a cortical layer, including a thin osmophilic layer at the surface; the cortex is underlain by a spongiform medial layer and the innermost basal layer comprises a palisade of fine striations (Endo & Wyss, 1992). Although the three cuticle layers are also expressed through early J3 (Endo & Wyss,

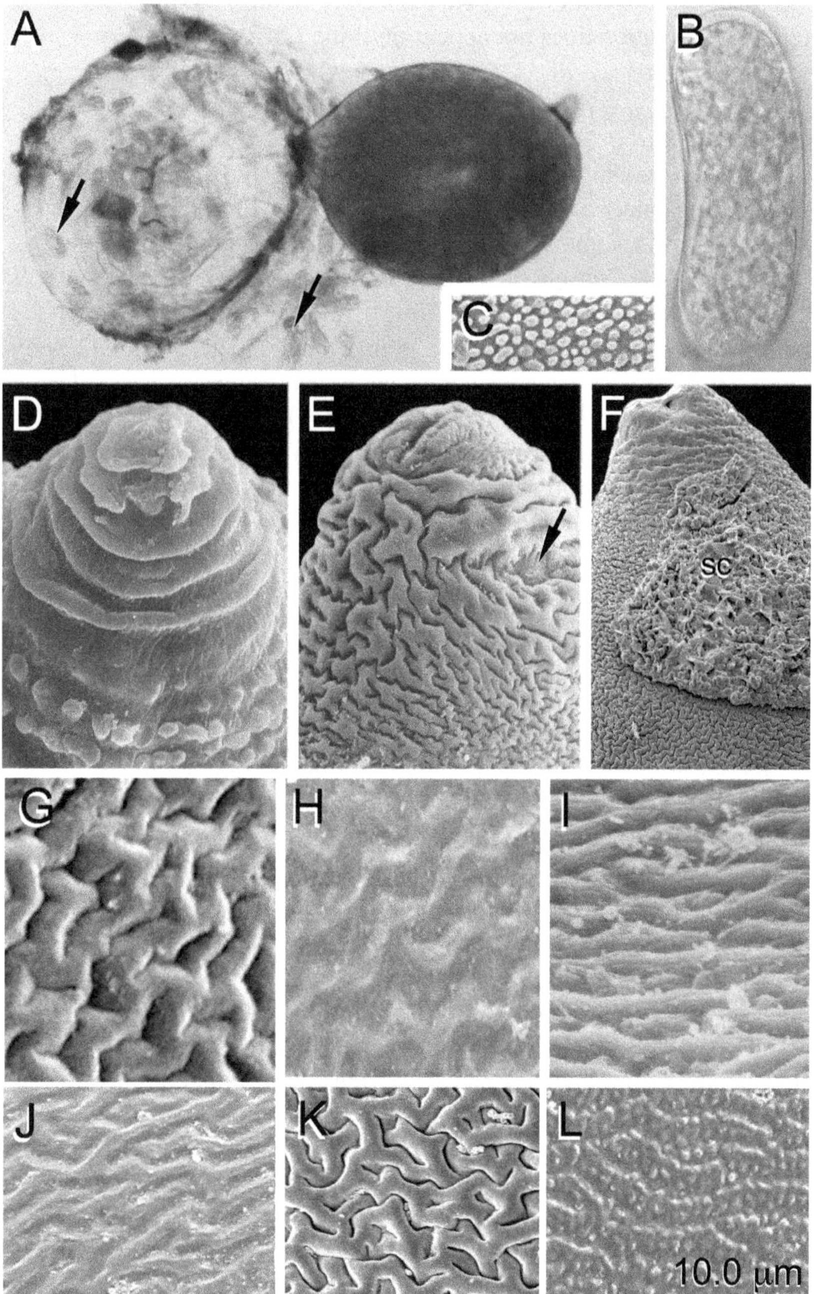
A
B
C
D
E
F
sc
G
H
I
J
K
L
10.0 μm

1992), their further persistence in subsequent sedentary juvenile stages is uncertain; clearly the basic pattern of layering is lost or highly modified in the adult female. In sedentary J2 and J3 there may be a distinctive fibrillar exudate, secreted by the epidermis, external to the osmophilic region, which is probably consistent with the subcrystalline layer of adult females (Endo & Wyss, 1992).

The J2 cuticle is superficially annulated, the annulation being interrupted by longitudinal lines (grooves) comprising the two fields, one on each lateral region. Variations occur among the species with respect to the number of lateral lines (three or four), areolation of lateral field and the position of the anterior and posterior end of particular lines (Fig. 9H, I). Generally in J2 the lateral lines terminate only a few annuli posterior to the phasmid openings. Lateral lines do not persist beyond the early sedentary stages (Cordero, 1989) and annulation clearly diminishes with development. Raski (1950) noted that, by the J4, juveniles developing into females are instead covered by irregular longitudinal markings. He also observed that in J4 females annulations are only retained in the head region. The process and regulation of transition through the juvenile stages from annulation to longitudinal folds, and then to the complex surface patterns in adult females, as well as to renewed annulation in the adult male, are unknown.

An important diagnostic feature of the surface cuticle concerning the labial patterns of J2 and their potential for defining taxa was first noted by Stone (1975). These patterns are interpreted as modifications from a basic scheme consisting of a labial disc surrounded by six lips, two lateral, two subdorsal and two subventral, all of nearly equal

Fig. 6. *Female (A), cysts (D-L) and eggs (B, C) of cyst nematodes. C-L: SEM. A: Mature female of* Heterodera schachtii *including an egg-filled gelatinous matrix. Black arrows indicate eggs within and outside egg mass; B: Egg with developing embryonic juvenile; C: Egg surface of* Cactodera cacti*; D: Anterior end of female of* Globodera tabacum*; E: Terminal region of* H. schachtii. *Arrow indicates position of anus; F: Terminal region of* C. cacti *including subcrystalline layer (sc); G: Zigzag cuticular pattern of mid-body of* H. glycines*; H: Lace-like cuticular pattern of mid-body of* Punctodera punctata*; I: Striated cuticular pattern of mid-body of* P. punctata*; J: Cuticular pattern anterior to mid-body in* Globodera tabacum*; K: Rugose cuticular pattern at mid-body in* G. tabacum*; L: Cuticular pattern posterior to mid-body in* G. tabacum. *(Scale bar applies to G-L.)*

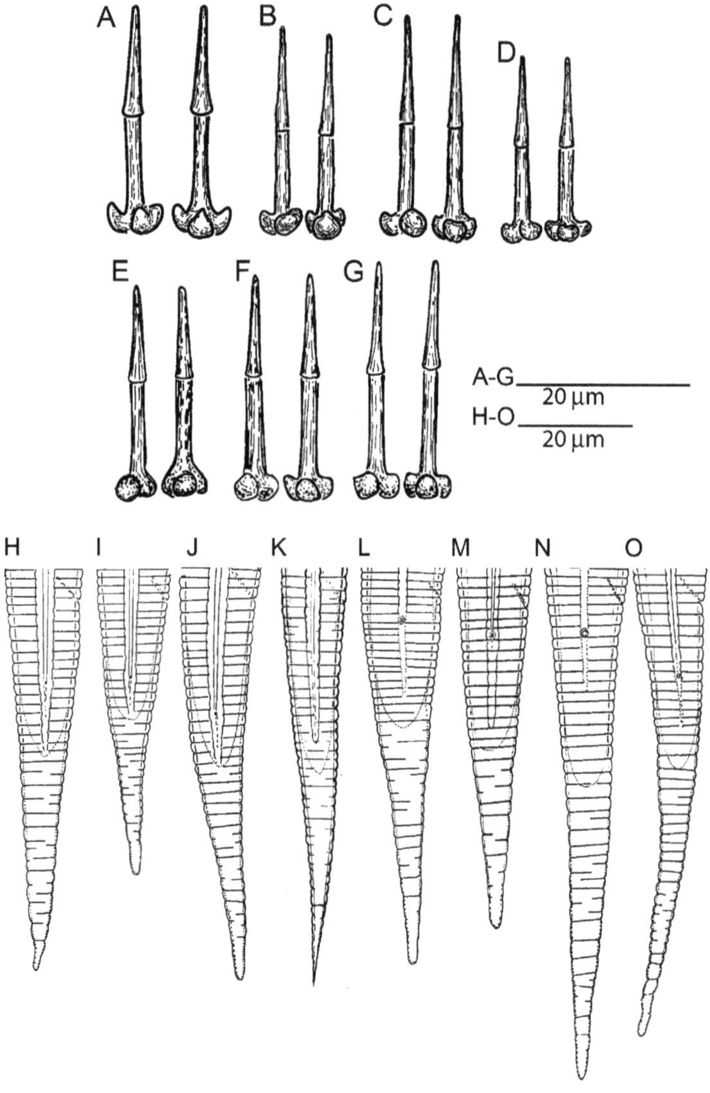

Fig. 7. *Examples of diversity of stylet and tail form of second-stage juvenile (J2) of cyst nematode species A-G, stylets (left is lateral; right is dorsal view). A:* Heterodera mani*; B:* H. glycines*; C:* H. trifolii*; D:* H. zeae*; E:* Globodera rostochiensis*; F:* G. pallida*; G:* G. tabacum*; H-O: Tails (lateral view). H:* H. goettingiana*; I:* H. schachtii*; J:* H. trifolii*; K:* H. spinicauda*; L:* H. avenae*; M:* H. hordecalis*; N:* H. ustinovi*; O:* H. bifenestra. *A-D after Wouts and Baldwin (1998), E-F after Baldwin and Mundo-Ocampo (1991), H-O after Wouts and Weischer (1977), Sharma (1998).*

size (Othman, 1985; Othman *et al.*, 1988; Baldwin & Schouest, 1990; Baldwin, 1992) (Fig. 9A). Although many cyst nematode J2 retain this basic pattern, in many other species this pattern is modified, often by fusion of adjacent submedial lips with one another and with the labial disc. Sometimes the submedial lips further fuse with the anterior labial annuli (Fig. 9B, C). In addition, the pattern may be further modified by changes in labial shape, and particularly by the dorsoventral elongation of the submedial relative to the lateral lips. The J2 labial pattern is also variable among species by the number of additional annuli (complete or interrupted) on the slightly offset labial region of the head (Fig. 9A-C). Changes in the labial region of J3 and J4 have not been described but, relative to the J2, the labial region is somewhat modified in adult males and extensively modified in adult females.

Typically, the labial pattern has a full complement of sensilla expressed as a ring of six inner labial papillae (with openings surrounding the stoma) and four cephalic papillae (occurring as slight elevations), and a pair of lateral amphids expressed on the lips as a pair of oval openings (Endo, 1980) (Figs 3C; 9E; 10A). In addition, four outer labial sensilla were described by Endo (1980), but these are not clearly expressed at the surface and, notably and alternately, six outer labial sensilla are widely conserved among Nematoda. However, sensory structures, as expressed on surface patterns, do not appear to vary or have diagnostic value among cyst nematode species. The internal structure of the amphids includes a cuticle-lined canal enclosing seven receptors, and a sheath cell enclosing two neurons, the finger cells, that terminate in an elaborate pattern of microvilli (Fig. 10B) (Endo, 1980), which are presumed to detect thermal cues (Mori & Ohshima, 1995). The extent to which labial patterns and associated sensilla vary among sedentary juvenile stages is not known.

The labial framework of J2 of Heteroderinae (Figs 3, 5A, B) resembles that of many Hoplolaiminae with its rounded shape, slightly offset appearance and heavy sclerotisation. This sclerotisation is specifically the cuticular lining of the cheilostom, including a hexaradiate framework in which anteriorly a cuticular vestibule and vestibule extension form a conduit that serves as a guiding apparatus for the stylet through its point of protrusion at the stoma (Baldwin *et al.*, 2004). The vestibule is a hub from which six blade-like radii extend to the body wall cuticle, organising the tissues of the head into two lateral and four submedial segments of approximately equal size (Fig. 3B). The radii and hub of the framework may vary among species in robustness of the sclerotised

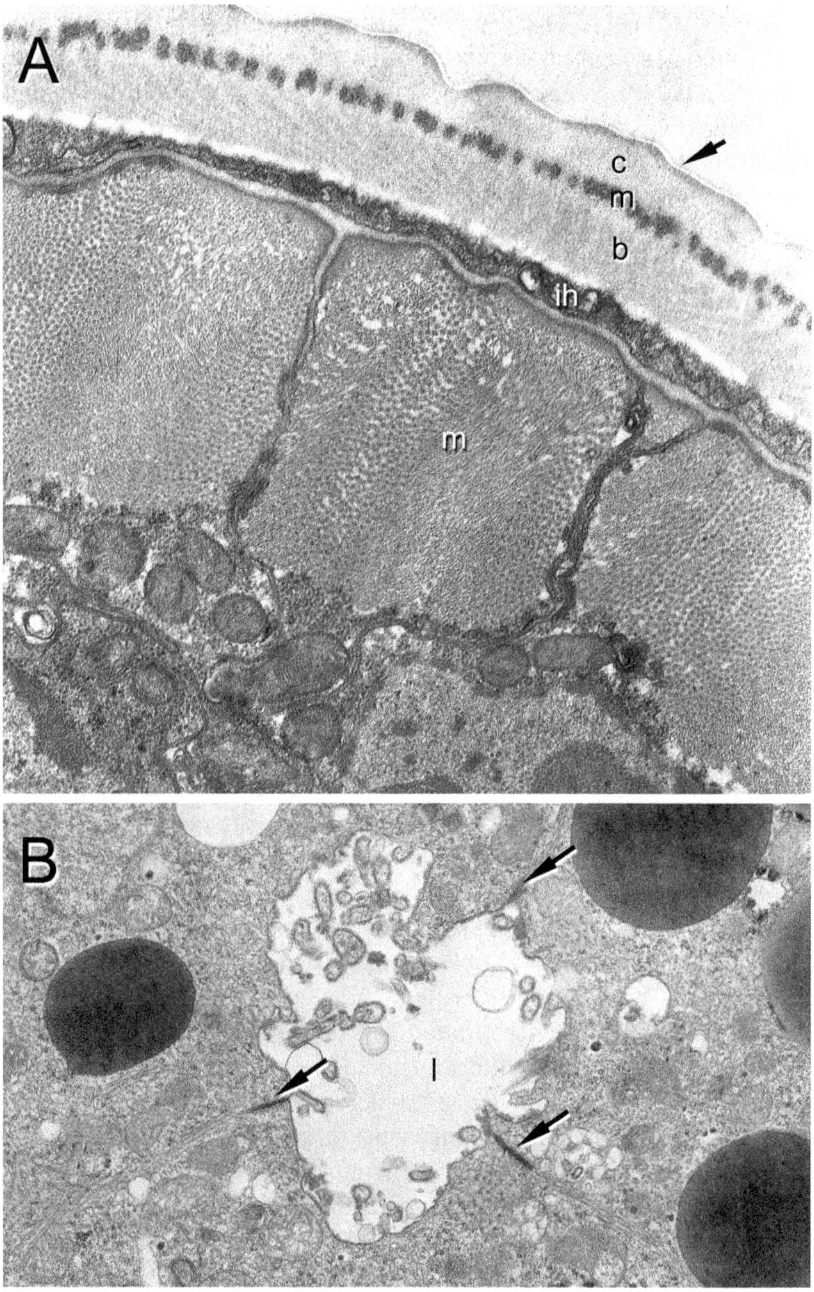
A
c
m
b
ih
m
B
l

region. The J2 stylet is robust, generally varying in length from 16 to 28 μm, with length typically being diagnostic for a species. The stylet of juveniles and adults, consisting of an anterior tapering cone, cylindrical shaft and posteriorly three large basal knobs, also varies among species by the size and details of the shape of the knobs (Figs 3A, B; 4A, B; 5A, B; 7) (Wouts & Weischer, 1977; Hirschmann & Triantaphyllou, 1979). The J2 stylet knobs, for example, may be distinctively rounded or anchor-shaped and may project anteriorly or slope posteriorly (Hirschmann, 1956; Wouts & Weischer, 1977; Mulvey & Golden, 1983). Strongly projected stylet knobs in the J2 distinguish *H. mani* from other species of the *Avenae* group, and anchor-shaped knobs in J2 of *G. pallida* are distinct from *G. rostochiensis*. However, this feature is subject to intraspecific variation and its use for identification generally requires the assessment of many individuals.

The size and shape of stylet knobs is functionally related to the pharyngeal-based stylet protractor muscles and the points of attachment of these muscles on the knobs. The stylet appears at the first moult and, in addition to the J2, it is also expressed in the sedentary J3 and at least the female J4, although in these stages it is described as "less robust" and having basal swellings in lieu of distinct knobs (Raski, 1950). This reduction in the knobs may be functional, considering that once infection is established the role of the stylet, with respect to invasion of the root and penetration, is reduced. Endo (1985) describes the development of the stylet during the second moult, including formation of the basal swellings in relation to stylet protractor muscles from a cell described as an arcade-like A4. It is likely that the A4 cell is homologous to the M1 marginal cell of the rhabditid or cephalobid stegostom, as in other Tylenchida (Baldwin *et al.*, 2004; Bumbarger *et al.*, 2006; Ragsdale *et al.*, 2008).

The pharynx of J2, including the corpus, isthmus and gland lobe, nearly fills the diameter of the body cavity, and variation in size, shape

Fig. 8. *Transmission electron micrographs of second-stage juvenile of* Heterodera glycines. *A: Transverse section of body wall including surface glycocalyx (arrow). Abbreviations: c = cortex of cuticle, b = basal striated layer of cuticle, ih = interchordal hypodermis, m (black) = medial layer of cuticle, m (white) = contractile region of somatic muscle; B: Intestinal epithelium including lumen (l). Arrows indicate junctional complexes at cell margins. After Endo (1988).*

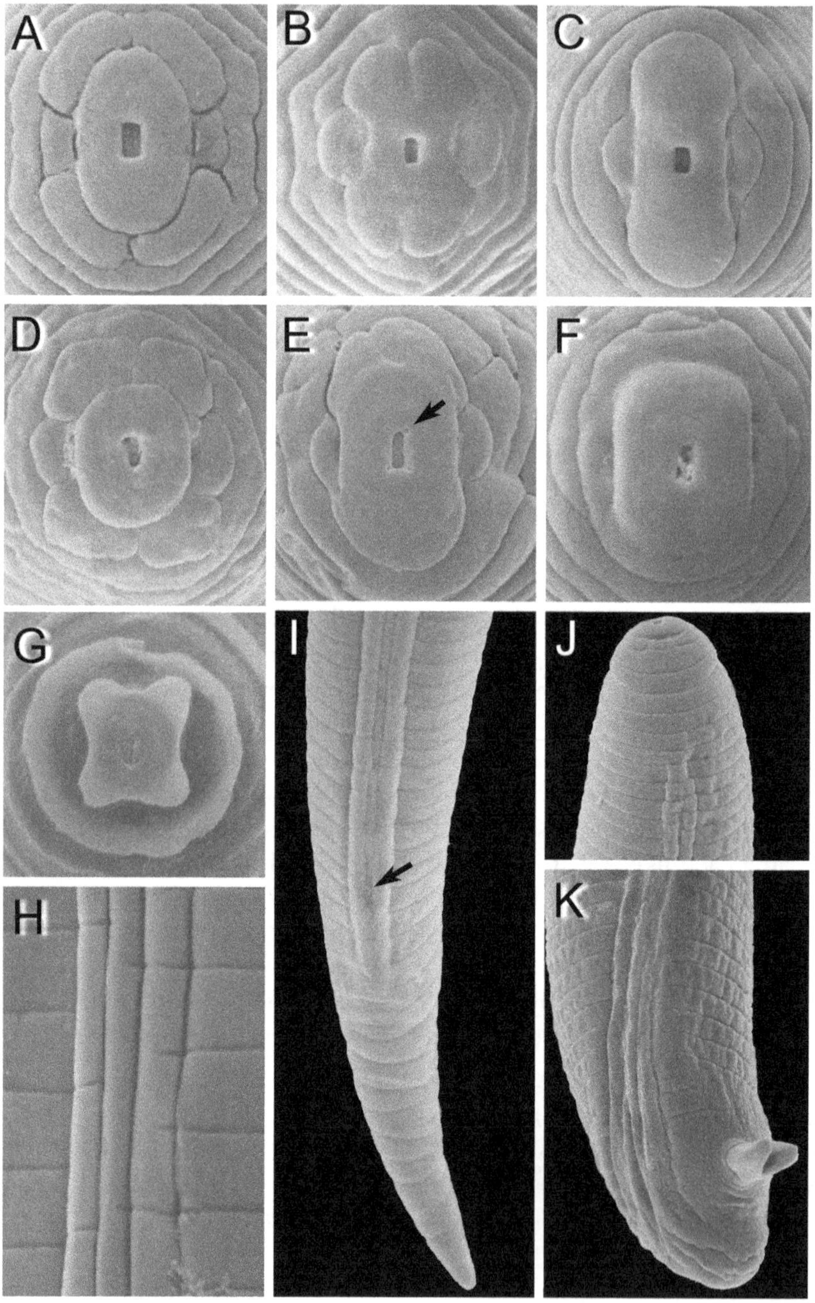
A
B
C
D
E
F
G
I
J
H
K

and orientation is not usually diagnostic among cyst nematodes (Fig. 5A, B). The pharynx of the J2 has been elegantly described with TEM (Endo, 1984). The procorpus largely comprises the noncontractile regions of the stylet protractor muscles, and dorsally it includes an ampulla as the anterior extension of the dorsal gland process. The cuticularised dorsal gland duct joins the pharyngeal lumen from a base that extends from the ampulla (Figs 3A; 5A, B), and the position of this junction relative to the base of the stylet knobs (4-12 μm) may vary among species (Hirschmann, 1956; Wouts & Baldwin, 1988). Endo (1984) describes "constraining muscles" at the junction of the procorpus and median bulb. The median bulb is characterised by radial muscles and a triradiate lumen and also includes the ampulla and ducts of two subventral glands. The isthmus, with a triradiate lumen and radial muscles and additional constraining muscles, primarily comprises gland processes and leads posteriorly to the gland lobe that is dominated anteriorly by the dorsal gland (Endo, 1984). In contrast to the male, all the glands appear to be robust and actively secreting in the J2. While not confirmed with TEM, in general the pharynx is similar among the feeding juvenile stages in its basic components, but with differences with respect to the median bulb that increases in size in J3, J4 and the adult female, as well as the proportions of the isthmus and components of the gland lobe. The median bulb in male juveniles tends to be smaller than in female juveniles of equivalent stage (Raski, 1950). Using TEM, Endo (1984) describes changes from sedentary J2 to J3 in the morphology of secretory products of the subventral gland. As noted in general for heteroderids, the three pharyngeal glands overlap ventrally and subventrally, and the lumen of the J2 pharynx joins with that of the intestine through closely apposed cell membranes; the transition from the intestine to the cuticle-lined rectum involves similar adjacent membranes. A border of tightly packed intestinal microvilli, as reported for other nematodes (Borgonie

Fig. 9. *SEM of cyst nematodes. A-C:* En face *lip patterns of second-stage juvenile; D-F:* En face *lip patterns of male. A:* Cactodera cacti*; B:* Heterodera glycines*; C:* H. avenae*; D:* H. cruciferae*; E:* H. fici. *Arrow indicates orifice of inner labial papillae; F:* H. avenae*; G:* En face *view of female* H. glycines*; H: Lateral field of second-stage juvenile of* Punctodera chalcoensis*; I: Lateral view of tail of second-stage juvenile of* H. fici. *Arrow indicates phasmid opening; J: Lateral view of head region of male* Betulodera betulae*; K: Ventrolateral view of tail region of* B. betulae.

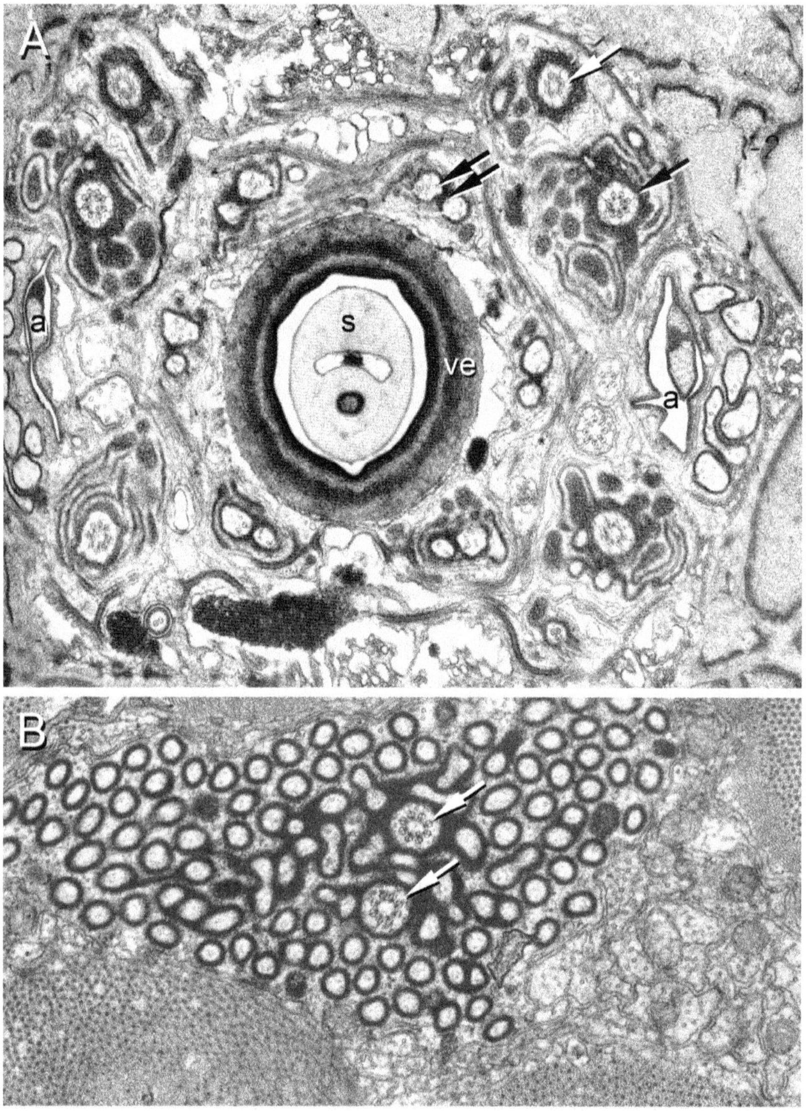

Fig. 10. *TEM of transverse sections of anterior region of second-stage juvenile of* Heterodera glycines *(after Endo (1980)); A: Region of vestibule extension (ve) including stylet (s) cone and amphid (a) duct. Single black arrow indicates neuron of inner labial sensilla, single white arrow indicates neuron of cephalic sensilla and double black arrow indicates pair of neurons of inner labial sensilla; B: Amphid showing ciliary region (arrows) and surrounding microvilli of neurons of the finger cells.*

et al., 1995), has not been observed in cyst nematodes and indeed appears to be absent or sparse and irregularly formed (Endo, 1988) in other Tylenchida (Fig. 8B) (Endo *et al.*, 1997). Following the second moult the rectum remains well developed in J3, and at this stage is reportedly larger in male juveniles. In J4, the rectum, but not the anus, is visible late in the stage (Raski, 1950), finally converging with the reproductive system with the development of accessory structures.

Whereas the tail regions of several cystoid Heteroderinae have pronounced ampulla in the phasmid duct of the infective J2, most cyst nematodes lack such ampulla and the phasmids appear as pores on each lateral region and anterior to the middle of the tail (Figs 5A; 9I). Nevertheless, phasmid openings in J2 vary among species, ranging from obscure in some (*e.g.*, *Heterodera oryzae*) to pronounced in others (*e.g.*, *H. sacchari*, *H. avenae*, *H. hordecalis* and *H. ustinovi*) (Luc & Merny, 1963; Wouts & Weischer, 1977; Luc, 1986b; Wouts & Baldwin, 1998). In J2, the position of phasmid openings relative to the anus and tail tip may be diagnostic for species. Whether phasmids are present in additional juvenile stages is mostly not known, although Carta and Baldwin (1990) demonstrated that phasmids deteriorate in post-penetration J2 and are lost beginning with the J3 and J4 in developing males of some cystoid and cyst nematodes.

Apart from the sensory structures, the nervous system of juvenile stages is known primarily at the level of the light microscope, including the position of the nerve ring (Fig. 5B), the assumption being that the system is consistent throughout life stages and among species (Fig. 4A, B). Notably, nerve commissures traverse the interchordal hypodermis and, with the light microscope, these are interpreted in J2 and males (Baldwin & Hirschmann, 1975b) as cephalids, hemizonid, hemizonion and caudalids (Fig. 3A) (Hirschmann, 1956). The positions of these commissures, particularly in relation to the excretory pore or rectum, have little if any diagnostic value (Wouts & Weischer, 1977). Similarly, the excretory system, including a cuticle-lined duct, cell body and a single lateral canal, probably varies little among species, stages or sex. Nevertheless, the position of the excretory opening in relation to other structures such as the pharynx may have some taxonomic importance in J2 and adult stages (Fig. 5A-C).

The reproductive system in migratory newly hatched J2 is an oval primordium located near the centre of the body and comprises two germ cells enclosed in an epithelial sheath that includes two cap cells (Fig. 5A,

D). By the time the J2 establishes a feeding site and becomes sedentary, germ cells of the primordium undergo multiple divisions, but it is not until the J3 female that the two gonad branches are evident, forming a 'V' with their distal tips extending anteriorly (didelphic/prodelphic). In conjunction with the V configuration, the point of connection migrates posteriorly during the third moult and the J4, and at the final moult it establishes an attachment to the vulva (Raski, 1950; Shahina & Maqbool, 1989). By contrast, in male juveniles a single anteriorly directed gonad branch is evident by J3 and is aligned immediately anterior to the rectum in J4 although the connections and organisation of the cloaca and accessory structures are not apparent until the final moult (Raski, 1950).

MALE ADULT

Males, although rare or absent in parthenogenetic forms, are common for most species of cyst nematodes (Fig. 2). They may be found in soil during their life of about 10 days (Evans, 1970) but they are more often on roots near young females, including within the egg sac that is attached to the female's posterior terminus (Burrows & Stone, 1985). Often several males are found with a single female and both sexes can mate several times (Green *et al.*, 1970). Many morphological features of cyst males are shared with males of other taxa that have females swollen at the stage in development when they are receptive for mating (*e.g.*, *Meloidogyne*); these similarities are functionally convergent and specifically include shape of the male, longitudinally twisted body, lack of caudal alae and spicules positioned far posterior to facilitate the male wrapping around the swollen female for mating (Figs 4C; 9K; 11A, B, D). The vermiform males, even within a particular cyst species, are variable in length (900-1600 μm), suggesting that size is affected by environmental variables that may include nutritional availability and crowding during development. This variability limits the value of all but a few male characters for species diagnosis (Wouts & Baldwin, 1998).

The body wall of males and the external layer of cuticle, as viewed by TEM, include three layers similar to those of the J2 (Baldwin & Hirschmann, 1975a) (Fig. 8A). As is the case for J2, the male cuticle is annulated and has three or four lateral lines, sometimes an areolated lateral field, depending on the species. Labial patterns of males are often considered phylogenetically and diagnostically informative but these patterns tend to be more variable within populations than, for example,

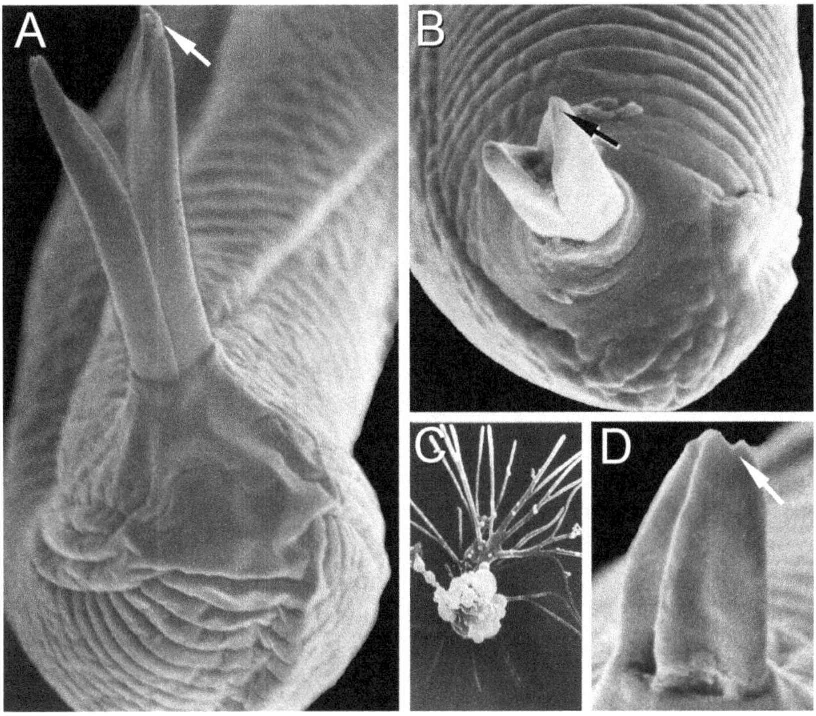

Fig. 11. *SEM of male cyst nematodes. A:* Cactodera eremica *showing protracted spicules with bifid tip; B:* Punctodera punctata *showing protracted spicules with smooth tip. Arrow indicates position of orifice to innervation; C: Sperm of* Heterodera schachtii *(after Cares & Baldwin, 1994); D: Partly protracted spicules of* H. avenae *showing bifid tip (arrow).*

patterns of juveniles and this variability somewhat limits diagnostic value (*e.g.*, Baldwin & Bell, 1985) (Fig. 9D-F). In relation to the basic hexaradiate pattern, male patterns are typically less derived and less dorsoventrally elongate than some of their J2 counterparts. Submedial lips are often fused with one another and with the labial disc but, in contrast to submedial lips in J2 of many species, apparently they are not likely to fuse with adjacent head annuli. Posterior to the labial disc, head annuli are often highly irregular and interrupted (Othman *et al.*, 1988). Labial patterns of males have a complement of sensilla equivalent to those of J2 and, in addition to the four outer labial sensilla reported in J2, two additional outer labial sensilla are noted in the lateral region that do not extend to the anterior surface (Baldwin & Hirschmann, 1975b).

Amphids resemble those of the J2, including the pair of receptors that terminate with microvilli, as finger cells, in each sheath cell (Fig. 10B). Male amphidial openings are oval, in contrast to those of *Meloidogyne* that appear as a refractive slit-like ampulla designated by Chitwood (1949a) as "amphidial cheeks" and used as a diagnostic feature to separate the males of the two genera. In cyst nematode males, the labial framework and stylet, while larger, are otherwise invariant in structure from those of J2 (Baldwin & Hirschmann, 1975a); the position of the dorsal gland orifice in relation to the stylet is also similar to that of J2. General innervation of males, excluding that of the copulatory region, apparently varies little from that of J2 including the position of hemizonion, hemizonid and similar interchordal commissures.

The pharynx of male cyst nematodes includes a distinct corpus, isthmus and gland lobe, with a membranous junction to the intestine and at the posterior end of the intestine, *i.e.*, similar to that of the J2. The completeness of the digestive system is remarkable considering that the male apparently does not feed as a migratory adult. This contrasts with the male of *Meloidogyne* which, with a similar biology, has a degraded digestive system where the isthmus is poorly defined, the dorsal gland is lost, the intestine is modified without a clear lumen and there is a large caecum that extends anteriorly adjacent to the median bulb (Baldwin *et al.*, 1977).

The reproductive system is a single outstretched testis that extends anteriorly from a *vas deferens*. The *vas deferens* joins posteriorly with a cloacal duct; further posteriorly the duct slightly enlarges, it is cuticle lined, and it ends as an opening to the exterior. The cloaca is shared by the reproductive and digestive system and encloses the terminal portion of a pair of cuticularised retractable spicules that are lodged within the cloacal duct (Figs 4C; 9K; 11A, B, D). The dorsal wall of the cloacal duct is thickened and protrudes between the spicules, forming a gubernaculum (Fig. 4A). The proximal end of the spicules consists of a manubrium and a broad shaft, to which two protractor muscles attach. The shaft extends posteriorly in a blade that is curved in cross section. The concave part of the blade of the two spicules face one another so that together they form a conduit through which sperm pass (Fig. 11A, B). The spicules vary in length (20-45 μm) among species and the distal tip may be expressed as a flat or rounded, bifid or tridentate spicule tip (Fig. 11A, B, D) (Clark *et al.*, 1973; Wouts & Baldwin, 1998). Near the tip each spicule has an orifice leading to a canal enclosing a sensory receptor. The spicules typically protrude through a sheath (often expressed as a cloacal *tubus*) that is terminal or nearly terminal with the location varying slightly among species, both with respect to position of the sheath and the transition of body annulation associated with the smooth-surfaced sheath (Othman *et al.*, 1988). Luc *et al.* (1988) define *Globodera* males as lacking a cloacal *tubus*. Although the shapes of spicules have been proposed for use as a diagnostic character for some species (*e.g.*, *Globodera*) (Behrens, 1972), more detailed studies revealed a considerable range of intraspecific variation in spicule shape (Manduric *et al.*, 2004). Mature sperm of Heteroderinae, including cyst nematodes, are amoeboid; the surface is dense with filopodia (Fig. 11C). However, preliminary studies suggest that sperm are distinctive among taxa with respect to symmetry, size and details of fine structure and development (Shepherd *et al.*, 1974; Walsh & Shepherd, 1983; Cares & Baldwin, 1995).

Female adult and cyst

Females have a number of features that distinguish genera and species, but cysts may be the stage richest in diagnostic characters. Furthermore, cysts are the stage often most readily available (Hesling, 1978). Females and cysts may be obtained directly from infected roots and cysts may also be recovered from soil by a range of techniques (Appendix I). Females in roots are often firmly attached by a cement-like material that encapsulates the head and neck region.

Whereas all cyst nematode females include a narrow anterior neck and swollen body, the female shape varies considerably in details that may be diagnostic (Figs 12, 13, 14A-C). This includes the presence or absence of a terminal posterior protrusion, the vulval cone, that determines if the female is globose when the cone is absent, as in *Globodera*, *Punctodera*, *Dolichodera and Paradolichodera*, or lemon-shaped where the cone is present, as in *Heterodera*. The shape and prominence of the vulval cone differs among *Heterodera* species and ranges from prominent and rather narrow in *H. schachtii* and *H. trifolii* to short and nearly insignificant as, for example, in *H. cruciferae* and *H. carotae*. In the latter case, the prominence may be so small that the cyst, rather than lemon-shaped, appears nearly spheroidal or ovoid (Hesling, 1978). Finer details of profile, including the shape of the cone, (Hesling, 1978) can aid identification of species such as *H. cardiolata*

Fig. 12. *Cysts. A:* Globodera pallida*; B:* G. tabacum*; C:* Punctodera chalcoensis*; D:* Betulodera betulae*; E:* Cactodera weissi*; F:* Heterodera schachtii. *(Original.) (Scale bar = 1 mm.)*

or *C. estonica* (Golden, 1986). Additional variation in shape includes a very small cone as in some *Cactodera* and *Betulodera* species (Fig. 14A-C). The formation and proportional size of the cone is affected by the relative elasticity of the body wall as the female enlarges, and this may be affected by development in relation to cuticle composition, layering and thickness, as well as attachment points and arrangement of vaginal muscles (Figs 15, 16) (Cordero & Baldwin, 1990, 1991; Cordero *et al.*,

Fig. 13. *Cysts. A:* Heterodera australis*; B:* H. filipjevi*; C:* H. latipons*; D:* H. glycines*; E:* H. mothi*; F:* H. sacchari. *(Original.) (Scale bar = 1 mm.)*

1991). Thus, size and shape, while reflecting underlying taxon-specific genetics, are variable because they are also subject to environmental factors including nutrition, age, host, physical pressure and crowding. In some cases, reliability of female/cyst shape and size in diagnostics can be improved by considering across-species comparisons of the largest individuals of a population (Hirschmann, 1956; Thorne, 1961). Regardless of limitations, specific trends in size of females and cysts are of some diagnostic value in combination with other characters. For

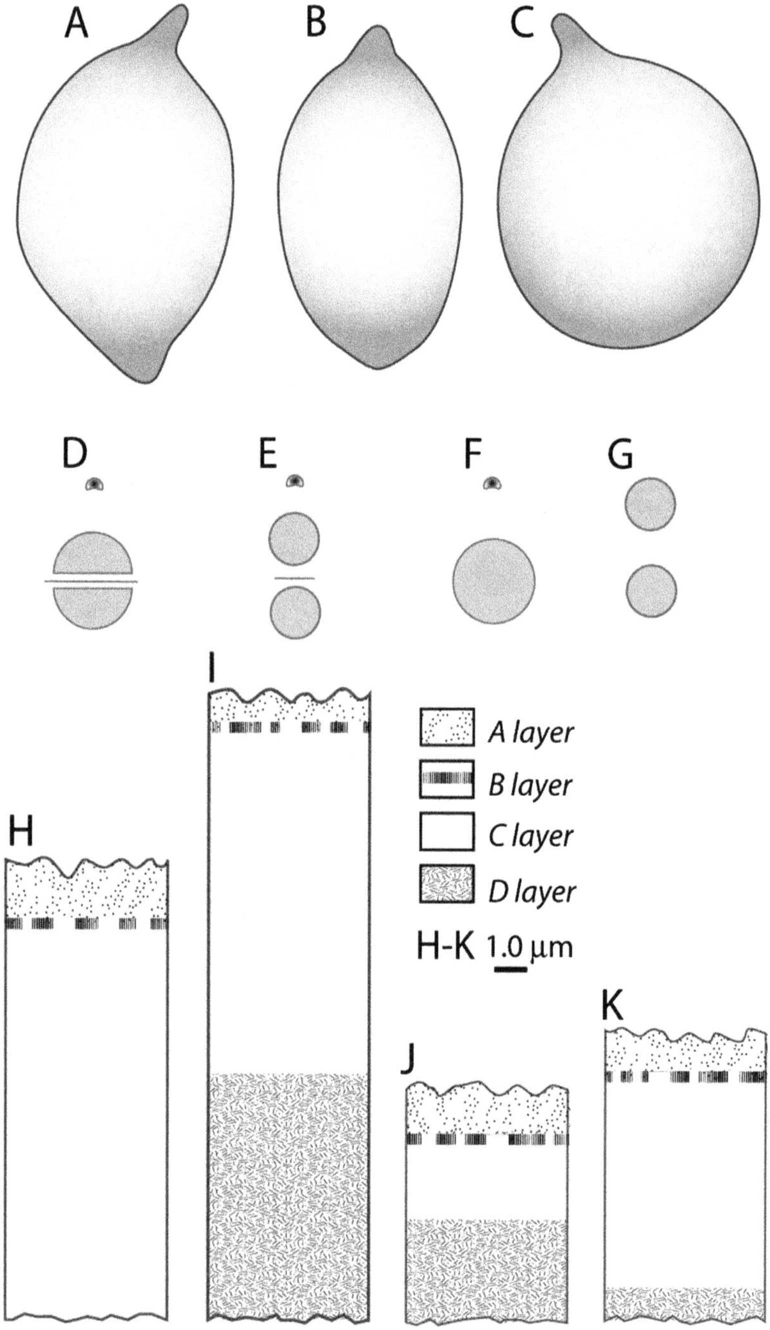
A
B
C
D
E
F
G
H
I
J
K
A layer
B layer
C layer
D layer
H-K 1.0 µm

example, females and cysts of some species, including *H. cruciferae*, *H. carotae* and *H. urticae*, tend to be relatively small, whereas cysts of *H. trifolii*, *H. betae*, *H. arenaria* and *H. avenae* are usually larger.

The female body wall is modified from that of the J2. In conjunction with a sedentary lifestyle, much of the somatic musculature of the female degrades and the body wall cuticle thickens, with the exception that some somatic muscles are retained and underlie the relatively thin body wall cuticle of the head and neck region. Although the outer cortical layers and striated layers of the cuticle are apparently retained from the J2, an additional broad C-layer occurs basal to the striations in all cyst nematodes (Fig. 14H, I) (Shepherd *et al.*, 1972). *Heterodera* is distinguished from *Globodera*, *Cactodera* and *Punctodera* by the presence of an additional inner cuticular layer, D, of specifically orientated fibrils (Fig. 14I) (Shepherd *et al.*, 1972; Cliff & Baldwin, 1985). A D-layer may be present even where the overall cuticle is relatively thin, as is the case in *Cactodera* or *Punctodera* (Cliff & Baldwin, 1985). Subsequent to this distinction, it was discovered that *Heterodera* also can develop a D-layer but that, in contrast to *Globodera*, this occurs very late in development. It was further discovered that layering varies with body region (Cordero, 1989; Cordero & Baldwin, 1990). Overall cuticle thickness apparently varies among species. For example, a diagnostic feature for *Globodera mali* is the presence of a thin and transparent cuticle which persists in the cyst. Regardless of overall layering, the cuticle structure is distinctively thin-walled and consists of a loose mesh of fibres in a discrete area of the posterior region of the body. In this area, the cuticle ruptures at cyst maturity forming taxon-specific apertures, the fenestrae. These aperatures may

---

Fig. 14. *Diagrammatic representation of female cyst nematodes (A-C, overall shape; D-G, fenestral region; H-K, layering of body wall cuticle. Redrawn from Cliff and Baldwin (1985)). A: Including substantial terminal cone; B: Including reduced terminal cone; C: Lacking terminal cone; D: Ambifenestrate; E: Bifenestrate; F: Circumfenestrae; broken line indicates position of vulva, anal region not fenestrate; G: Circumfenestrae; broken line indicates position of vulva, anal region fenestrate; H: Body wall cuticle of* Heterodera schachtii, *including A, B, and C layers; I: Body wall cuticle of* Globodera rostochiensis, *including A, B, C and D layers; J: Body wall cuticle of* Cactodera cacti, *including A, B, C and D layers; K: Body wall cuticle of* Punctodera punctata, *including A, B, C and D layers.*

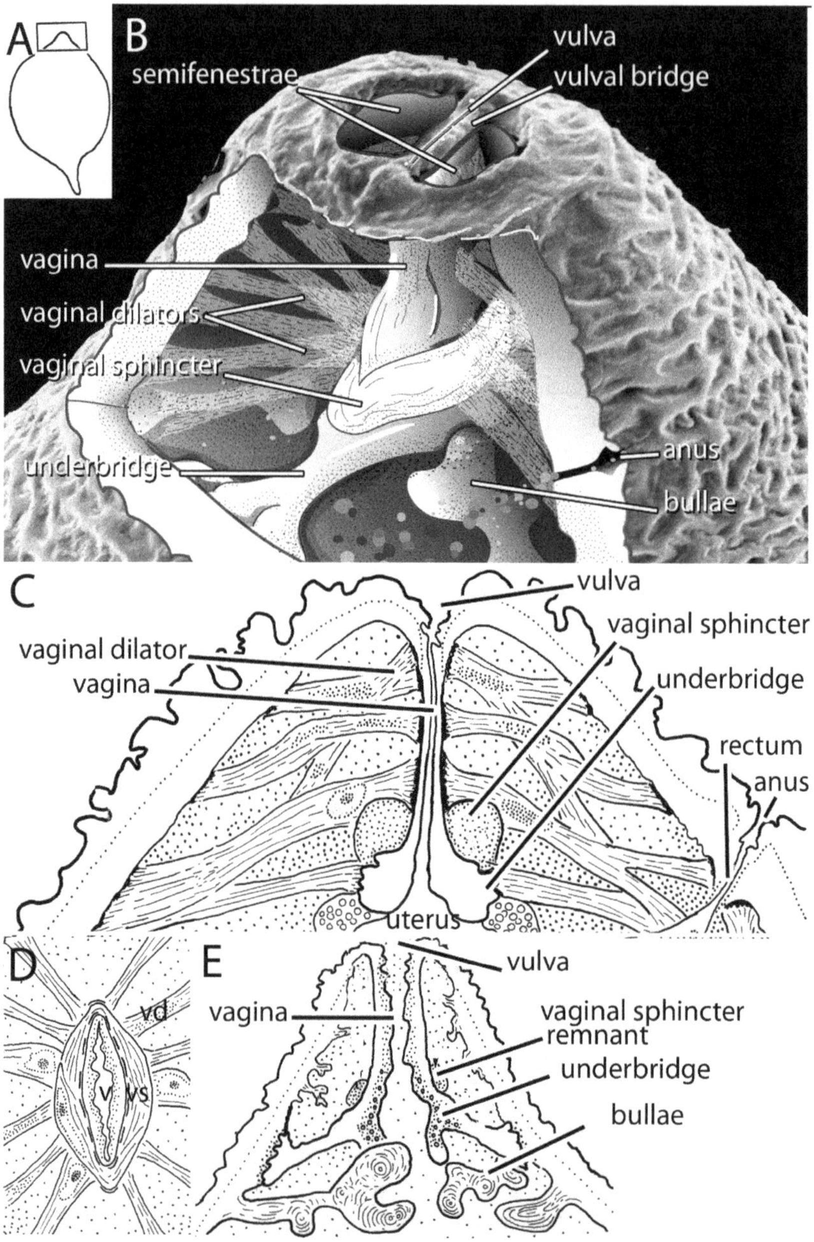
A
B
vulva
semifenestrae
vulval bridge
vagina
vaginal dilators
vaginal sphincter
underbridge
anus
bullae
C
vulva
vaginal sphincter
vaginal dilator
vagina
underbridge
rectum
anus
uterus
D
vd
v
vs
E
vulva
vagina
vaginal sphincter remnant
underbridge
bullae

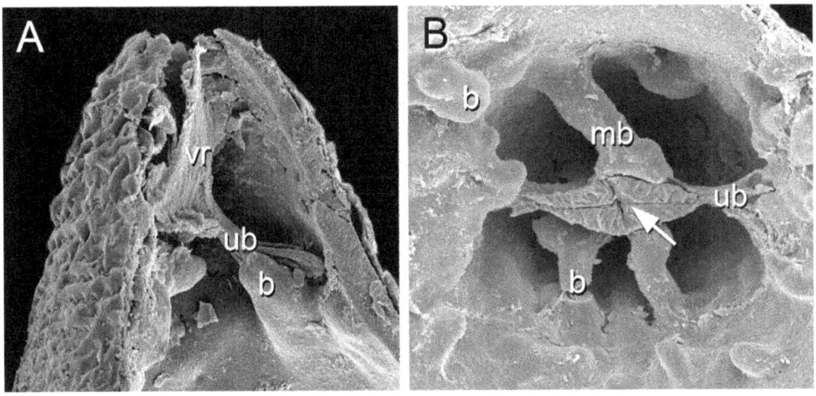

Fig. 16. *SEM of cyst cone of* Heterodera schachtii. *A: Near lateral view with cutaway showing internal structures including vulval remnant (vr), underbridge (ub) and bullae (b); B: View from inside cyst orientated toward terminus showing Mulvey's bridge (mb). Arrow indicates remnant of base of vagina lining.*

allow the exit of J2 from the cyst. On the outside of the cuticle we have noted that a subcrystalline layer may be present and, while variable, its occurrence has long been recognised as somewhat taxon-specific (Fig. 6F) (Liebscher, 1892; Fuchs, 1911; Kirjanova, 1969). Some have suggested that the subcrystalline layer might be protective and that it consists of two sublayers, with the outer layer fracturing into pyramidal blocks presenting an irregular polygonal appearance (Hesling, 1978). There has been considerable speculation concerning the source of the subcrystalline layer, with earlier suggestions that it might at least partially be formed from cast cuticles (Franklin, 1951). The specific association with certain taxa is consistent with the layer persisting in aseptic culture (Zunke, 1986; Cordero, 1989), in contrast to a hypothesis

Fig. 15. *Drawings depicting basic structures of terminal region of cyst nematodes. A: Overview of female. Box indicates terminal region as enlarged in B and C; B: Ventrolateral 3D view with cut away showing internal structures. Some features that may be exclusive to either female or cyst stages are combined to show a more complete range of possible structures; C: Right lateral view of* Heterodera schachtii*; D: Transverse view through vulva (v) at level of vaginal sphincter (vs) and vaginal dilators (vd) of* H. schachtii*; E: Dorsal view of mature cyst. C-E after Cordero and Baldwin (1991).*

that the layer is generated by a symbiotic fungus that feeds on metabolic products of the cyst (Brown *et al.*, 1971).

Female labial patterns, as viewed by SEM, apparently vary little among species of cyst nematode. Typically the labial disc is squarish and offset above a disc apparently comprised of fused lips (Fig. 9G). Generally, additional annulations or a pattern of tubercles (Franklin, 1939, 1951; Othman *et al.*, 1988) occur near the head, regardless of more complex patterns that may occur posteriorly and throughout most of the body of females and cysts (Fig. 6D). These surface patterns may be relatively taxon-specific, ranging from parallel transverse ridges approximating striations, to a more typical zigzag pattern defined by angular incisures and that may vary from coarse to fine among taxa (Fig. 6D-L). Applying the distinctive features of these patterns in diagnostics is promising, but there are difficulties because among the named patterns there is a range of intermediate forms described by inconsistent and imprecise terminology, including rugose, reticulate or lace-like (Fig. 6G-L) (Jones, 1952; Hesling, 1978; Baldwin & Mundo-Ocampo, 1991). Patterns may be further altered by natural variation within a population, between regions of the same body and by age of the individual. For example, a striking range of pattern variation occurs among the mid-body and adjacent anterior and posterior regions of cysts of *G. tabacum* (Fig. 6J-L). Nevertheless, zigzag and related mid-body patterns apparently differ among species in the form and scale of the pattern (Othman *et al.*, 1988). Beyond surface textures, cuticle patterns may further vary among species by the presence or absence of subsurface punctations (Franklin, 1939, 1951) that can often be attributed to spots of dense deposits or radial columns in the subsurface layer of the cuticle (Jones, 1952; Hesling, 1978; Baldwin & Mundo-Ocampo, 1991). While mid-body surface patterns generally vary too much within species for diagnostic purposes, they can be useful, for example, in distinguishing those few cyst species that retain annulation (examples in *Punctodera* and *Cactodera*) or with unique features, such as the longitudinal lines that distinguish *G. millefolii* from other *Globodera* spp.

Cuticular patterns on the posterior or terminal region of females and cysts are typically modified from those of the rest of the body (Fig. 19). Details of these cuticular characters, together with some structures underlying the cuticle, have long been recognised for their diagnostic value (Figs 15-19) (Franklin, 1951; Cobb & Taylor, 1953; Oostenbrink & den Ouden, 1954; Cooper, 1955; Hesling, 1978; Golden, 1986).

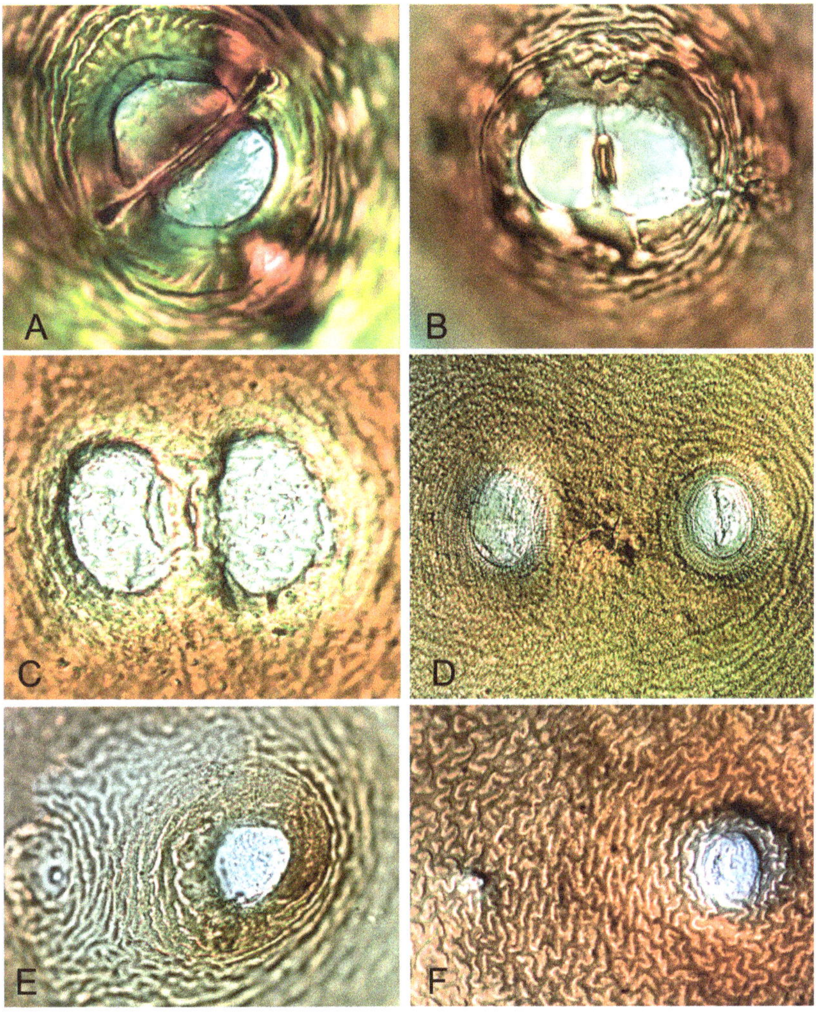

Fig. 17. *Vulval plate. A:* Heterodera trifolii *(ambifenestrate); B:* H. pratensis *(bifenestrate); C:* Heterodera *sp. (bifenestrate); D:* Punctodera punctata *(circumfenestrate); E:* Cactodera radicale *(circumfenestrate); F:* Globodera rostochiensis *(circumfenestrate). (Courtesy of V.N. Chizhov, Russia.)*

Often the characters and terminology differ between round cyst species of *Globodera*, *Punctodera*, *Dolichodera* and *Paradolichodera* (Golden, 1986), that lack a cone, and lemon-shaped species of *Heterodera*, *Cactodera* and *Betulodera*. Here the terminal region of groups of round and lemon-shaped females and cysts are discussed separately:

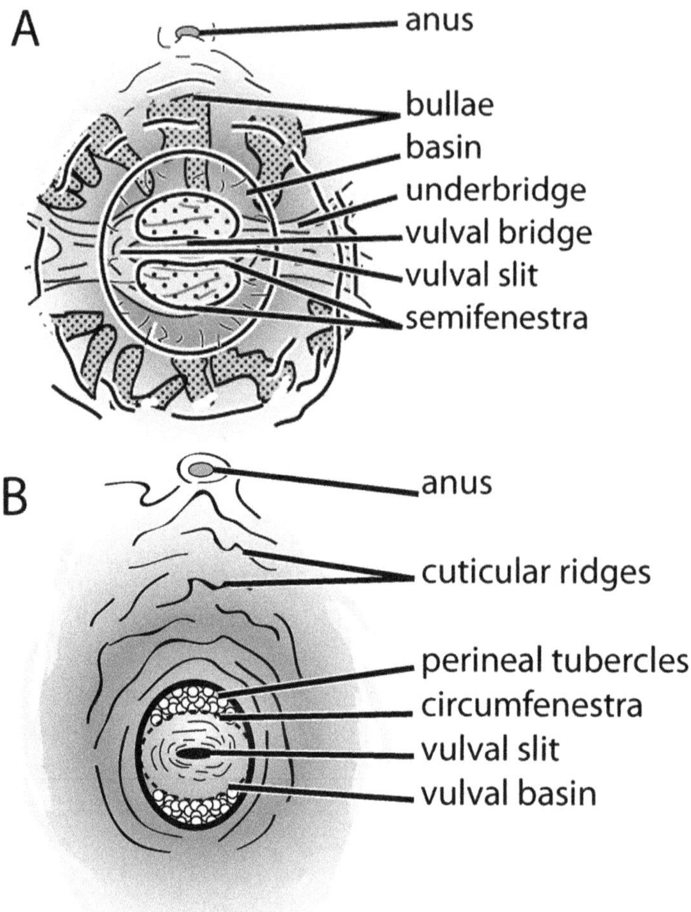

Fig. 18. *Terminal view of tail region of young cysts (composite of species for genus) showing typical diagnostic features and associated terminology. A: Cone of* Heterodera *(ambifenestrate type); B: Terminal region of* Globodera *(circumfenestrate type). After Mulvey (1973).*

i) Globose females and cysts (Figs 12A-D; 14C). In round cyst nematodes the terminal region, including the vulval-anal area is continuous with the contour of the female and cyst (Figs 14C; 18B). This region encompasses the vulval slit and the length of this slit may be informative for diagnostics. The vulval slit is surrounded by an area of thin cuticle and, as the female matures, the region is sometimes sunken to form a cavity called the vulval basin in which the margin is often defined by an

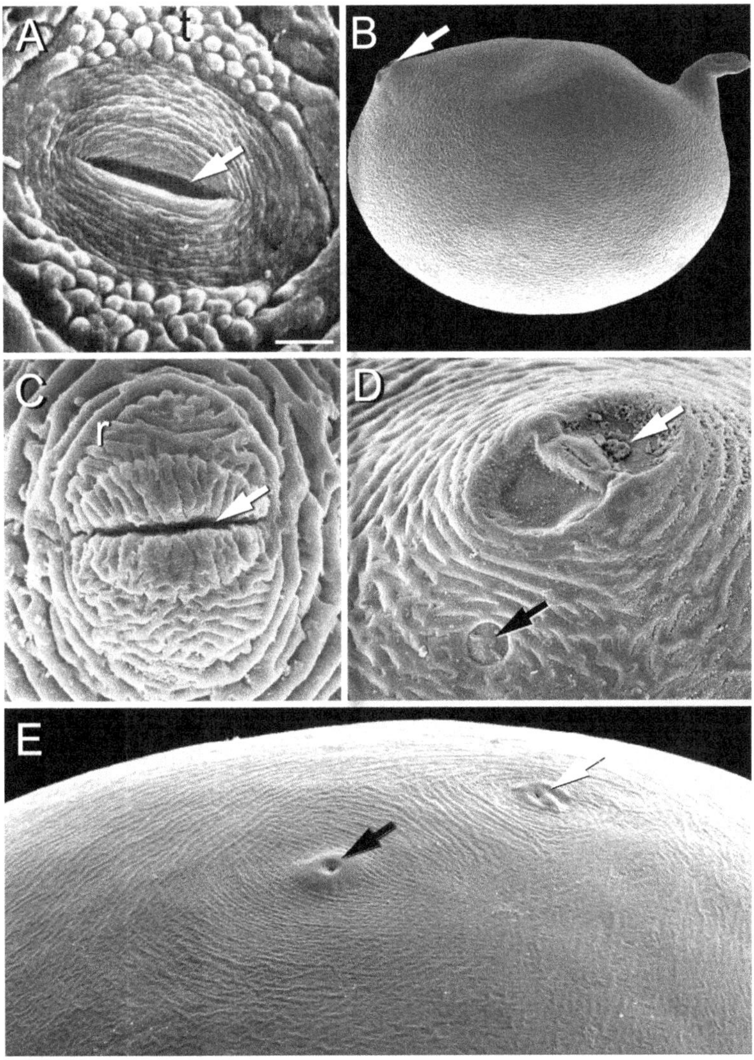

Fig. 19. *SEM of female cyst nematodes. A: Vulval region of mature* Globodera tabacum virginiae *(Mulvey, 1973). Abbreviation: t = tubercles defining crescent region of vulval basin. Arrow indicates vulval slit; B: Entire female of* Heterodera glycines. *Arrow indicates fenestral region shown in D; C: Vulval region of young* H. glycines, *showing rivulets (r) on each side of the vulval slit (arrow); D: Vulval and anal region of mature* H. glycines. *White arrow indicates fenestral region on ventral side of vulval slit, black arrow indicates anus; E: Terminal region showing vulval (white arrow) and anal (black arrow) region of mature* Punctodera chalcoensis.

elevated rim. As the cyst matures, the region of thin cuticle, the fenestra, ruptures resulting in a single opening, the circumfenestrae (Fig. 14F). The area of the fenestra is more or less equal to that of the vulval basin (Stone, 1986) and the shape and size of the circumfenestrae may be diagnostic. The area near the margins of the circumfenestrae may have distinctive cuticular patterns. Specifically, some *Globodera* species have perineal tubercles (Figs 18B; 19A) tightly packed within a pair of anterior (ventral) and posterior (dorsal) areas called vulval crescents adjacent to the margins of the fenestra. These may be diagnostic by presence or absence and, if present, in size and details (Figs 18B; 19A).

Dorsal to the circumfenestra is the anus and, in species of *Punctodera*, the region encompassing the anus develops a second fenestra (Figs 14G; 17D; 19E). Particularly in *Globodera*, the distance from the edge of circumfenestra to the anus and also the ratio obtained by dividing this measurement by the length of the fenestra (Granek's ratio defined by Hesling, 1973), while variable, may be useful in diagnostics. This ratio contrasts slightly with that originally proposed by Granek (1955) as the distance from the anus to the nearest edge of the vulval opening divided by the diameter of the vulval opening. In round cyst nematodes the cuticular surface pattern between the vulva and anus, the perineum or perineal area, may be diagnostic. In *Globodera*, the pattern of this region is described as maze-like, reticulate or with parallel cuticular ridges (Fig. 18B). These patterns, including the number of cuticular ridges between the anus and vulva, have been considered to be diagnostic for some species (Stone, 1973a; Golden, 1986) (Table 14).

ii) Lemon-shaped females and cysts (Figs 13, 15, 18A). In lemon-shaped cysts of *Heterodera*, *Cactodera* and *Betulodera* it has been noted that the posterior region is set off from the contour of the rest of the body as a cone, and that this protrusion varies among species from prominent to very small. The vulva or vulval slit is situated at the terminus of the cone and the anus is a small orifice on the dorsal side of the cone. In *Heterodera*, the vulval slit is bordered on each side by a narrow band of cuticle and together the pair of bands form the vulval bridge (Fig. 15A). The width and length of the vulval bridge are used in diagnostics among *Heterodera* species. For example, the vulval bridge varies from between 2 to 4 μm wide for most *Heterodera* species compared with 20 μm and more for *H. hordecalis* and *H. latipons*. Also, the length of the vulval slit is a key character for identification of *Heterodera*. Species with the shortest (*ca* 6 μm)

(Fig. 17B, C) vulval slit distinguish *H. aucklandica* and *H. spinicaudata* whereas the longest vulval slit (64-65 μm) (Fig. 17A) characterises *H. leuceilyma* and *H. amygdali*. In most lemon-shaped cysts the vulval slit and vulval bridge are flanked on the dorsal and ventral regions by semifenestrae characterised in females by thin cuticle. In mature cysts the semifenestrae rupture, while typically retaining the vulval bridge (Figs 15B; 18A). Within *Heterodera*, two types of semifenestration are further distinguished: ambifenestrate, in which each semifenestra is semicircular (Figs 14D; 17A) and bifenestrate, in which each semifenestra is circular (Figs 14E; 17B, C). Fenestral length and width are used for species identification. In old ambifenestrate cysts where semifenestrae are initially separated by a thin vulval bridge (*H. urticae*, *H. carotae* and others), the bridge may eventually be ruptured so that the fenestrae appear together as a single, roughly circular, hole. Under such circumstances the opening may be misinterpreted as circumfenestrate, but with close examination it is possible to detect remains of the bridge and infer the original ambifenestrate pattern (Hesling, 1978). Among lemon-shaped cysts some species are exceptions in they do not form seminfenestral openings. For example, cysts of the *Afenestrata* group of *Heterodera* do not develop fenestral openings. Furthermore, species of *Cactodera* and *Betulodera*, regardless of the presence of at least a small cone, instead develop circumfenestra (Figs 14F; 18B).

As for *Globodera* (see above), in most *Heterodera* species, the fenestral region is situated in a sunken region defined by Cooper (1955) as a basin and demarcated by slightly elevated cuticle. The rim of the vulval basin may include, in some species of *Heterodera*, wrinkle-like rivulets on the dorsal and ventral regions (Cooper, 1955; Hesling, 1978) (Fig. 19C).

During cyst formation in some *Heterodera* species, the cuticle may thicken at the proximal end of the vagina. From the vagina it is attached laterally by forked or branched projections to adjacent sides of the body wall cuticle; these connections comprise the underbridge (Figs 15, 16, 18A). The underbridge is an internal structure of the terminal cone and is located beneath the vulval bridge as viewed in an excised terminal cone mounted on end (Figs 15, 16, 18A). Presence, absence, length, shape and position of the underbridge are all important diagnostic features for species of *Heterodera* (Mulvey, 1974). For example, species of the *Schachtii* group have a strongly developed and pigmented underbridge, whereas species from the *Goettingiana* group have a slight and fragile underbridge that is vulnerable to damage during preparation of the vulval cone for microscopic examination. In *H. sacchari*, *H. leuceilyma* and *H. goldeni*, the underbridge includes prominent finger-like projections that help to distinguish these species. The position of the underbridge relative to the surface can be diagnostic, as is the case for *H. filipjevi* where it is relatively close to the surface (posterior terminus) by comparison with the much deeper position (45-54 μm) from the surface in *H. leuceilyma*. Consistent with the underbridge, Raski (1950) noted sclerotised structures which extend from the vagina to the body wall and that persist in the cyst. In some cases the vagina connecting the two bridges may persist with longitudinal markings (ridges or folds) as a vulval remnant or "sheaf" (Fig. 16A) (Golden, 1986). A structure called Mulvey's bridge is described in newly-formed cysts of some species (*H. trifolii* and several others) as occurring at right angles to the underbridge, but this too is believed to be persistent remnants of musculature (Fig. 16B) and it does not persist in old cysts (Mulvey, 1959; Hesling, 1978; Cordero, 1989).

In some species, as the cyst matures, large deposits originating from the hypodermis may extend into the body cavity and darken with age. These persist as bullae (Figs 15, 16, 18A), particularly in those species that also have an underbridge (Cooper, 1955; Hesling, 1978; Cordero, 1989). Bullae, first described and named by Cooper (1955), are prominent, single or branched knob-like or finger-like structures. They vary in size, shape, position and pattern, and may have diagnostic value (Franklin, 1951; Wouts & Baldwin, 1998). For example, in *H. avenae* and other species of the *H. avenae* species complex, bullae are clustered in the distal end of the cone, whereas in *H. schachtii* and other species of the *Schachtii* group they tend to be scattered in the vicinity of the underbridge (Hesling, 1978). Four finger-like bullae are typically diagnostic for *H. zeae*. However, due to intraspecific variation, bullae are described as "present or absent" for *H. latipons* and *H. bifenestra*. While not specifically designated as bullae, dark knob-like structures occurring at the ends of the vulval bridge are termed vulval knobs by Sharma and Swarup (1984) and these have been found in species of the *H. cardiolata* complex.

Often compared to bullae are other small tooth-like pigmented structures, the vulval denticles (*Cactodera*, *Betulodera*, *H. oryzae*, *H. urticae*, *H. mothi* and others). These have been demonstrated to be remnants of vulval musculature (Golden & Raski, 1977; Cordero, 1989;

Cordero *et al.*, 1991), as may also be the case for vulval bodies (Wilson, 1968; Mulvey, 1973). Whereas vulval denticles and vulval bodies are sometimes used in diagnostics, it is crucial to recognise their limitations for this purpose considering their source and the fact that they differ fundamentally from small bullae.

By definition, cysts are tanned and the pigmentation and transitions of colour as they mature is diagnostic (Figs 12, 13). The pigment is the result of polyphenoloxidase tanning (Ellenby, 1946; Awan & Hominick, 1982) and may first appear as brown globules (Jones, 1950). Franklin (1951) was among the first to note that cyst colour at maturity may vary among species from light tan to nearly black. In *G. rostochiensis*, for example, the cysts as they mature are said to pass through a golden stage to brown, whereas in *G. pallida* they change directly to brown. Franklin (1951) suggested that the yellow phase of some *Globodera* species may result from pigmentation of body contents, implying that this yellow colour is subsequently obscured by darker pigmentation in the body wall. As a diagnostic tool, cyst colour transformations are not limited to *Globodera*. For example, females of *H. betae* change from white through yellow to brown, whereas females of *H. schachtii* do not have a distinct yellow colour. This distinction has practical diagnostic value because both species parasitise sugar beet. Notably, a yellow phase has been reported in conjunction with cyst development for *H. goettingiana*, *C. cacti* and *H. trifolii* (Wouts & Baldwin, 1998).

With the exception of stylet length, shape of the stylet knobs and position of the dorsal gland orifice, the digestive system of females apparently varies little among species of cyst nematode. Details of the stylet knobs may be species-specific, but often is not practical for diagnostics because of difficulty in recovering the stylet and observing it in a consistent orientation (Wouts & Baldwin, 1998). The stylet leads to a pharynx distinguished by a large median bulb and gland lobe but, in contrast to males and infective J2, there is little fine structural information to elucidate the details of this region (Fig. 4B). Considering the food source, the intestine may function primarily as a storage organ, consistent with Raski's (1950) observation that the rectum is not well developed in the adult female. With the vulva considered terminal, the anus occurs on the vulval lip defined as dorsal. In females, little is known of the sensory and secretory-excretory systems and how these might change relative to immature stages.

The reproductive system, while diagnostic with respect to surface structures, is otherwise not known to vary among cyst nematode species. The vulva leads to two long, coiled, reproductive tracts through a dorsoventrally flattened, cuticle-lined, vagina (Fig. 15). The region is highly muscular with muscles from the vagina extending to the body wall. Notably, in *Heterodera* these radiate as 48 dilator muscles (four levels with six muscles on each side) along the length of the vagina, whereas there are fewer and they are less developed in *Cactodera* (Cordero & Baldwin, 1991; Cordero *et al.*, 1991). The differences in musculature appear to be congruent with females laying some eggs into an egg sac *vs* retaining (muscular reduced) most eggs through the cyst stage. Near the junction with the uterus and just posterior to the underbridge, the vagina is surrounded by a *sphincter vaginae* muscle (Fig. 15) and a remnant of this may persist in the cyst (Fig. 15E). Proximally the tracts consist of a muscular uterus followed by an oviduct including the spermatheca, and distally the ovary consists of a long growth zone and relatively short germinal zone (Triantaphyllou & Hirschmann, 1962). The extent to which eggs are laid *vs* retained in the body varies among species, but where eggs are laid they are protected in an egg sac. Even species that do not produce an egg-containing sac may produce small quantities of gelatinous exudate (Hesling, 1978; Turner & Rowe, 2006), and in young females this exudate may be occupied by one or more males. The source of this exudate appears to be near the junction of the vagina with the uterus, although no specific secretory cells have been observed.

DIMENSIONS FOR MORPHOLOGICAL IDENTIFICATION

Morphology remains a key tool for descriptions and diagnostics of cyst nematode species. Beyond the morphological features described above there are ratios and measurement that have become part of the tradition of descriptions and some of these are listed below.

a – body length / greatest body diam. (male, J2) or L/W ratio (cyst, female, egg)

b – body length / distance from anterior end to junction of pharynx and intestine (male, J2)

b′ – body length / distance from anterior end to posterior end of pharyngeal glands (male, J2)

c – body length / tail length (male, J2)
c′ – tail length / body diam. at anus (male, J2)
DGO – distance from anterior end to orifice of dorsal gland opening (female, male, J2)
Granek's ratio – distance from the edge of fenestra to the anus divided by the length of the fenestra (*Globodera* cyst vulval cone)
T – distance from cloaca to anterior part of testis × 100 / body length (male)

Nematology Monographs & Perspectives, 2010, Vol. 8A, 43-70

Chapter 2

Biology and evolution

Life cycle

The life cycle of cyst nematodes comprises the egg, four juvenile stages and the adults (Fig. 2). The first-stage juvenile is limited to the egg; the J2 hatches from the egg while within the cyst or the gelatinous matrix that comprises the egg sac. Hatching is the result of a complex interplay between external signals and the internal readiness of the J2 to emerge. Hatching of large numbers of J2 for many cyst nematode species occurs in response to host root diffusates. Some species, for example, *G. pallida*, *G. rostochiensis*, *H. carotae*, *H. cruciferae*, *H. goettingiana* and *H. humuli*, show a high level of dependence on host root diffusate for hatching to occur, whereas *H. avenae*, *H. cajani*, *H. schachtii* and *H. sorghi* can hatch in large numbers in the absence of any stimulus (Perry, 2002). Those J2 that hatch within the cyst leave the cyst *via* either the open fenestral region or the opening created at the neck where the female's head has deteriorated or otherwise broken away. The J2 are then attracted toward the host root, responding to gradients of a variety of stimuli. Some of the gradients that exist around roots include CO_2, amino acids, pH and sugars; these are likely to act as general, non-specific attractants for long-distance migration of nematodes in the soil (Perry, 1997). Once in the root area, the J2 responds to other specific attractants in host root diffusate to locate a host root and orientate to a preferred invasion site (Perry, 1997, 2002). Once at the invasion site, a J2 enters the root epidermis using repeated thrusts of its protrusible, pointed stylet. Penetration usually occurs in the zone of elongation behind the growing root tip (Fig. 20A). The J2 migrates intracellularly and crosses the cortical tissue toward the vascular cylinder. Cortical cell walls are disrupted by a combination of the mechanical action of highly coordinated stylet thrusts and the action of enzymes secreted by the nematode (Wyss & Grundler, 1992; Davis *et al.*, 2004). Reaching the pericycle, the J2 carefully penetrates a single

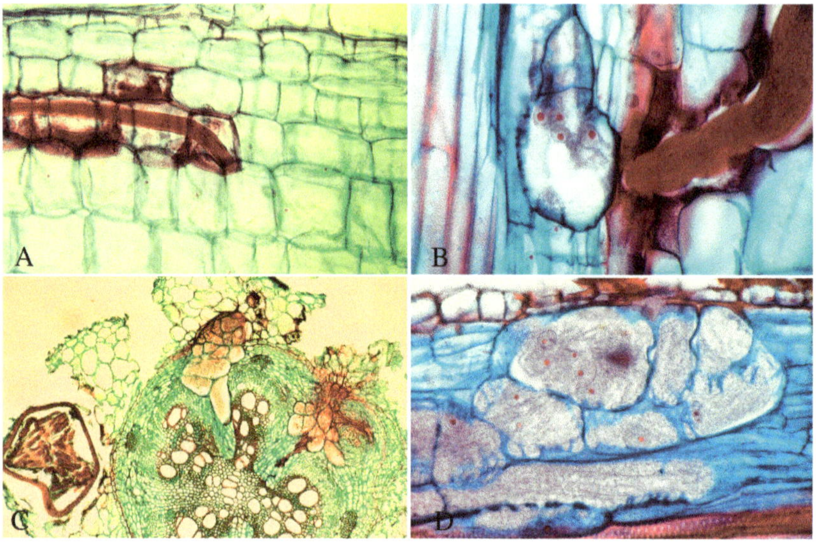

Fig. 20. *A: Histological section of soybean root showing intracortical penetration by an infective juvenile of soybean cyst nematode at 1 day after inoculation; B: Longitudinal section of soybean root showing a third-stage juvenile of soybean cyst nematode at its feeding site and a syncytium. The syncytium, which functions as a food reservoir, has dense cytoplasm and multiple nuclei that result from dissolution of cell walls; C: Cross section of susceptible soybean cultivar at 40 days after inoculation showing sectors of root occupied by syncytia. Some syncytia have been replaced by parenchymatous tissue; D: Extensive syncytium induced by a third-stage juvenile of soybean cyst nematode at 10 days after inoculation. Note cell wall fragments dispersed throughout cytoplasm. After Endo (1998).*

cell with its stylet and then injects secretions of the dorsal pharyngeal gland into this cell. These secretions mediate a massive reprogramming of cell development and induce the formation of a large, multinucleate, feeding structure, a syncytium, which is formed by dissolution of walls initially separating adjacent cells (Fig. 20B-D). This syncytium acts as a nutrient sink. The syncytium is highly metabolically active and characterised by multiple enlarged nuclei with nucleoli, dense cytoplasm and a proliferation of mitochondria, endoplasmic reticulum and plastids. The syncytia develop ingrowths in the cell wall boundary adjacent to the conducting tissues (Fig. 20D). The ingrowths are associated with increased surface area of the adjacent plasmalemma, and thus they significantly enhance the capacity of the syncytium to take up nutrients

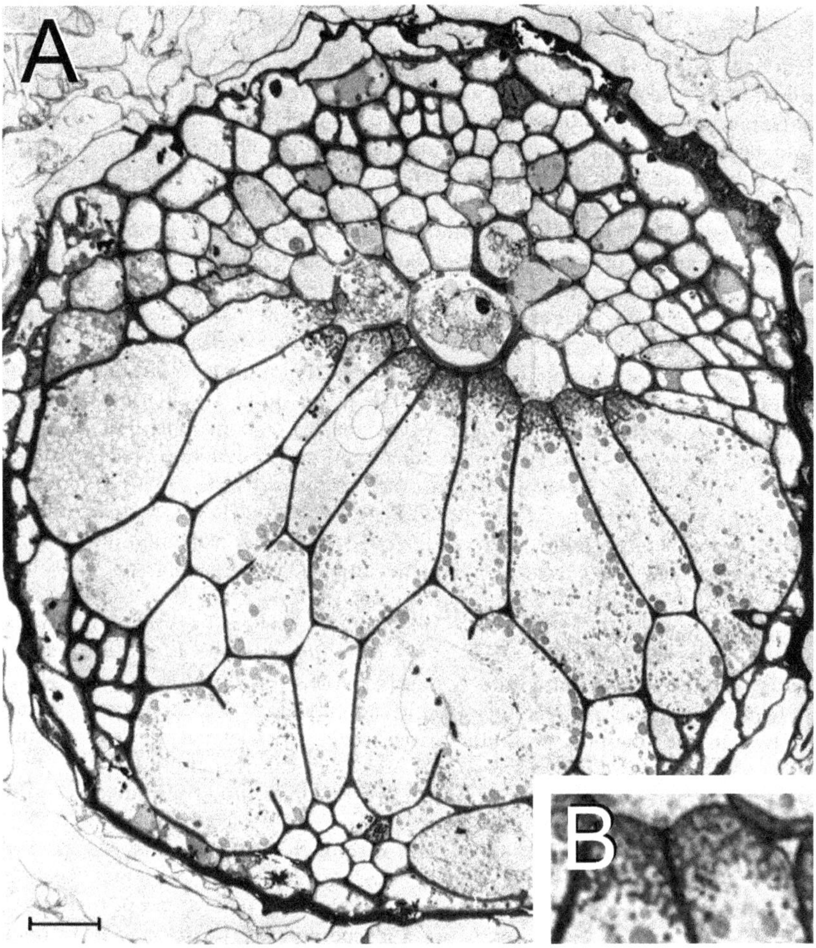

Fig. 21. *Syncytial host response (A) including enlargement showing wall ingrowths (B). After Wyss* et al. *(1984).*

through the apoplast to the symplast of the syncytium (Fig. 21). After establishing a feeding site, the J2 undergoes three moults to reach the adult stage.

Cyst nematodes are sexually dimorphic. Juveniles destined to become males feed until the end of the J3 stage, at which point a motile, vermiform male develops within the J4 cuticle. After emerging from the cuticle and leaving the root, males are attracted to females that exude sex pheromones. Many males may surround each female and multiple mating occurs. Males have short lives, remaining active for a few days. By feeding on the roots the J2 female becomes swollen and develops into a mature adult, which ruptures the surface of the root with the posterior portion of its body, thereby exposing its vulva for copulation and egg deposition. After mating, eggs develop and contain maturing embryos. In most species most eggs are retained within the female that then dies. The cuticle of the female body is tanned by a polyphenoloxidase to form the tough, leather-like, cyst wall. Females are usually capable of producing up to 500-700 eggs. Females of some species of *Heterodera* can produce a large gelatinous egg sac in which some eggs are deposited. Eggs in the gelatinous matrix, which consists primarily of carbohydrate, may be partially protected by the presence of antimicrobial compounds such as chitinase and polyphenoloxidase. The life cycle from egg to egg varies depending on temperature but is usually completed in about 30 days. The number of generations per year also varies from one to several and depends on species (Lilley *et al.*, 2005; Turner & Rowe, 2006). In some species, eggs within the cyst are especially persistent when conditions are not favourable for plant growth and infection. Under these conditions they can persist in soil for many years. The incidence of diapause varies greatly among species and between populations of the same species from different climatic areas. The combination of diapause and quiescence enables *G. rostochiensis*, for example, to persist in the soil for more than 20 years in the absence of a host plant and biological antagonists (Wright & Perry, 2006).

Symptoms of infection

Field symptoms of severe cyst nematode infection often include reduced growth, stunting, wilting, chlorosis of leaves, and reduction of quality and quantity of yield (Figs 22, 23). Above-ground symptoms of damage are not unique to cyst nematodes. They could be mistaken for nutrient deficiencies, drought stress, herbicide injury, or other plant diseases, and often can remain undetected for several years. Symptoms of infection are expressed more fully when mineral nutrition is limited or root growth is otherwise impeded. Areas of severely affected and symptomatic plants are often round or elliptical in shape (Fig. 23). Severely infected plants may die before flowering. Most below-ground symptoms of cyst nematode damage are also not unique. Infected

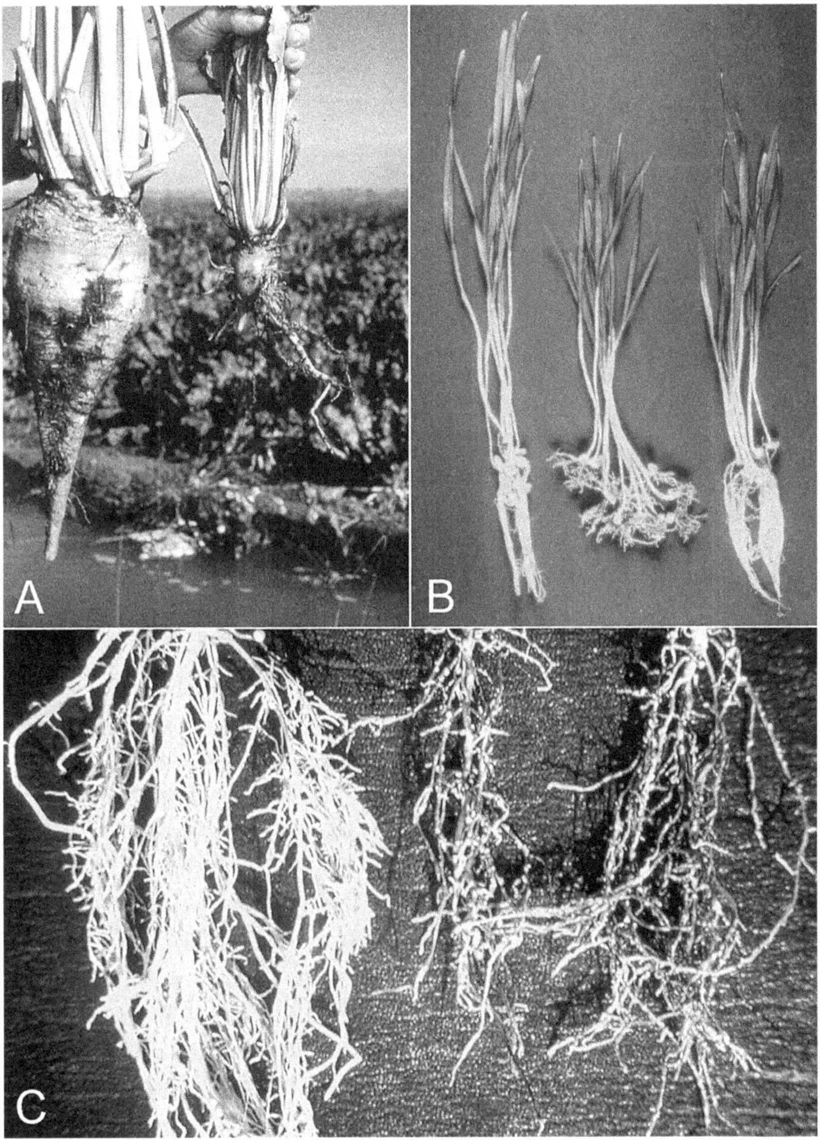

Fig. 22. *Symptoms of infection. A: Healthy (left) and infected (right) sugar beet plants by sugar beet cyst nematode; B: Healthy (left) and infected (centre and right) wheat seedlings by the cereal cyst nematode; C: Healthy (left) and infected (centre and right) soybean root systems by the soybean cyst nematode. Courtesy of CDFA, USA.*

Fig. 23. *Infected fields with cyst nematodes. A: Soybean field infected with soybean cyst nematodes with uneven plant height and slight chlorosis (courtesy of G. Tylka, USA); B: Sugar beet field infected with the sugar beet cyst nematodes; C: Cereals infected with the Filipjev cereal cyst nematode; D: Potato field infected with the golden cyst nematode; E: Cabbage field infected with cabbage cyst nematode. Courtesy of V.N. Chizhov, Russia.*

roots are dwarfed or stunted (Fig. 22). Extensive branching and local thickening of roots can also be observed. Cyst nematode infections may also make the roots more susceptible to attacks by other soil-borne plant pathogens. Many abiotic factors, for example, fertility, pH, soil type and organic matter content, influence nematode population development and damage severity.

Cytogenetics and reproduction

Most cyst nematodes are amphimictic, *i.e.*, their populations consist of both males and females which reproduce sexually. The sexual fate first becomes apparent at the end of the parasitic J2 phase. The sex determination mechanism of cyst nematodes remains unknown. It has been suggested that sex expression is genetically controlled but might be modified by environmental influences. Observation of unbalanced sex ratios in populations under different growth conditions suggests an environmental influence on sex determination. Under unfavourable conditions the proportion of males increases, whereas under favourable conditions the majority of individuals develop into females.

Several experimental studies with cyst nematode species showed their ability to produce interspecific and infertile hybrids. Successful hybridisation was observed between *H. schachtii* and *H. glycines* and between *G. pallida* and *G. rostochiensis* (Miller, 1983). Hybridisation between some species of cyst nematode appears to be a natural, but very rare phenomenon.

Cyst nematodes are telogonic – within a gonad a series of increasingly developing gametes can be viewed at stages conducive to determining chromosome number. A chromosome number of $n = 9$ is thus interpreted as conserved among cyst nematodes with $2n = 18$, characterising species that reproduce by amphimixis. Additional species are classified by Triantaphyllou and Hirschmann (1980) into groups apparently representing three independent lines of evolution toward parthenogenesis. In two of these lines, involving the *H. trifolii* species complex (*H. galeopsidis*, *H. betae* and others) and *H. sacchari*, parthenogenesis is of the mitotic type and is associated with high chromosome numbers suggesting triploidy ($3n = 27$), tetraploidy ($4n = 36$), and aneuploidy ($2n$ or $3n = 24, 26, 28, 32, 34$). In a third line, represented by *B. betulae*, parthenogenesis is meiotic, this species being cytologically distinct from all others by the presence of a haploid chromosome number of n = 12 or 13 and males are rare (Triantaphyllou & Hirschmann, 1980).

Origin and phylogeny

The relationship of Heteroderinae to other Tylenchida has long been the subject of speculation; several authors have reviewed the evolution and phylogeny of cyst nematodes (Golden, 1986; Stone, 1986; Krall, 1990; Baldwin, 1992; Siddiqi, 2000). Several classifications follow Chitwood (1949a) and Chitwood *et al.* (1956), implying a unique common ancestor between cyst and root-knot nematodes, but an alternative hypothesis is that Heteroderinae have a unique common ancestor with Hoplolaimidae. Paramonov (1967) was the first to propose that cyst nematodes might represent a particular phylogenetic lineage of hoplolaimids on the basis of feeding adaptations on host plant roots that appear to suggest a transition of this group to sedentary endoparasitism. This idea was later supported by Wouts and Sher (1971) and Krall and Krall (1978). They viewed the hypothesis of a common ancestor with Hoplolaimidae as strengthened by shared characters, including the enlarged phasmid ampulla, the longitudinal lines of the labial region and a relatively short tail. Additional evidence included the cuticular structure in females, features of nurse cells induced in infected plants, and cytogenetic information that was argued to support further the hypothesis of a common ancestor between Heteroderinae and Hoplolaimidae, while also rejecting a sister relationship between Heteroderinae and Meloidogyninae (Baldwin, 1992). Molecular phylogeny also strongly supports a close relationship between Heteroderinae and Hoplolaimidae that excludes Meloidogyinae (Baldwin *et al.*, 2004; Subbotin *et al.*, 2006).

Stone (1975) was one of the first to suggest an evolutionary scheme for species groups of *Heterodera* and genera of Heteroderinae based on six types of head morphology of J2, as revealed by SEM. Hesling (1978) presented a modified version of this tree. From *Meloidodera* at the base, two branches were proposed, one leading to *Cryphodera* and *Zelandodera* and another leading sequentially to *Atalodera*, *Globodera*, *Cactodera* and *Heterodera*. *Sarisodera* and *H. punctata* (= *Punctodera*) were shown to branch separately from *Atalodera*. Later Stone (1979) published a modified evolutionary scheme for cyst nematodes based on morphological (J2 labial patterns, vulval morphology) and host range data. *Globodera* was regarded as retaining the primitive hexaradiate labial pattern, differing markedly from others in cyst morphology. It was also suggested that *Globodera* co-evolved as an independent branch with hosts of Solanaceae and Compositae. In addition, *Punctodera* was considered to have evolved as a branch independent from other lineages of cyst nematodes. Stone (1979) suggested that the main lineage of cyst nematodes comprised *Heterodera* and that it co-evolved on dicotyledon hosts. The ancestor was regarded as a *H. cruciferae*-like form from which species groups of *Heterodera* designated *Betulae* (*Betulodera*), *Fici*, *Humuli*, *Schachtii*, *Cruciferae*, *Cacti* (*Cactodera*), *Avenae*, *Oryzae* and *Graminis* were derived.

Krall and Krall (1978) believed that several lineages leading to modern groups and species of heteroderids were derived from a hypothetical ancestor similar to modern hoplolaimids. They considered that the most ancestral group might have been characterised by cysts that were lemon-shaped and ambifenestrate, similar to modern representatives of the *Schachtii* group. From this main *Protoschachtii* stem they proposed that several lineages evolved: *Punctodera*, *Cactodera* + *Globodera* and *Heterodera*. Krall and Krall (1978) further suggested that *Punctodera* was the most ancestral lineage within the cyst nematodes. Bifenestrate cyst nematodes of *Bidera* (the *Avenae* group) were considered to be highly specialised and derived separately from other *Heterodera*.

Early ideas about evolution of Heteroderinae were limited by lack of rigorous phylogenetic methodology such as cladistics based on Hennig (1966). These methods, already in wide use for other groups of organisms, only later began to find applications in phytonematology (Ferris, 1979, 1985; Coomans, 1983). Ferris (1979) was the first to employ a cladistic approach to heteroderid phylogeny. She presented arguments to establish polarity of character states of five characters and used these to infer phylogenetic relationships among seven species groups (including several genera). She considered the circumfenestrate cyst as expressed in genera *Globodera*, *Punctodera* and *Cactodera* to be derived relative to the bi- or ambi-fenestrate cyst as expressed in the *Avenae* group. She suggested that the derived condition of the vulval slit (elongate) and labial patterns derived by elongation and fusion provided a basis to propose that the *Humuli*, *Goettingiana* and *Schachtii* groups successively shared a common hypothetical ancestor (Ferris, 1979) to which the *Avenae* group is sister. Ferris (1979,

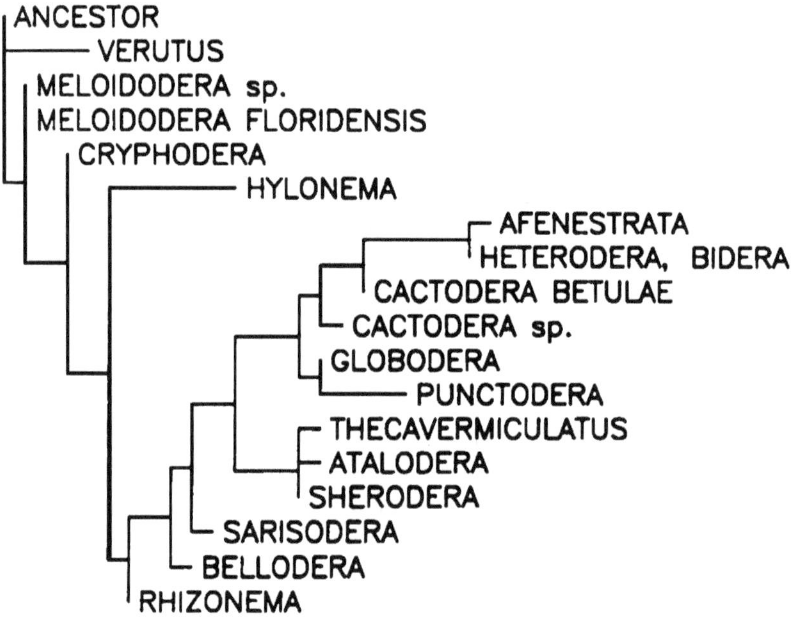

Fig. 24. *Tree based on PAUP parsimony analysis of nine morphological and nurse cell responses of cyst and non-cyst-forming nematodes. After Baldwin and Schouest (1990).*

1985) also developed a matrix of eight polarised characters that, in combination with biogeographic data, further elucidated relationships among genera of cyst and non-cyst heteroderids. Baldwin and Schouest (1990) expanded upon cladistic analysis of genera of Heteroderinae as a basis for revised classification (Fig. 24). Their approach was the first to use computer-based algorithms reflecting various models of phylogenetics. Baldwin and Schouest (1990) were also the first to introduce to heteroderid phylogenetics a range of novel characters and character states, including some from transmission electron microscopy (TEM) applied to detailed female, cyst, male and J2 morphology and details of nurse cell feeding structures in hosts (Baldwin, 1985, 1992; Baldwin & Schouest, 1990). They also expanded upon earlier treatment of SEM labial patterns of the J2 as a single character and instead treated them as several independent characters reflecting various aspects of fusion relative to the plesiomorphic labial disc surrounded by six separate lips; in addition, they considered some aspects of male labial patterns. On the basis of this analysis they strengthened evidence for a single ancestor for all cyst-forming genera. Some aspects of the analyses also suggested as separate clades *Thecavermiculatus* + *Atalodera* + *Sherodera* and *Hylonema* + *Rhizonema* + *Sarisodera* + *Bellodera*. Furthermore, the work provided support for arguments that *Meloidodera*, *Cryphodera*, *Zelandodera* and *Verutus* branched near the base of the tree. Baldwin and his Ph.D. students, Mundo-Ocampo, Cordero, Carta, Cares and Othman, further tested heteroderid phylogenies through SEM and TEM studies of characters of labial patterns, cyst cones, phasmids and sperm (Mundo-Ocampo & Baldwin, 1984; Othman *et al.*, 1988; Carta *et al.*, 1990; Cordero *et al.*, 1991; Cares & Baldwin, 1995).

Wouts (1985) developed a phylogenetic classification of the family Heteroderidae in which a putative cladogram consisted of six main lineages corresponding to subfamilies Verutinae, Meloidoderinae, Cryphoderinae, Heteroderinae, Ataloderinae and Punctoderinae; the subfamilies were composed of 17 valid genera (Fig. 25). According to Wouts (1985), the hypothetical heteroderid ancestor developed from a nematode that resembled modern members of the family Hoplolaimidae and which was possibly closely related to *Rotylenchulus.* The genus *Verutus*, characterised by a very large equatorial vulval slit, was considered the most primitive genus within heteroderids. The genus *Meloidodera* was considered to be further derived by a reduction in the length of the vulva. Genera, interpreted as developing later, exhibited a subterminal vulval slit and progressive loss of annulation of the female cuticle. In this process, four evolutionary lines were said to have emerged: *i*) a posterior shift of the vulva and the formation of more or less distinct vulval lips gave rise to the genera *Zelandodera* and *Cryphodera*; *ii*) changes in the labial configuration of the J2 gave rise to the genera *Hylonema*, *Afrodera*, *Heterodera* and *Bidera*; *iii*) changes in the composition of the female cuticle resulted in the genera *Thecavermiculatus*, *Atalodera*, *Sherodera*, *Sarisodera* and *Bellodera*; and *iv*) reduction in vulval slit size led to the development of the genera *Dolichodera*, *Globodera*, *Cactodera* and *Punctodera.* Wouts (1985) placed *Punctodera* at the apex of the Punctoderinae and considered the genus relatively derived. Wouts's phylogeny proposes independent development of cysts in different lineages.

New insight into the phylogeny of cyst nematodes by analysis of ribosomal RNA gene sequences has been provided by Ferris (1998), Ferris *et al.* (1999a, b), Sabo *et al.* (2001), Subbotin *et al.* (2001a,

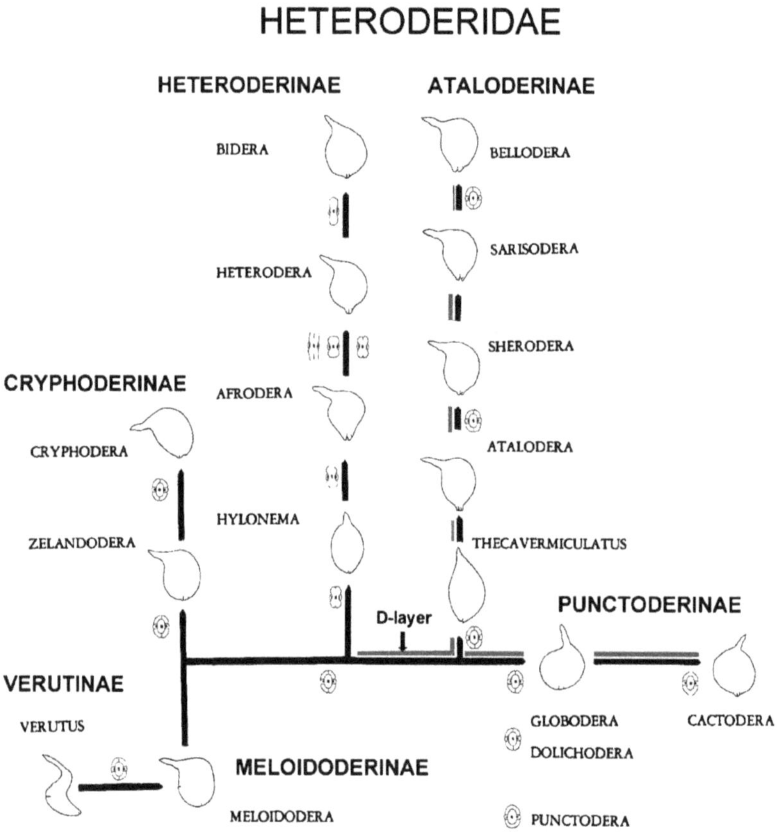

Fig. 25. *Evolutionary development of body and cuticle of female and lip region of J2 in subfamilies of the Heteroderidae. After Wouts (1985) with modifications.*

2006), Tanha Maafi *et al.* (2003) and Sturhan *et al.* (2007). Molecular data support the hypothesis of Wouts (1985) that *Verutus* branches basal to, and as a sister to the clade of all other heteroderids (Fig. 26). Molecular phylogenies also strongly support monophyly of *Cactodera*, *Punctodera*, *Betulodera*, *Globodera* and *Paradolichodera*, and they resolve some of the basal relationships within *Heterodera*, usually placing the *Goettingiana* group as branching closest to the origins of the cyst clade (Fig. 27). Molecular data also support many aspects of Stone's (1979) hypotheses, including recognition of the *Schachtii*, *Goettingiana*, *Avenae*, *Cyperi* and *Humuli* groups of *Heterodera*, albeit with species composition modified within each group (Fig. 27).

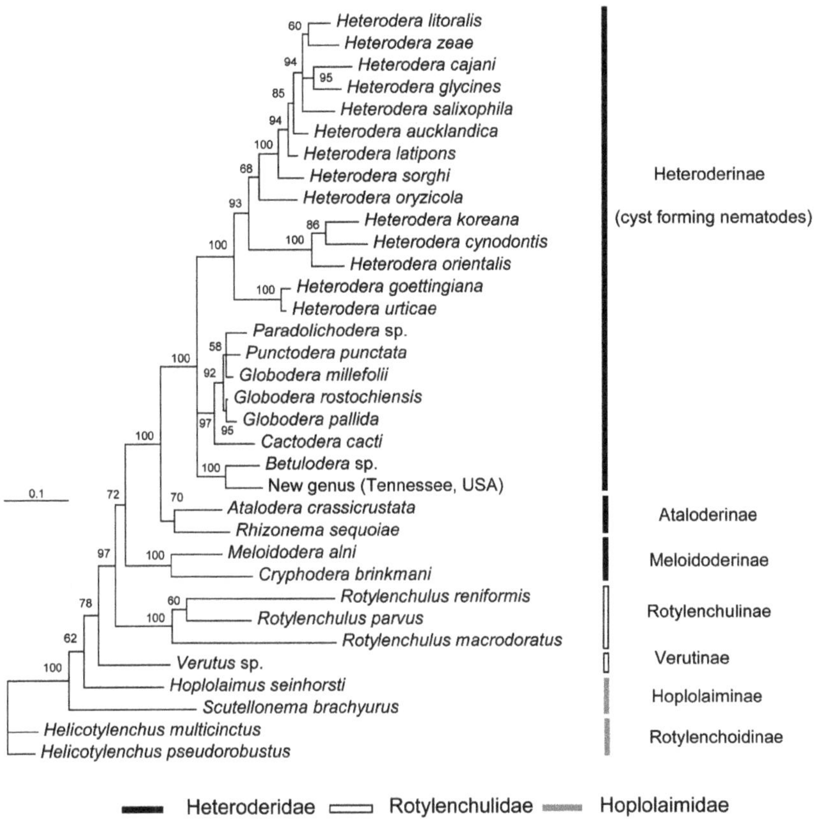

Fig. 26. *Phylogenetic relationships between Heteroderidae, Rotylenchulidae and Hoplolaimidae: Bayesian 50% majority rule consensus tree from two runs as inferred from D2 and D3 sequences of rRNA gene alignment under the GTR + I + G model.*

Some inconsistencies between molecular phylogeny and earlier-proposed phylogenetic hypotheses or groupings may be attributed to symplesiomorphies as well as homoplastic (*i.e.*, convergent) evolution. For example, according to our data a bifenestral vulval cone developed independently at least three times during the evolution of cyst nematodes. Likewise, the presence of three incisures in the lateral field of the J2 seems to have arisen independently twice. Finally, a short vulval slit apparently evolved independently three times in the *Avenae* group of *Heterodera*, as well as in *H. bifenestra* and *H. turcomanica*.

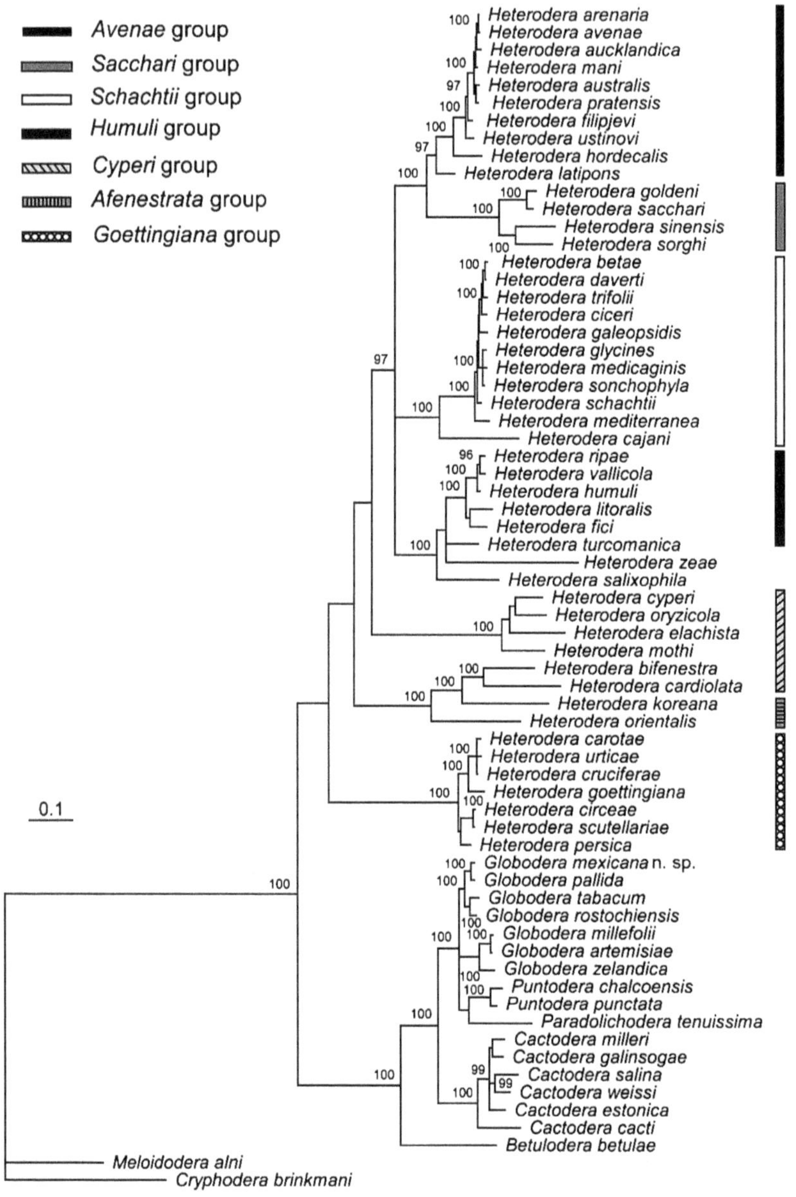

Fig. 27. *Phylogenetic relationships among Heteroderinae: Bayesian 50% majority rule consensus tree from two runs as inferred from ITS1-5.8S-ITS2 sequences of rRNA gene alignment under the GTR + I + G model. Posterior probabilities more than 70% are given for appropriate clades.*

Co-evolution with plants

Considering taxonomy according to extant plant classification (Angiosperm Phylogeny Group, 2003), cyst nematode species have been described from the type plant hosts belonging to the following orders: the genus *Heterodera* (80 species) – Poales (42); Fabales (10), Caryophyllales (5), Rosales (5), Lamiales (4), Mapighiales (3), Apiales (2), Brassicales (2), Urticales (2), Saxifragales (1), Myrtales (1), Fagales (1), Sapindales (1), Asterales (1); the genus *Globodera* (10 species) – Solanales (6), Asterales (2), Rosales (1), Myrtales (1); the genus *Cactodera* (13 species) – Caryophyllales (10), Asterales (1), Poales (1); the genus *Punctodera* (4 species) – Poales (4); the genus *Dolichodera* (1 species) – Poales (1); the genus *Paradolichodera* (1 species) – Poales (1); the genus *Betulodera* (1 species) – Fagales (1). Thus, almost half of known cyst nematode species have been described from plants in the order Poales, one of the largest orders of flowering plants.

Krall and Krall (1970, 1973, 1978) and Krall (1990) argued the importance of a comparative ecological approach to study the evolution of cyst nematodes. They proposed a hypothesis of co-evolution of heteroderids using concepts of Grossheim (1945) who suggested that the evolution of flowering plants or angiosperms can be presented in three stages of development. On this basis Krall and Krall (1978) proposed that among heteroderid species parasitising angiosperms those that are characterised by a wider host range (lower degree of trophic specialisation) are less highly evolved relative to those species that have a more restricted host range. They noted that no cyst nematode species has been found on gymnosperms or on the group of angiosperms interpreted by Grossheim as in the first stage of their phylogenetic development; this is the Protanthophyta, including Magnoliales, Ranunculales and others. Grossheim designated the second stage of angiosperm development as Mesanthophyta, including large orders of flowering plants, and these, while including families phylogenetically distant from one another, nevertheless are parasitised by certain cyst nematodes. Grossheim considered the third most specialised and highest stage of development of flowering plants to be the Hypsanthophyta, including grasses, sedges and many others. These are infected by cyst nematode species that have relatively limited host ranges that are often limited to a single plant genus or family (Krall, 1990). Additional aspects of hypotheses of co-evolution of cyst nematodes and their plant hosts are developed and discussed by Stone (1979, 1985) and Sturhan (2000).

Through reconstruction of molecular phylogeny of heteroderids, Subbotin *et al.* (2001a) further evaluated hypotheses of co-evolution of heteroderids with their hosts as proposed by Krall and Krall (1978) and Stone (1979). Molecular data tend to support the view that each of the different heteroderid clades are linked to hosts from a single family or group of closely related plant families. Nevertheless, molecular evidence suggests that some cyst species have secondarily colonised plant hosts unrelated to the host families of the cyst nematode's sister species.

An updated phylogenetic tree including mapping of host plant orders is given in Figure 28. Outgroups for this analysis are the genera *Meloidodera* and *Cryphodera*. These were previously proposed as ancestors for cyst nematodes (Krall & Krall, 1978; Wouts, 1985) and subsequently confirmed as such by molecular phylogeny (Subbotin *et al.*, 2006). The species of these genera parasitise a rather wide spectrum of plants, including gymnosperms and angiosperms. Indeed, the ability of some species of *Meloidodera* and *Cryphodera* to parasitise *Pinus* species and for additional sedentary Heteroderinae such as *Sarisodera sequoiae* to parasitise *Sequoia sempervirens* (D. Don) Endl. is consistent with the argument of Krall and Krall (1978) that suggests an ancient origin of these nematode genera. Molecular phylogenetics further suggest that within cyst nematodes that form a circumfenestral cone there are several distinct lineages: *i*) more basally derived *Betulodera* having hosts from Fagales; *ii*) *Cactodera* associating with Caryophyllales (Amaranthaceae, Caryophyllaceae, Portulacaceae) and, perhaps, relatively recently with Asterales and Poales; *iii*) *Punctodera* and *Paradolichodera* parasitising Poales; *iv*) *Globodera* having hosts from Asterales and Myrtales; and *v*) *Globodera* parasitising Solanales.

Six lineages are evident within *Heterodera* – three of them with monocotyledons and three with dicotyledons. The *Goettingiana* group of *Heterodera* includes species parasitising plants belonging to various taxonomic groups. Based on analysis of ITS-rRNA gene sequences for some species parasitising Brassicales and Apiales, Myrtales and Lamiales, it is plausible to hypothesise that some of these species are associated with changes in host specialisation relatively recently, jumping to phylogenetically unrelated but ecologically similar host plants. Several lineages of *Heterodera* are associated with Poales. In some phy-

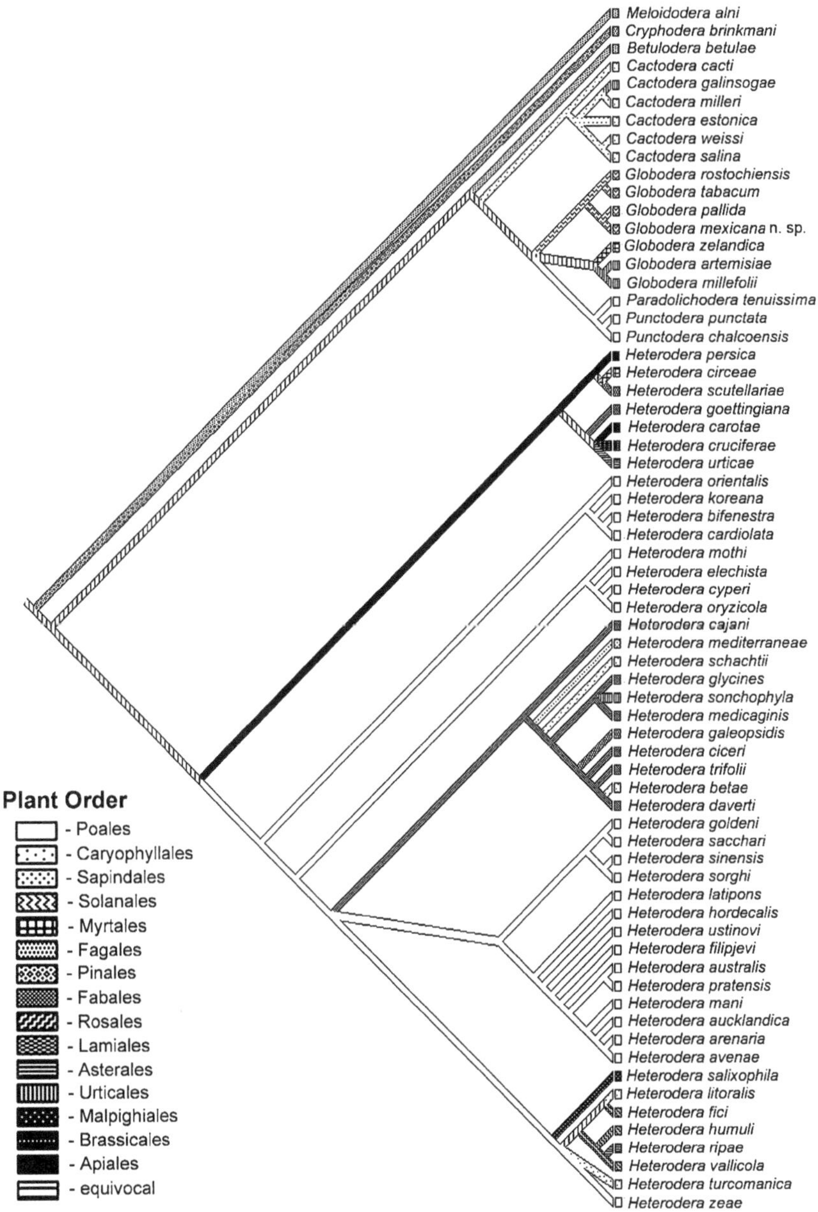

Fig. 28. *Phylogenetic relationships among Heteroderinae as inferred from Bayesian inference analysis with the mapping of host plant orders.*

logenetic trees (Subbotin *et al.*, 2001a; Tanha Maafi, 2003) the *Cyperi* groups form a clade with the *Afenestrata* group, *H. cardiolata* and *H. bifenestra*. All these species, except for *H. bifenestra*, were found associated with plants from the families Cyperales and Poaceae (subfamilies Panicoideae, Chloridoideae, Bambusoideae). Another lineage associated with Poales includes species from the *Avenae* group that parasitise Poaceae from the subfamily Pooideae and from the *Sacchari* group that parasitise plants from subfamilies Panicoideae and Arundinoideae.

The tropical species *H. cajani*, distributed in India, Pakistan and Egypt, and attacking agricultural crops from the Fabaceae, shares a common ancestor with species from the *H. schachtii* group *sensu stricto*. We therefore suggest that the *Schachtii* group primarily co-evolved with plants from the order Fabales and secondarily colonised Lamiales, Caryophyllales, and Asterales. Another lineage includes species of the *Humuli* group associated with hosts from Caryophyllales, Urticales, and Rosales, as well as *H. salixophila* and *H. zeae* parasitising plants from Malpighiales and Poales, respectively. The position of *H. zeae* within *Heterodera* needs to be more carefully tested with several gene datasets. A clearer understanding of co-evolution of Heteroderinae is dependent on more complete resolution of phylogenies of nematodes as well as of the plant species with which they have evolved.

Biogeography

Cyst nematodes are adapted to a variety of climates and are distributed worldwide. Most cyst nematodes occur in colder areas, including temperate regions and higher elevations of the tropics where temperatures remain at or below 20°C. However, several *Heterodera* and most *Cactodera* species occur in the subtropics and tropics tolerating average temperatures well above 20°C. Molecular phylogenetics suggests an early divergence between warm climate tropical and temperate cyst nematode species (Subbotin *et al.*, 2001a).

Considering present knowledge on distribution of cyst nematodes, the following pattern can be constructed. The highest biodiversity of *Heterodera* occurs in Asia, where 53 species have been found, followed by Europe with 34, North and Central America with 22, Africa with 18, Oceania with nine and South America with six species (Table 2). By contrast, *Globodera* is distributed as follows: Europe,

Table 2. *Occurrence of* Heterodera *species in each geographic region.*

Europe (34)	Africa (18)	Asia (53)	North and Central America (22)	South America (6)	Oceania (9)
H. arenaria	*H. africana*	*H. agrostis*	*H. avenae*	*H. avenae*	*H. aucklandica*
H. aucklandica	*H. avenae*	*H. australis*	*H. canadensis*	*H. axonopi*	*H. australis*
H. avenae	*H. cajani*	*H. amygdali*	*H. carotae*	*H. fici*	*H. avenae*
H. betae	*H. cardiolata*	*H. avenae*	*H. cruciferae*	*H. glycines*	*H. cruciferae*
H. bifenestra	*H. cruciferae*	*H. bamboosi*	*H. cyperi*	*H. schachtii*	*H. fici*
H. carotae	*H. daverti*	*H. bergeniae*	*H. fici*	*H. trifolii*	*H. graminis*
H. ciceri	*H. fici*	*H. cajani*	*H. filipjevi*		*H. litoralis*
H. circeae	*H. gambiensis*	*H. cardiolata*	*H. glycines*		*H. schachtii*
H. cruciferae	*H. glycines*	*H. carotae*	*H. goettingiana*		*H. trifolii*
H. cyperi	*H. goettingiana*	*H. ciceri*	*H. graminophila*		
H. daverti	*H. goldeni*	*H. cruciferae*	*H. humuli*		
H. fici	*H. hordecalis*	*H. delvii*	*H. koreana*		
H. filipjevi	*H. humuli*	*H. elachista*	*H. latipons*		
H. galeopsidis	*H. latipons*	*H. fici*	*H. lespedezae*		
H. glycines	*H. oryzae*	*H. filipjevi*	*H. leuceilyma*		
H. goettingiana	*H. sacchari*	*H. galeopsidis*	*H. longicolla*		
H. hordecalis	*H. schachtii*	*H. glycines*	*H. mani*		
H. humuli	*H. trifolii*	*H. glycyrrhizae*	*H. mothi*		
H. latipons		*H. goettingiana*	*H. schachtii*		
H. mani		*H. goldeni*	*H. trifolii*		
H. medicaginis		*H. hordecalis*	*H. ustinovi*		
H. mediterranea		*H. humuli*	*H. zeae*		
H. pratensis		*H. johanseni*			
H. ripae		*H. kirjanovae*			
H. rosii		*H. koreana*			
H. salixophila		*H. medicaginis*			
H. schachtii		*H. mediterranea*			
H. scutellariae		*H. menthae*			
H. sonchophila		*H. mothi*			
H. spinicaudata		*H. orientalis*			
H. trifolii		*H. oryzae*			
H. urticae		*H. oryzicola*			
H. ustinovi		*H. pakistanensis*			
H. zeae		*H. persica*			
		H. phragmitidis			
		H. plantaginis			
		H. raskii			
		H. ripae			
		H. riparia			
		H. sacchari			
		H. saccharophila			
		H. schachtii			
		H. sinensis			
		H. skohensis			
		H. sorghi			
		H. spiraeae			
		H. swarupi			
		H. trifolii			
		H. turangae			
		H. turcomanica			
		H. uzbekistanica			
		H. vallicola			
		H. zeae			

Table 3. *Occurrence of* Globodera, Cactodera, Dolichodera, Paradolichodera, Betulodera *and* Punctodera *species in each geographic region.*

Europe	Africa	Asia	North and Central America	South America	Oceania
			Globodera		
G. artemisiae	*G. rostochiensis*	*G. artemisiae*	*G. bravoae*	*G. leptonepia*	*G. rostochiensis*
G. rostochiensis	*G. pallida*	*G. rostochiensis*	*G. rostochiensis*	*G. rostochiensis*	*G. pallida*
G. pallida	*G. tabacum*	*G. pallida*	*G. pallida*	*G. pallida*	*G. zelandica*
G. millefolii		*G. mali*	*G. mexicana*	*G. tabacum*	
			sp. n.		
G. tabacum		*G. tabacum*	*G. tabacum*		
			Cactodera		
C. cacti	*C. amaranthi*	*C. cacti*	*C. acnidae*	*C. amaranthi*	*C. cacti*
C. estonica	*C. cacti*	*C. estonica*	*C. amaranthi*	*C. cacti*	
C. radicale		*C. thornei*	*C. cacti*		
			C. eremica		
			C. estonica		
			C. evansi		
			C. galinsogae		
			C. milleri		
			C. rosae		
			C. salina		
			C. thornei		
			C. weissi		
			Dolichodera		
			D. fluvialis		
			Paradolichodera		
					P. tenuissima
			Betulodera		
			B. betulae		
			Punctodera		
P. punctata		*P. punctata*	*P. punctata*		
P. stonei			*P. chalcoensis*		
			P. matadorensis		

and North and Central America and Asia each with five species, South America with four, Oceania and Africa each with three. This includes *G. rostochiensis*, *G. pallida* and *G. tabacum*, which are reported from almost all continents. For *Cactodera*, 12 species were reported in North and Central America, three species in Europe and Asia, two species in Africa and South America and one in Oceania (Table 3). Notably, *C. cacti* has been included for many countries because it has been found in glasshouses, although it has been reported under natural conditions only from Central and South America. The monospecific genera *Dolichodera* and *Betulodera* have been reported from North and

Central America, whereas *Paradolichodera* is only known from Oceania. Three species of the genus *Punctodera* have been found in North and Central America, two in Europe and one in Asia (Table 3).

Although no fossils of cyst nematodes are known, biogeography of cyst nematodes, together with their host plants and molecular data, may provide some evidence on the approximate time and areas of origin of cyst nematodes. Krall and Krall (1978) stated that the centre of origin for any organism group is likely to be the extant area with the highest species diversity. On this basis they suggested that the present Mediterranean and Middle Asia regions could be argued to be the centre of origin for the genus *Heterodera* and that the New World is most likely the centre of origin for *Cactodera.* Ferris (1979) believed that *Punctodera* originated from a Laurasian parasite of dicotyledons and *Cactodera* and *Heterodera* were suggested as seeming to have both Laurasian and Gondwanaland lines from Pangaea. She further argued development of the *Avenae*, *Goettingiana* and *Schachtii* groups in Laurasia and the *Humuli* group in Gondwanaland. Within *Globodera*, most species apparently fall within two groups: those that parasitise Solanaceae concentrated in the New World, and those that parasitise Compositae indigenous to Europe. Ferris (1979) proposed Laurasian radiation of *Globodera* and migration south from North America. This point of view has been supported by Picard *et al.* (2008), who applied a molecular clock approach on the ITS phylogeny and estimated *Cactodera* with *Punctodera* + *Globodera* divergence at 173-130 mya and divergence of the South American and Laurasian *Globodera* occurring between 80-60 mya.

Stone (1979) suggested that *Globodera* might have originated in Gondwanaland and remained on the part of the supercontinent that later became South America, a proposal in contrast to a scenario in which the ancestors of European species were suggested to have been carried northwards when fragments of Gondwanaland encountered Laurasia. Based on sources of resistance in tuberous *Solanum* in north-west Argentina, Stone (1979) hypothesised that the origin of the potato cyst species *G. rostochiensis* and *G. pallida* was in that region and suggested subsequent migration northward in conjunction with development of additional species. Grenier *et al.* (2010) suggest that the uplift of the Andes Mountains has triggered a variety of adaptive biotic radiations for Solanaceous plant-parasitic nematodes and allowed the emergence of cold-adapted biota that may have played a crucial role in biodiversification along the cordillera and in *Globodera* speciation. Molecular clocks proposed for *Globodera* suggest that *G. pallida* and *G. rostochiensis* species could have emerged as early as 15-18 million years ago, which represents a time-span compatible with the potato section's divergence (Picard *et al.*, 2008; Grenier *et al.*, 2010).

Native to South America, the potato cyst nematode *Globodera pallida* is one of the principal pests of Andean potato crops and is also an important global pest following its introduction to Europe, Africa, North America, Asia and Oceania. After studying mitochondrial sequences and microsatellite loci from populations of *G. pallida* from Peru and European countries, Plantard *et al.* (2008) showed a clear south to north phylogeographic pattern in Peruvian populations and were able to identify the origin of western European populations with high accuracy. They are all derived from a single restricted area in the extreme south of Peru, located in a triangle of 200 km side length between the north shore of Lake Titicaca and the city of Cusco (Fig. 29). However, the question of the number of introductions remains uncertain because these data could support two alternative hypotheses of origin of western European populations. Either one unique introduction, if possible intra-population haplotype diversity for *cyt*b gene is taken into account, or double or even multiple introductions, if the results of only phylogenetic analysis are considered.

Nematode distribution data combined with newly developing data on the origin of monocotyledon and dicotyledon hosts continues to require more complete surveys and analysis. Increasing support for worldwide biosurveys will be a critical resource toward consideration of coevolution and vicariance biogeography for testing phylogenies of Heteroderinae genera (Baldwin, 1992).

Pathotypes

A number of studies have shown intraspecific variation in relation to nematode behaviour on hosts. The following definitions are now widely accepted. 'Host races' are biological races of a nematode species distinguished by inherited ability or inability to successfully parasitise certain species of plant. 'Pathotypes' are known as biological races

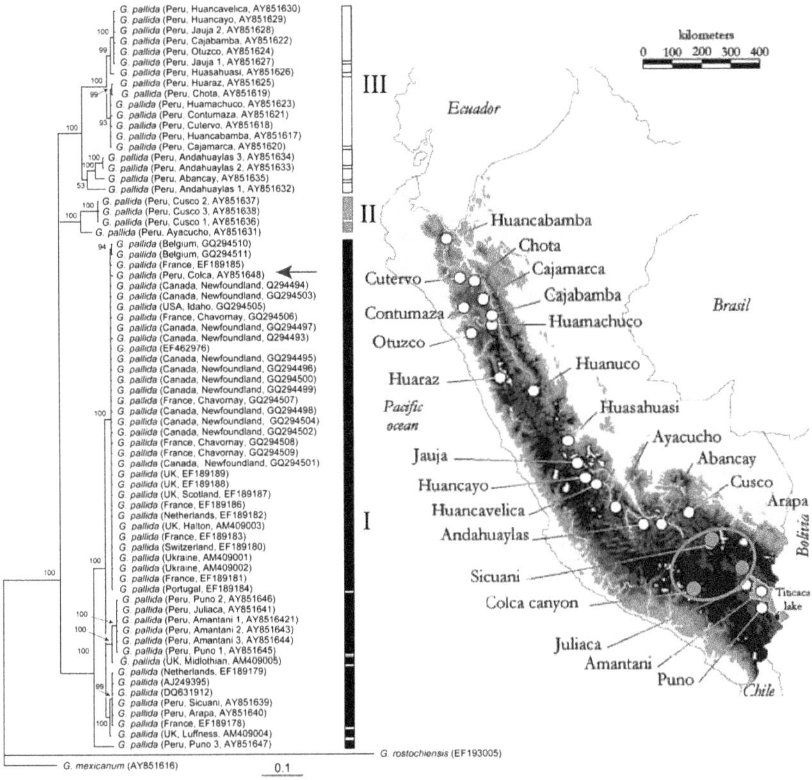

Fig. 29. *Phylogenetic tree of the mitochondrial haplotypes of* Globodera pallida sensu lato *found in Europe, USA, Canada and Peru with a map illustrating the location of the populations sampled in Peru. After Plantard* et al. *(2008) and Madani* et al. *(2010). Area of origin of European populations is indicated by a circle. Arrow indicates the position of sample from Colca, Peru, within the clade containing European populations.*

distinguished from others of the same species by inherited ability or inability to reproduces on specific lines of a host plant species. Pathotypes may exist within a host race. Host races and pathotypes have become an important part of an infraspecific classification of some cyst nematodes. They have been recognised and classified for *G. pallida*, *G. rostochiensis*, *H. avenae*, *H. australis*, *H. cajani*, *H. filipjevi*, *H. hordecalis*, *H. schachtii* and *H. zeae*. As sources of resistance to other cyst nematode species are discovered and utilised, other pathotype complexes are likely to be found (Stone, 1977). Problems of pathotype and race definition and application for nematodes have been discussed in several articles (Kort *et al.*, 1977; Sturhan, 1985; Niblack *et al.*, 2002).

Potato cyst nematodes are classified into pathotypes on the basis of their ability to multiply on different potato clones. However, their pathotype status is still a matter of debate (Phillips & Trudgill, 1983; Mugniéry *et al.*, 1989). Two pathotype schemes for potato cyst nematodes have been suggested. Kort *et al.* (1977) proposed the International Scheme for *Globodera* pathotypes in which, among the European populations, five pathotypes within *G. rostochiensis* and three within *G. pallida* were recognised. A second approach, the Latin American Scheme (Canto & Scurrah, 1977), recognised among the South American populations of *G. rostochiensis* and *G. pallida* four and seven pathotypes, respectively. The distribution of pathotypes of both species according to the Latin American Scheme for South American countries was re-analysed by Franco *et al.* (1998) (Table 4). As noted by Bakker *et al.* (1993), these schemes are not able to reflect all of the genetic diversity introduced into Europe.

The first description of pathotypes of *H. avenae* was given by Andersen (1959) and, in earlier classifications of this species, Arabic numerals (1, 2, 3, *etc.*), letters (A, B, C, *etc.*), and Roman numerals (I, II, III, *etc.*) were used to designate the diversity (Andersen & Andersen, 1982). The pathotype scheme for the cereal cyst nematode *H. avenae* and related species *H. australis* and *H. filipjevi* is based on their capacity to reproduce on barley cultivars with different resistance genes, as given in Table 5. The International Cereal Cyst Nematode Test Assortments consist of 12 barley, six oat and six wheat lines, called differentials. Pathotypes of *H. avenae* belonging to groups 1 and 2 are the most numerous and widely distributed in Europe, North Africa, America and Asia (Andersen & Andersen, 1982; Al Hazmi *et al.*, 2001; Mokabli *et al.*, 2002; Smiley *et al.*, 2007). Only one pathotype of the Australian cereal cyst nematode seems likely for South Australia and Victoria (O'Brien & Fisher, 1979). Two pathotypes are distinguished for *H. filipjevi*. Several pathotypes, which are different from known ones and are not listed in Table 5, were found in China (Yuan *et al.*, 2010).

A race test for *H. glycines* was first developed based on variable comparative development of females on four differential soybean lines. This test was proposed by a group of nematologists and soybean breeders (Golden *et al.*, 1970; Riggs & Schmitt, 1988) to distinguish 16 races of *H. glycines* (Table 6). However, the scheme was shown to describe

Table 4. *Pathotype groups of the potato cyst nematodes,* Globodera rostochiensis *and* G. pallida.

Species and accession	Species												
	Globodera species:	*G. rostochiensis*					*G. pallida*						
	Virulence group:[a]	Ro1		Ro3		Ro5	Pa1		–	–	–	Pa2/3	
	European pathotypes:[b]	Ro1	Ro4	Ro2	Ro3	Ro5	Pa1		–	–	Pa2	Pa3	–
	South. American pathotypes:[c]	R1A	R1B	R2A	R3A	–	P1A	P1B	P2A	P3A	P4A	P5A	P6A
	Ploidy, Resistance gene												
Solanum tuberosum ssp. *tuberosum*	4x, (minor)	+/–	+	+	+	+	+	+	+	+	+	+	+
S. tuberosum spp. *andigena* CPC 1673	4x, H1 on chromosome 5	–	–	+	+	+	+	”	“	“	+	+	“
S. kurtzianum KTT 60.21.19	2x, K1 K2 A & B	–	(+)	–	(+)	(+)	+	+	+	+	+	+	+
S. vernei GLKS 58.1642.4	2x, Quantitative	–	+	–	–	+	+	+	+	–	+	+	+
S. vernei Vt 62.33.3	2x, Quantitative		–	–	–	+	–	+	–	–	–	+	+
Ex. *S. multidissectum* hybrid P55/7	2x, 1 + polygenes H2	+	+	+	+	+	–	–/+	–	+	+	+	+
S. t. ssp. *andigena*	H3 + polygenes	+	“	“	“	“	(–)	“	“	“	(–)	(–)	“
CIP 280090.10	Quantitative	“	“	“	“	“	“	“	“	“	–	–	+
S. vernei hybrid 69.1377/94	2x, Polygenes	–	–	–	–	–	–	“	“	“	–	–	“
S. vernei hybrid 63.346/19	2x, Polygenes	–	–	–	–	–	+	“	“	“	+	+	“
S. spegazzinii	2x, Fa = H1	–	–	+	+	+	“	“	“	“	“	“	“
S. spegazzinni	2x, Fb + 2 minor	+	–	+	–	–	“	“	“	“	“	“	“
	Glo1 on chromosome 7	(–)	+	+	+	+	“	“	“	“	“	“	“

*Adapted from Cook and Noel (2002), [a] – Trudgill (1985); [b] – Kort *et al.* (1977); [c] – Canto-Saenz and de Scurrah (1977); + = compatible interaction : nematode multiplication, potato susceptible; – = incompatible interaction : nematode no multiplication, potato resistant; () = partial or uncertain interaction; “ = no information.

Table 5. *Pathotype groups of the cereal cyst nematodes,* Heterodera avenae, H. australis[*] *and* H. filipjevi.

Species	*H. avenae*								*H. australis*	*H. filipjevi*	
Pathotype group	Ha1 group							Ha2 group	Ha3 group		
Pathotype Different species and cultivar	Ha11	Ha21	Ha31	Ha41	Ha51	Ha61	Ha71	Ha12	Ha13	Ha23	Ha33
Barley											
Varde	+	“	“	+	“	+	+	+	+	+	+
Emir	+	+	“	+	“	–	+	+	+	+	+
Ortolan/Drost	–	–	–	–	–	–	–	+	+	+	+
Morocco	–	–	–	–	–	–	–	–	–	–	–
Siri	–	–	–	+	+	+	–	–	+	+	+
KVL 191	–	–	–	“	+	+	+	–	“	“	“
Bajo Aragon	–	“	“	–	“	–	–	–	+	+	+
Herta	+	+	–	“	–	“	–	+	+	“	“
Martin 403	–	“	“	–	“	–	–	–	–	+	+
Dalmatische	(–)	“	“	+	“	–	(+)	+	+	(–)	+
La Estanzuela	“	“	“	“	“	“	+	“	“	(–)	“
Hartian 43	–	“	“	“	“	“	–	–	“	–	+
Oat											
Nidar	+	“	“	(+)	“	+	–	+	+	+	+
Sol II	+	–	–	–	–	+	–	+	+	+	+
Pura Hybrid BS1	–	–	“	–	–	–	–	–	+	–	+
Avena sterilis 1376	–	–	“	–	–	–	–	–	–	–	–
Silva	(–)	“	“	–	“	(–)	–	(–)	(–)	(–)	+
IGV.H. 72–646	–	“	“	–	“	–	–	–	+	+	+
Wheat											
Capa	+	+	“	+	“	+	+	+	+	+	+
AUS10894	–	“	“	–	“	–	+	–	(–)	+	+
Loros	–	–	“	–	“	(–)	[illegible]	–	(–)	+	+
Psathias	“	“	“	+	“	“	“	+	+	+	–
Iskamish K-2-light	+	“	“	–	“	(–)	“	+	+	+	+

[*] Adapted from Cook and Rivoal (1998) and Turner and Rowe (2006); + = susceptible; – = resistant (<5% new females compared to numbers on susceptible control); () = intermediate; “ = no observation.

Table 6. *Races of the soybean cyst nematode,* Heterodera glycines, *according to the race determination scheme of Riggs and Schmitt (1988).*

Soybean differential	Race															
	1	2	3	4	5	6	7	8	9	10	11	12	13	14	15	16
Pickett	–	+	–	+	+	+	–	–	+	+	–	–	–	+	+	–
Peking	–	+	–	+	–	–	–	–	+	–	+	+	+	+	–	+
PI 88788	+	+	–	+	+	–	+	–	–	–	+	–	–	–	+	+
PI 90763	–	–	–	+	–	–	+	+	–	+	–	+	–	+	+	+

– = resistant (female index < 10% cultivar Lee); + = susceptible (female index > 10% that of susceptible control cultivar Lee).

inadequately the extensive genetic diversity of *H. glycines* populations existing in areas of soybean production (Epps & Duclos, 1970), and it was noted that the classification could be expanded to include many more groups (or types) by increasing the number of differentials. Thus, Riggs *et al.* (1981) distinguished up to 25 'races' on 12 resistant soybean lines, demonstrating that the set of four differentials used by Golden *et al.* (1970) did not account for all the variability in host compatibility of *H. glycines* (Niblack *et al.*, 2002). Niblack *et al.* (2002) proposed an HG Type Test to describe population variation better and to expand the flexibility of the race classification system, with evaluation also based on development of adult females. The HG Type system first uses three of the four resistant soybean genotypes currently recommended as indicator hosts (Peking, PI 548402, PI 88788 and PI 90763). This latter scheme is open-ended (additional soybean lines can be added) and easily adaptable to different areas (Table 7). To identify the HG Type of an *H. glycines* population, a bioassay is conducted according to specific recommendations, and the population is named according to the number associated with soybean indicator lines, or sources of resistance, on which it is virulent (Niblack *et al.*, 2006). Virulence is measured as a Female Index (FI) calculated as following: $(N_X/N_S) \times 100$, where N_X = average number of females on cultivar "x" and N_S = average number of females on standard susceptible cultivar Lee 74. HG Types should be reported not only by number but also with the FI values for each number (Table 8) (Niblack *et al.*, 2006). The "–" following the HG Type designation indicates that the test was incomplete because less than the full complement of seven HG Type indicator lines were included in the test.

Table 7. *Indicator lines for HG Type classification of genetically diverse populations of* Heterodera glycines. *After Niblack* et al. *(2006).*

Number	Indicator line
1	PI 548402 (Peking)
2	PI 88788
3	PI 90763
4	PI 437654
5	PI 209332
6	PI 89772
7	PI 548316 (Cloud)

Table 8. *HG Types for several populations of* Heterodera glycines *collected in southern Illinois. After Niblack* et al. *(2006).*

Sample number	Mean no. females on Lee	Female indices			
		PI 548402	PI 88788	PI 437654	HG Type
14	163	7	332	0	2–
52	403	3	30	0	2–
57	124	7	8	0	0
59	140	19	13	0	1.2–

Nematology Monographs & Perspectives, 2010, Vol. 8A, 71-80

Chapter 3

Methods for applying electron microscopy

Transmission electron microscopy (TEM)

TEM has proved to be a valuable tool for understanding detailed morphology of heteroderids, including males, juveniles, females and cysts (Shepherd *et al.*, 1972; Baldwin & Hirschmann, 1973, 1975a, b, 1976; Baldwin *et al.*, 1977; Endo, 1980, 1983, 1984, 1985, 1998; Baldwin, 1982, 1983, 1985; Cliff & Baldwin, 1985; Carta & Baldwin, 1990; Cordero & Baldwin, 1990, 1991; Cordero *et al.*, 1991; Cares & Baldwin, 1994a, b, 1995) as well as in elucidating feeding sites and other aspects of host parasite relationships (Mundo-Ocampo & Baldwin, 1983a, b, c, 1984, 1992; Endo, 1991). The level of TEM detail generally has not had practical diagnostic applications, although in some cases first observing details with TEM then provides a basis to interpret accurately corresponding features at the level of the light microscope. For example, TEM provided a clearer understanding of cyst males relative to those of root-knot males with respect to the latter having unique slit-like amphidial openings, the loss of a dorsal gland, unusual tripartite gland openings and a caecum, all characters subsequently distinguishable with the light microscope once TEM provided initial insight. TEM has also been invaluable in interpreting and testing homology of certain diagnostic characters otherwise subject to misinterpretation (*e.g.*, bullae and denticles), in addressing questions of structure as they relate to function and biology and in interpreting character evolution and phylogeny (Baldwin & Schouest, 1990; Cordero & Baldwin, 1990, 1991; Cordero *et al.*, 1991).

TEM methods vary in detail with the goals and subjects (*e.g.*, J2, males, females, cysts) of investigation (Fig. 30). Can the goals be achieved with a few sections by sampling a few strategic regions as in the case for determining body wall cuticle layers in females? Conversely, do the goals require resolution and serial sections as needed for 3D reconstruction of the stylet or sensory organs? Regardless of variables

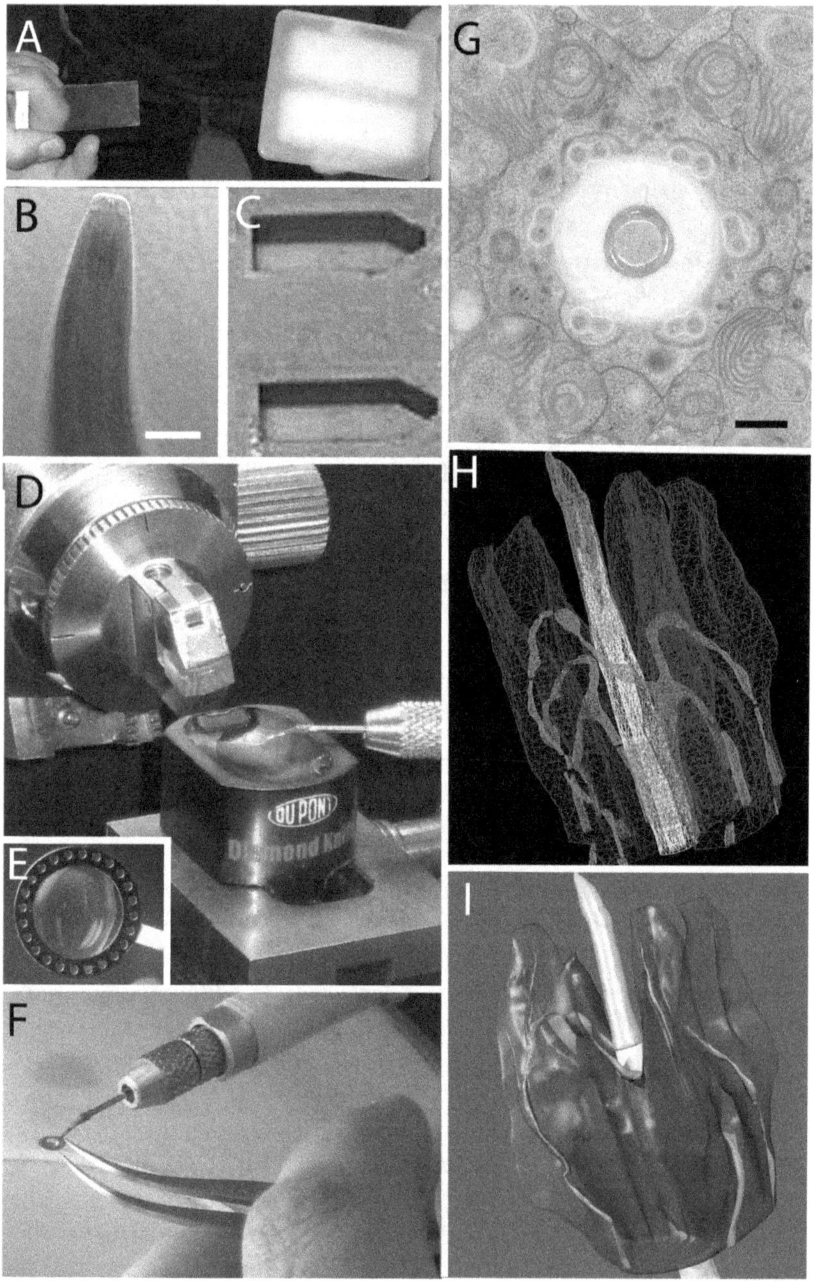
A
B
C
D
E
F
G
H
I
DU PONT
Diamond Knife

in relation to specific goals, the basic steps are likely to include: *i*) killing and fixation; *ii*) dehydration; *iii*) infiltration and embedding; *iv*) polymerisation; *v*) block trimming; *vi*) thin sectioning; *vii*) mounting; *viii*) heavy metal differential staining; *ix*) coating; *x*) recording and photography; and *xi*) reconstruction/spatial interpretation. Herein each step is summarised as it applies to the more demanding steps required for 3D reconstruction with notes added suggesting how the methods can be simplified where the requirements are less rigorous. Although to date the emerging methods of 3D reconstruction have been primarily applied to taxa other than heteroderids (Bumbarger *et al.*, 2006, 2007, 2009; Ragsdale *et al.*, 2008, 2009), they offer great potential for understanding cyst nematode comparative morphology, homologies and developmental biology.

KILLING AND FIXATION

A long-standing challenge in TEM of all stages of cyst nematodes is in achieving adequate fixation. There may be multiple reasons, but often the problem is attributed to relative impermeability of the cuticle; that is, the slow penetration of fixatives allows catabolic processes and a consequent deterioration of fine structures. Often, initially cutting the fresh specimen in buffer or fixative speeds fixation, but the action can result in distortion, exacerbated by releasing the pressure of the hydrostatic skeleton. Both cut and whole specimens can be difficult to handle safely in the presence of toxic fumes from fixatives, although generally this can be addressed

Fig. 30. *Steps in preparing specimens for transmission electron microscopy. A: Moulds (right) for preparing specimens in an epoxy 'slide' for light microscope viewing and orientation in blocks suitable for sectioning; B: Specimen fixed in osmium tetroxide as viewed in epoxy slide. Scale bar = 10 μm; C: Moulds used for re-embedding specimen from slide in to a form suitable for trimming and sectioning on ultramicrotome; D: Specimen and diamond knife mounted on ultramicrotome; ribbons of sections being picked up by loop; E: Enlarged view of loop used for transferring ribbons of sections; F: Ribbons of sections being transferred from loop to film-covered slot grid; G: TEM transverse section through vestibule region of a tylenchid (*Aphelenchus avenae*; after Ragsdale* et al.*, 2009) photographed with a CCD camera and assembled as a montage. Scale bar = 10 μm; H: Wireframe rendition of stylet region as reconstructed from serial TEM sections (Ragsdale* et al.*, 2008); I: 3D depiction as developed from serial TEM sections (compliments of Ragsdale).*

by a system in which the air around the dissecting microscope can be directly evacuated into a fume hood. Handling can also be aided by processing specimens in porous containers (Bumbarger *et al.*, 2006) and/or embedding specimens or specimen pieces in blocks of 3% water agar in which fixation, dehydration and infiltration can proceed (Dolinski & Baldwin, 2003). A range of fixatives are suitable for TEM of heteroderids and those most suitable may vary with the goals of the project; for example, if the emphasis is on resolving cell membranes, fixation with only osmium tetroxide may be used, whereas in studies where resolution of cytoplasm is a priority, it may prove valuable to first fix in aldehyde, such as glutaraldehyde, or aldehyde mixtures, such as Karnovsky's (1965) solution. Fixative protocols further vary by the buffers in which they may be delivered as well as by agents that may be added to regulate osmolarity (Baldwin & Eddleman, 1992; Zhang & Baldwin, 1999; Dolinski & Baldwin, 2003). In recent years, technologies have provided approaches that largely overcome problems of catabolism and osmolarity, issues that historically have been limiting for the use of TEM for some stages of heteroderids. This involves placing specimens in a custom designed stainless steel chamber (Bumbarger *et al.*, 2006) and rapid freezing in liquid nitrogen under high pressure. This is followed by chemical fixation with osmium tetroxide by freeze substitution in an autofreeze substitution apparatus (Bumbarger *et al.*, 2006) – thus catabolism is stopped by freezing and transition, without thawing, to chemical fixation. Chemical fixation is followed by rinsing, typically with several exchanges of buffer solution, to remove any residual fixative.

Dehydration

Removing all water from the sample is essential to the following steps of infiltration and embedding. In addition to removing water it is important to complete the process in a solvent suitable for infiltration with epoxy, and this varies with the epoxy being used. Typically, dehydration is through a series from aqueous solutions of buffer to 100% ethanol or acetone. Several changes in 100% strength of the final solvent are needed to be certain that dehydration is complete. One must also be certain that the final solvents are indeed 100%, recognising the hydroscopic properties of many such materials. Although complete dehydration is essential there is some concern that specimens left more than a day in high concentrations of solvents might result, even in fixed specimens, in some fine structure being leached.

INFILTRATION AND EMBEDDING

Infiltration with epoxy resin is conducted by introducing dehydrated specimens into gradually increasing concentrations of epoxy solution in solvent. The principle is ultimately to replace all of the solvent with epoxy and this is accomplished not only by exchanges of solution with increasingly higher concentrations of the solvent but in conjunction with gradual evaporation of solvent. Examples of commonly used epoxy resins for TEM include Spurr's, Durcupan and Epon 812. The resin used may have an impact on final contrast of images and the formulation may affect the final properties of the epoxy (hardness, plasticity) that ideally should match the hardness of the specimen. This is especially a consideration where hard cuticular structures are being investigated including, for example, stylets or spicules. Generally it is advisable to evaluate each batch of mixed epoxy by polymerising an empty test block prior to using it to embed specimens.

POLYMERISATION

Specimens are polymerised in epoxy within flat embedding plates. Moulds for these plates can be those that provide a pyramidal end for orientating the specimen ('stepped' microtome embedding mould); for more precise orientation the slide-shaped mould (Giammara & Hanker, 1986) is recommended (Fig. 30A-C). Specimens mounted in these slides can then be recorded under light microscopy *via* through-focus video (De Ley & Bert, 2002) to provide a record of orientation that is an invaluable aid to subsequent reconstruction and interpretation (Fig. 30B). The specimen is then cut from the flat mould and re-embedded and again polymerised with additional epoxy in a stepped embedding mould. Methods for polymerising epoxies vary, but most commonly this is accomplished by heating in an oven at about 70°C.

BLOCK TRIMMING, MOUNTING AND SECTIONING

In order to obtain good results, the faces of the epoxy blocks containing the nematode specimens must be trimmed to a trapezoid shape with sloping sides. The quality of the sections is usually enhanced by working from a face only slightly larger than the specimen. Although a number of approaches are available, it is advisable to trim the block under a dissecting microscope by using a single-edged razor blade and, with a shaving technique, removing thin layers of epoxy at a time. Following razor trimming often the block face can be further trimmed and smoothed with an old diamond knife after mounting the block on the ultramicrotome. For mounting, the position of the specimen as recorded in the through-focus video (see above) can help inform accurate orientation. Subsequent sectioning generally requires a sharp, clean, diamond knife with a flawless edge. During sectioning the thickness of each section is estimated based on interference colour with sections *ca* 75 nm thick generally ideal for resolving most structures. A record of the thickness of each section and any missing sections is kept since these will be a valuable reference for subsequent reconstruction. Ribbons of sections floated on to filtered water are mounted on Synaptek® wide slot grids that were previously coated with a thin plastic film such as Formvar® or Pioloform® (Fig. 30D-F). The ribbons are picked up in a loop where the surface tension of the water forms a film across the loop in which the ribbon floats. Under the dissecting microscope the ribbon is then orientated on the slot grid and allowed to dry. In contrast to mesh grids, a slot grid avoids having crossbars that potentially obstruct valuable sections; furthermore, a wide slot facilitates orientation of sections.

Heavy metal differential staining

TEM images are created by the contrast of differential staining using heavy metals through which electrons are restrained from passing. Typically, these include solutions of uranyl acetate and lead citrate. In some cases uranyl acetate is introduced prior to sectioning such as fixation in a 1% solution at the time of osmium tetroxide fixation (Bumbarger *et al.*, 2007, 2009). However, this does not preclude secondary staining of sections on grids by floating upon or submerging grids in uranyl acetate in a saturated solution of ethanol or methanol, followed by careful rinsing in the solvent. A number of approaches are used to stain grids with lead citrate with perhaps that of Reynolds (1963) being most common. Furthermore, TEM supply companies provide a wide range of holders designed for staining grids. One of the greatest difficulties in staining, particularly with lead solutions, is to avoid unwanted residues that appear as 'dirt' on the TEM images. These are often products of reaction of the lead solution with carbon dioxide, and so various approaches such as temperature, pH, isolation from air (as in submerging rather than floating grids) and careful rinsing are considerations to minimise this artefact.

Coating

To stabilise sections under an electron beam it is often necessary to coat the grid with a thin film of carbon; such films of carbon are highly transparent to electrons and of very low contrast. A vacuum evaporator produces such a film on grids within a chamber by passing a current across two adjacent carbon rods under a high vacuum. This may be done to the grids coated with a plastic film (see above) either before or after the sections are mounted.

Recording and photography

The process of recording TEM images has been revolutionised by digital photography. For projects involving reconstruction, recording images on the electron microscope is added by photographing each serial section at the same magnification. To avoid digital camera limitations based on CCD size, it is best to create images based on a montage, for example, of the cross section of the entire specimen (Fig. 30G). In modern TEM digital camera systems this is usually achieved by software with automated montage capabilities internal to the digital camera.

Reconstruction for interpretation

Where serial sections are being used for 3D reconstruction, a crucial first step is to verify that all the images are in sequential order. This is achieved by reference to notes taken during sectioning, by the transition in the size of the sections and by reference to the through-focus video record. At this stage, sequential numbers are assigned to every image, allowing sufficient digits to accommodate the entire series (*e.g.*, for 220 images start with 001). The next step toward reconstruction is to obtain a converted and uniformly resized sequential series. For example, in the case of the Gatan TEM camera, images are provided in 32 bit DM3 (Gatan Digital Micrograph) format. The series of images must then be exported to TIFF by using Adobe Photoshop® or Image J® (free at rsbweb.nih.gov/ij/) or other software programs. It is possible, but not essential, to write software scripts that automatically convert an entire series of images. Once the images are converted and uninformative borders are removed, the image with the largest canvas size is located and borders are added to all other sections so that all the images of the series have identical dimensions; this is also easily facilitated by scripting. The next step toward 3D reconstruction is met through the IMOD Linux-based software package (Boulder Lab for 3D Electron Microscopy of Cells). The first goal of employing this software is to convert the series of TIFF images to a stack expressed as a single file in MRC format. In this format it is possible to move rapidly through the sections sequentially in a manner that is analogous to moving through frames of a movie. The IMOD software provides a system that is then used to align each image to the next, and then ultimately across all of the images in the stack, so that comparable parts of each section are orientated with sections above and below. With images aligned, the outline of each cell of interest on each section is outlined on an overlay in a separate file and the outline for each cell is coded with a distinct colour. By moving back and forth between sections it is possible to verify continuity across the image stack. With the outlines complete the software allows expression of each cell in three dimensions as a series of horizontal frames that can then be processed to a wireframe (Fig. 30H). The wireframes of each model can then be exported as 'object' files (vrml format) to Blender® (open source software, available at www.blender.org), which is an animation software package that allows refinement of the model as a 3D image or movie by designating light direction, shadow, texture, transparency and, in the case of movies, movement of particular components such as allowing for visual dissection.

TEM holds great promise, beyond the descriptive studies of the past, for testing hypotheses of heteroderid biology, development functional morphology, character evolution and phylogeny. It also holds potential for a clearer understanding of mutants and transgenics relevant to understanding gene function. The combination of TEM methodology with computer technology, as well as the interface with other advances in microscopy including confocal approaches, provides a new arsenal of tools toward understanding cyst nematodes.

Scanning electron microscopy (SEM)

SEM has proved to be a valuable tool for understanding heteroderid morphological diversity and its significance in diagnostics and phylogenetics, specifically including labial patterns and lateral lines of J2 and males, J2 tail patterns including phasmid openings, body wall and cone patterns of females and cysts, egg patterns, and spicules (Clark *et al.*, 1973; Mulvey, 1973, 1974; Green, 1975; Stone, 1975; Mulvey & Golden, 1983; Baldwin & Bell, 1985; Othman & Baldwin, 1986; Othman *et al.*, 1986, 1988).

Procedures for preparation for SEM include fixation, dehydration, coating and viewing. Although there are a number of specific variations on details of preparation, the methods given here are those currently used for heteroderids at the Department of Nematology, University of California, Riverside, California, USA (Fig. 31).

After recovery, selection and cleaning of specimens (J2, male, female or cyst), the nematodes are concentrated in 12.5 ml distilled water within a 25 ml glass vial. The vial of specimens is heated by placing it in a

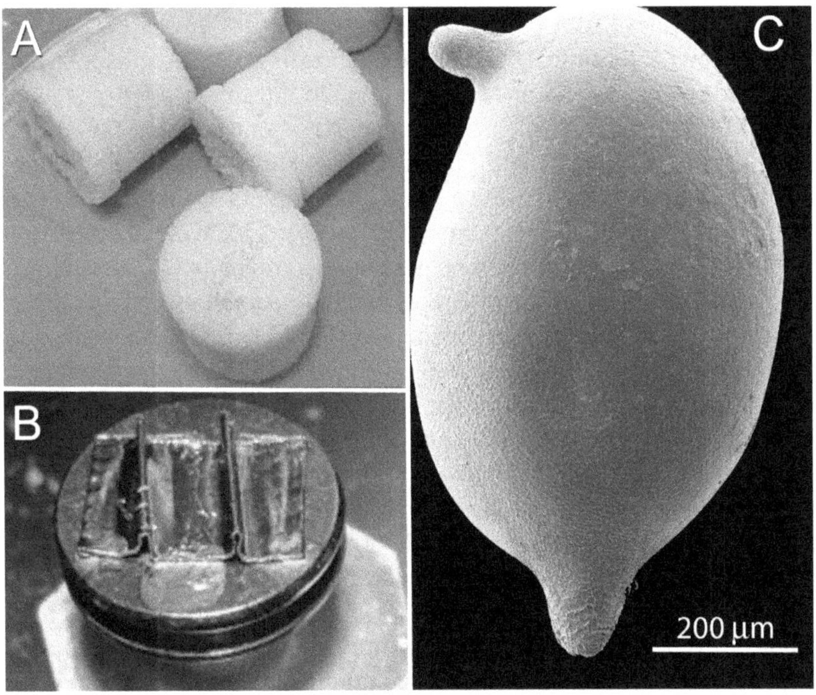

Fig. 31. *Steps in preparing specimens for scanning electron microscopy. A: Porous container used for fixing and drying specimens; B: SEM stubs upon which nematode specimens are mounted; C: Lateral view of* Heterodera schachtii *young cyst as viewed with SEM.*

hot water bath (55-60°C) and then the vial is immediately filled with 10% formalin solution in 0.1 M phosphate buffer (pH 7.0). After fixation the specimens are maintained at room temperature until further process. Alternative fixatives can include buffered FA (formalin and acetic acid), FAA (formalin, ethanol and acetic acid) or glutaraldehyde. Nematodes in the resulting 5.0% formalin dilution are left for a minimum of 24 h to several days at room temperature.

After primary fixation, nematodes are rinsed with several changes of 0.1 M phosphate buffer, then specimens are transferred by hand picking to a Beam® capsule or other convenient container suitable to retain the specimens under vacuum and high pressure.

Since fragile structures such as the labial region are particularly subject to distortion it is often helpful to increase rigidity by post-fixation for 2-12 h in 4.0% osmium tetroxide solution buffered with 0.1 M phosphate buffer. This is followed by rinsing with additional buffer within a 15 min period, followed by rinsing in distilled water. Dehydration is through a gradual transitional series of ethanol (not denatured) aqueous dilutions ranging from 20% through 100% absolute ethanol, specimens are left at each step for about 20 min; the entire process usually takes about 2 h. To avoid collapse with air drying specimens, drying is best done in a critical point dryer that during the process regulates temperature and pressure to avoid damage by surface tension of liquids. Dried specimens are mounted in suitable arrangements by positioning them on a face of a double sticking copper tape (available at http://www.tedpella.com/), which in turn is attached to aluminium stubs. If specimens are embedded in glycerin, they could be mounted on glass chips or overlapping a hair. Stubs with nematodes are coated with a 25 nm layer of gold palladium using a sputter coater.

It is best to examine and photograph specimens while they are freshly coated and, typically, nematodes are best examined at about 10 kV. Since the dried specimens are likely to be hydroscopic, if they must be stored it is advisable to do so in a sealed desiccation chamber.

SEM preparation of cyst nematodes often results in defectives and artefacts even where material is carefully processed – both care and persistence are required and these limitations often require preparation of a quantity of material to ensure success. Common defects include specimen collapse or distortion, precipitations on the nematode surface resulting from contaminated fixatives or buffers, inadequate rinsing, or deposits that occur due to problems during the sputter-coating process.

Nematology Monographs & Perspectives, 2010, Vol. 8A, 81-99

Chapter 4

Methods of biochemical and molecular diagnostics

Protein-based diagnostics

Biochemical approaches such as gel electrophoresis for separating protein and enzyme profiles have shown great potential for assisting in identification of cyst nematodes. Isoelectric focusing (IEF), a method for separating molecules which differ in their charge characteristics in a pH gradient, enables generating stable banding patterns and achieving resolution of proteins into sharp bands. Differences in banding patterns among species or populations and the presence of taxon-specific band-markers on the gel are used for diagnostics. The IEF protein profiles have been generated for many cyst nematode species (Table 9). The introduction of miniaturised electrophoretic systems, such as PhastSystem (Pharmacia Corp.), has made it possible to study small quantities of soluble proteins from a single cyst in a fully automatic process (Fig. 32A). This technique is currently used in many nematology laboratories for routine diagnostics of *G. pallida* and *G. rostochiensis*. The essential differences between IEF profiles obtained for these two

Table 9. *References for diagnostics of cyst nematodes using IEF of proteins.*

Species	References
Globodera artemisiae	Sumiya *et al.* (2002)
G. millefolii	Doneghan & McKenzie (1987)
G. pallida	Fleming & Marks (1982, 1983); Ohms & Heinicke (1983); Fox & Atkinson (1984); Marks & Fleming (1985); Fleming & Dolan (1986); Doneghan & McKenzie (1987); Santos *et al.* (1989, 1995); Jatala & Gavilano (1991); Karssen *et al.* (1995); Ibrahim *et al.* (2001); Sirca *et al.* (2003)

Table 9. *(Continued).*

Species	References
G. rostochiensis	Fleming & Marks (1982, 1983); Ohms & Heinicke (1983); Fox & Atkinson (1984); Marks & Fleming (1985); Fleming & Dolan (1986); Doneghan & McKenzie (1987); Santos *et al.* (1989, 1995); Jatala & Gavilano (1991); Karssen *et al.* (1995); Subbotin *et al.* (1999b); Ibrahim *et al.* (2001); Sumiya *et al.* (2002); Sirca *et al.* (2003); Tzortzakakis *et al.* (2004)
G. tabacum	Sumiya *et al.* (2002)
Heterodera arenaria	Holgado *et al.* (2004)
H. aucklandica	Sturhan & Rumpenhorst (1996); Gäbler *et al.* (2000)
H. australis	Subbotin *et al.* (2002)
H. avenae	Rumpenhorst (1985); Fleming & Dolan (1986); Shahina & Maqbool (1995); Sturhan & Rumpenhorst (1996); Subbotin *et al.* (1996, 2006); Gäbler *et al.* (2000); Holgado *et al.* (2004)
H. bifenestra	Rumpenhorst (1985)
H. cajani	Shahina & Maqbool (1995)
H. cardiolata	Shahina & Maqbool (1995)
H. carotae	Fleming & Dolan (1986)
H. cruciferae	Shahina & Maqbool (1995)
H. fici	Trindade *et al.* (1992)
H. filipjevi	Sturhan & Rumpenhorst (1996); Subbotin *et al.* (1996, 2002); Gäbler *et al.* (2000); Holgado *et al.* (2004)
H. glycines	Lawson *et al.* (1984); Radice *et al.* (1988)
H. goettingiana	Trindade *et al.* (1992); Shahina & Maqbool (1995)
H. hordecalis	Rumpenhorst (1985)
H. humuli	Fleming & Dolan (1986)
H. latipons	Rumpenhorst (1985)
H. mani	Rumpenhorst (1985); Fleming & Dolan (1986); Sturhan & Rumpenhorst (1996); Gäbler *et al.* (2000); Holgado *et al.* (2004)
H. pratensis	Gäbler *et al.* (2000)
H. sacchari	Shahina & Maqbool (1995)
H. schachtii	Fleming & Dolan (1986); Trindade *et al.* (1992); Shahina & Maqbool (1995)
H. sorghi	Shahina & Maqbool (1995)
H. ustinovi	Gäbler *et al.* (2000)
Punctodera punctata	Rumpenhorst (1985); Fleming & Dolan (1986)

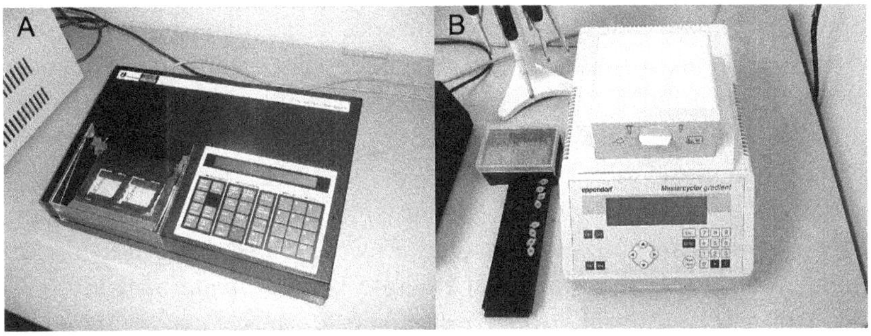

Fig. 32. *A: Pharmacia PhastSystem; B: Thermal cycler (general views).*

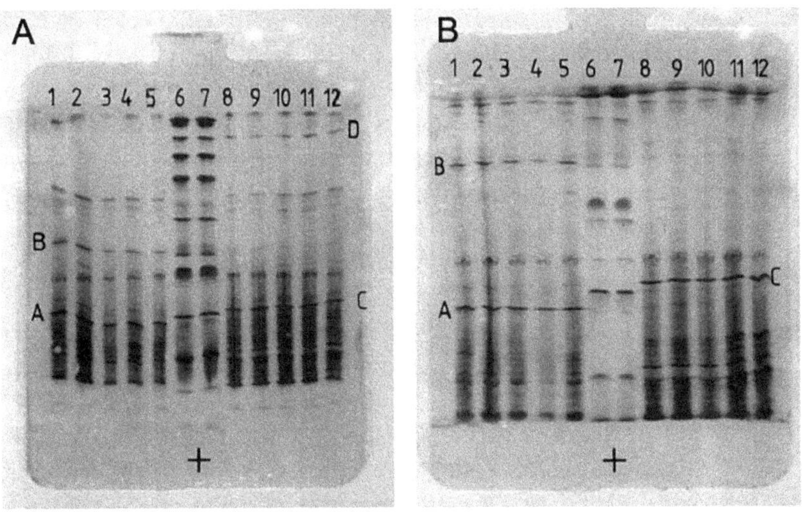

Fig. 33. *IEF patterns of* Globodera pallida *(lanes 1, 2, 3, 4, 5),* G. rostochiensis *(lanes 8, 9, 10, 11, 12), individual specimens and broad* pI *calibration kit (lanes 6 and 7). A: IEF on PhastGel 3-9; B: IEF on PhastGel 5-8, followed by silver staining. Specific bands are marked by capital letters. After Karssen* et al. *(1995).*

Globodera species are in several bands: *G. rostochiensis* shows a stable major band at p*I* 5.7 and a weaker band at p*I* 8.7; *G. pallida* shows the stable major band at p*I* 5.9 and a weaker band at p*I* 6.9 (Fig. 33). A protocol for diagnostics of potato cyst nematode species is given in Appendix III.

The application of enzyme staining technique for characterisation of a single protein or small subset of proteins on gel provides another approach to diagnostics. Isozymes analysis is also considered as one of the most effective biochemical techniques for comparing nematode species. Comparative analysis of esterase profiles was successfully applied to differentiate cyst nematode species by several research teams (Ganguly *et al.*, 1990; Nobbs *et al.*, 1992; Ibrahim & Rowe, 1995; Meher *et al.*, 1998). References to several of these studies are given in Table 10. Ibrahim and Rowe (1995) were the first to conduct comprehensive studies of esterase banding patterns using IEF and native gel electrophoresis to compare 19 species of *Heterodera* from diverse geographic origins. Twenty-six major bands of esterase activity were detected; some bands were common to two or more species, whereas others were species-specific. All 19 species of cyst nematode were separable based on esterase profiles. The gel with nonspecific esterase profiles for these species obtained by IEF is given in Figure 34. However, some studies revealed intraspecific variation in esterase patterns and showed that several esterase phenotypes can exist within the same species. This fact can potentially create some difficulties for diagnostic procedures, where all variants of given species have not yet been previously characterised and linked to that species. Presently, eight esterase electrophoretic phenotypes have been resolved for *H. glycines* (Noel & Liu, 1998), seven for *H. avenae* (Andrés *et al.*, 2001; Venkatesan *et al.*, 2004), three for *H. cajani* and *H. sorghi* (Venkatesan *et al.*, 2004) and two for *H. zeae* (Venkatesan *et al.*, 2004) and *H. filipjevi* (Andrés *et al.*, 2001). Several other enzymes, such as malate dehydrogenase, phosphoglucoisomerase, phosphoglucomutase and superoxide dismutase, showed some polymorphism among species and populations. These enzymes have also been proposed as reliable diagnostic markers for some cyst nematode species (Molinari *et al.*, 1996; Andrés *et al.*, 2001). Detailed protocols for preparation of nematodes for enzyme analysis, which include procedures for enzyme extraction, isoelectric focusing and polyacrylamide gel electrophoresis, have been described in several publications (Ibrahim & Perry, 1993; Ibrahim & Rowe, 1995; Karssen *et al.*, 1995; Andrés *et al.*, 2001).

The IEF of proteins is still considered to be a simple and efficient technique for nematode diagnostics; however, as with any technique, it has some weaknesses. The greatest disadvantage to this method is the relatively large quantity of nematode tissues that often is required for obtaining distinct protein or enzyme profiles. Other disadvantages are its sensitivity to methods of sample preparation and to specific development

Table 10. *References for diagnostics of cyst nematodes using esterase phenotypes.*

Species	References
Globodera pallida	Fox & Atkinson (1988); Zaheer *et al.* (1996)
G. rostochiensis	Fox & Atkinson (1988); Zaheer *et al.* (1996)
Heterodera arenaria	Ibrahim & Rowe (1995)
H. avenae	Berge *et al.* (1981); Rumpenhorst (1985); Bossis & Rivoal (1989); Ganguly *et al.* (1990); Ibrahim & Rowe (1995); Romero *et al.* (1996); Andrés *et al.* (2001); Bishnoi *et al.* (2004); Venkatesan *et al.* (2004)
H. bifenestra	Ibrahim & Rowe (1995)
H. cajani	Ganguly *et al.* (1990); Ibrahim & Rowe (1995); Meher *et al.* (1998); Shashi Singh *et al.* (1998); Venkatesan *et al.* (2004)
H. cardiolata	Ibrahim & Rowe (1995); Meher *et al.* (1998)
H. carotae	Ibrahim & Rowe (1995)
H. cruciferae	Ibrahim & Rowe (1995)
H. elachista	Nobbs *et al.* (1992); Ibrahim & Rowe (1995)
H. fici	Ibrahim & Rowe (1995)
H. filipjevi	Andrés *et al.* (2001); Bishnoi *et al.* (2004);
H. glycines	Dickson *et al.* (1971); Esbenshade & Triantaphyllou (1988); Lee *et al.* (1988); Radice *et al.* (1988); Ibrahim & Rowe (1995); Noel & Liu (1998)
H. goettingiana	Ibrahim & Rowe (1995)
H. hordecalis	Ibrahim & Rowe (1995)
H. latipons	Andrés *et al.* (2001)
H. mani	Rumpenhorst (1985); Ibrahim & Rowe (1995); Andrés *et al.* (2001)
H. oryzae	Nobbs *et al.* (1992); Ibrahim & Rowe (1995)
H. oryzicola	Nobbs *et al.* (1992)
H. sacchari	Nobbs *et al.* (1992); Ibrahim & Rowe (1995)
H. schachtii	Ibrahim & Rowe (1995)
H. sorghi	Ganguly *et al.* (1990); Ibrahim & Rowe (1995); Meher *et al.* (1998); Venkatesan *et al.* (2004)
H. trifolii	Ibrahim & Rowe (1995)
H. zeae	Ibrahim & Rowe (1995); Meher *et al.* (1998, 2004); Shashi Singh *et al.* (1998); Venkatesan *et al.* (2004); Correia & de Abrantes (2005)

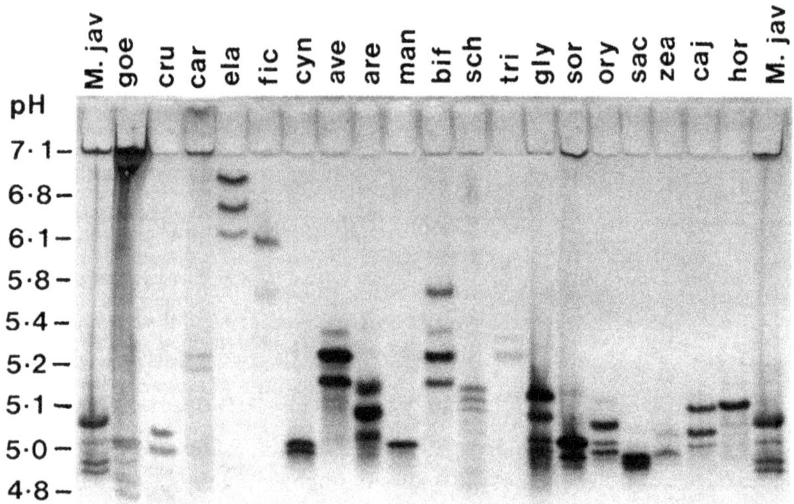

Fig. 34. *Comparative nonspecific esterase profiles obtained by isoelectric focusing of* Heterodera *spp.:* Heterodera goettingiana *(H. goe),* H. cruciferae *(H. cru),* H. carotae *(H. car),* H. elachista *(H. eta),* H. fici *(H. jici),* H. cardiolata *(H. cyn),* H. avenae *(H. ave),* H. arenaria *(H. are),* H. mani *(H. man),* H. bifenestra *(H. bif),* H. schachtii *(H. sch),* H. trifolii *(H. tri),* H. glycines *(H. gly),* H. sorghi *(H. sor),* H. oryzae *(H. ory),* H. sacchari *(H. sac),* H. zeae *(H. zea),* H. cajani *(H. caj),* H. hordecalis *(H. hor),* Meloidogyne javanica *(M. jav). pH range 4-7. After Ibrahim and Rowe (1995).*

stages of the target nematode. Samples prepared from juveniles and females can also generate different IEF profiles resulting from different protein expression in these development stages. Repeated freezing and thawing of samples may lead to denatured proteins, loss in functional activity sometimes inducing artefacts in banding patterns. Apparent loss of bands could be the result of the enzyme concentration falling below the detection sensitivity of the assay system.

DNA-based diagnostics

By comparison with the approaches discussed above, analysis of DNA for taxon diagnostics has several advantages. DNA does not rely on expressed products and therefore is not influenced by environmental conditions or developmental stage. The DNA diagnostics, being much more sensitive than any biochemical diagnostic technique, can be used with nanograms of DNA extracted from a single specimen or even from a fragment of an individual nematode (Appendix IV). For diagnostic purposes, various types of samples can potentially be used as total soil extracts that include a mixture of nematode species and infected plant materials. Dried, alcohol-fixed, formalin-fixed specimens or specimens embedded and mounted on permanent slide, can also be used for diagnostics, although with varying levels of consistency.

GENES USED FOR MOLECULAR DIAGNOSTICS

The main DNA regions targeted for diagnostics of cyst nematodes are nuclear ribosomal RNA genes. These include 18S, 28S and especially the internal transcribed spacer 1 (ITS1) and internal transcribed spacer 2 (ITS2), which are situated between 18S and 5.8S and 5.8S and 28S rRNA genes, respectively. The flanking 18S and 28S rRNA gene regions contain a rather conservative nucleotide sequences and thus universal primers can be designed and used for amplification of ITS regions. The ITS regions, being under a higher mutation rate than other rRNA regions, usually have sufficient variability in sequences to distinguish nematodes at the species level. Two nuclear protein coding genes, actin gene (Kovaleva *et al.*, 2005; Mundo-Ocampo *et al.*, 2008) and heat shock protein 90 gene (Skantar & Carta, 2005; Mundo-Ocampo *et al.*, 2008), have also been used for characterising cyst nematodes. The universal primers for rRNA genes and protein coding genes, which are currently used for diagnostics and the study of phylogenetic relationships for cyst nematodes, are given in Table 11 and Figure 35.

Genes of mitochondrial DNA (mtDNA), with their relatively higher rate of mutations relative to rRNA genes, have great potential for diagnostics of races and populations of cyst nematodes. Recent studies on sequencing and phylogenetic analysis of cytochrome *b* for *G. pallida* allowed the identification of several haplotypes within this species, and revealed the haplotype distribution of this species in South America and the native location of the *G. pallida* haplotypes presently introduced and distributed in Europe, USA and Canada (Plantard *et al.*, 2008; Madani *et al.*, 2010).

Targets for developing a diagnostic method can also be identified by screening random regions of the genome to find DNA fragment sequences that are unique to a particular taxon. This can be done using PCR-based techniques, such as random amplified polymorphic

Table 11. *Primer combinations used for molecular diagnostics and phylogenetic studies of cyst nematodes.*

Primer combination and code (direction)*	Primer sequence (5′-3′)	Amplified region	References
G18SU (f)	GCTTGCCTCAAAGATTAAGCC	18S rRNA	Blaxter *et al.* (1998)
R18Tyl1 (r)	GGTCCAAGAATTTCACCTCTC		Chizhov *et al.* (2006)
F18Tyl2 (f) R18Tyl2 (r)	CAGCCGCGGTAATTCCAGC CGGTGTGTACAAAGGGCAGG	18S rRNA	Chizhov *et al.* (2006)
18S (f)	TTGATTACGTCCCTGCCCTTT	ITS1-rRNA	Vrain *et al.* (1992)
rDNA1.58S (r)	GCCACCTAGTGAGCCGCGCA		Szalanski *et al.* (1997)
18S (f) 26S (r)	TTGATTACGTCCCTGCCCTTT TTTCACTCGCCGTTACTAAGG	ITS1-5.8S-ITS2 of rRNA	Vrain *et al.* (1992)
F194 (f) F195 (r)	CGTAACAAGGTAGCTGTAG TCCTCCGCTAAATGATATG	ITS1-5.8S-ITS2 of rRNA	Ferris *et al.* (1993)
TW81 (f) AB21 (r)	GTTTCCGTAGGTGAACCTGC ATATGCTTAAGTTCAGCGGGT	ITS1-5.8S-ITS2 of rRNA	Curran *et al.* (1994)
D2A (f) D3B (r)	ACAAGTACCGTGAGGGAAAGTTG TCGGAAGGAACCAGCTACTA	D2-D3 of 28S rRNA	Nunn (1992)
Actin717 (f) Actin1741 (r)	ATGGGACAGAAGGACTCG GGATGTCGATGTCGCACT	Actin gene	Kovaleva (unpubl.)
U831 (f) L1110 (r)	AAYAARACMAAGCCNTYTGGAC TCRCARTTVTCCATGATRAAVAC	Hsp90	Skantar & Carta (2005)

* f – forward, r – reverse.

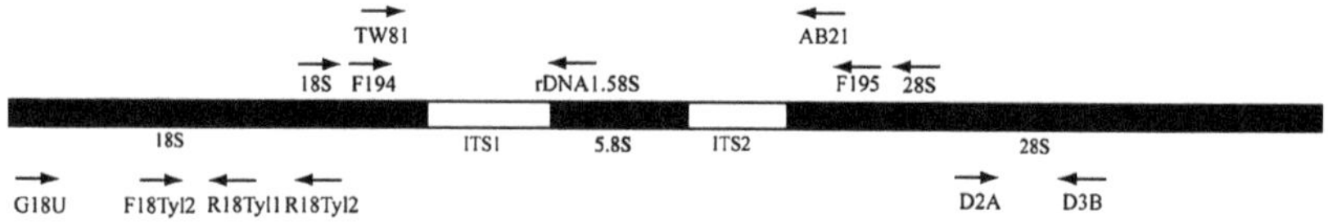

Fig. 35. *Structure of rRNA genes and positions of universal primers.*

DNA (RAPD) (Fig. 36) or amplified fragment length polymorphism (AFLP) (Fig. 37), which provide randomly generated fragments from the genome. These fragments are separated by gel electrophoresis and the patterns are compared for different taxa. Potential diagnostic and unique bands are extracted, cloned and sequenced and then used to design a specific sequence-characterised amplified region (SCAR).

Polymerase chain reaction (PCR)

Over the last decade there has been significant progress in molecular diagnostics of nematodes, due to the development and introduction of PCR. This method enables numerous copies to be obtained from a single or a few molecules extracted from organism DNA by chemical synthesis *in vitro*. The PCR method requires a DNA template (starting material) containing the region to be amplified, two oligonucleotide primers flanking and complementary to this target region, DNA polymerase and four deoxnucleotide triphosphates mixed in a buffer containing magnesium ions. PCR is performed in a tube in a thermocycler with programmed heating and cooling (Fig. 32B). The procedure consists of a succession of three steps determined by temperature conditions: *i*) template denaturation (94°C for 3-5 min); *ii*) primer annealing (55-60°C for 1-2 min); and *iii*) extension of DNA chain (72°C for 1-2 min). PCR is carried out for 30-40 cycles. The resulting amplified product is electrophoretically separated according to size on agarose or polyacrylamide gels and visualised using ethidium bromide or any other DNA-binding dyes. Once identified, target DNA generated by PCR amplification can further be characterised by various analyses, including restriction fragment length polymorphism (RFLP), dot blotting or sequencing, which gives full information on the order of the nucleotide bases in amplified fragment. Currently, several PCR-based methods are used for diagnostics of cyst nematodes: PCR-RFLP, PCR with specific primers and real-time PCR.

PCR-RFLP

Variation in sequences in PCR products generated from different samples can be revealed by restriction endonuclease digestion. The PCR-RFLP method is used to diagnose monospecific samples (one species in a tube) and it consists of two steps: *i*) amplification of the gene marker region using universal primers; and *ii*) digestion of the resulting

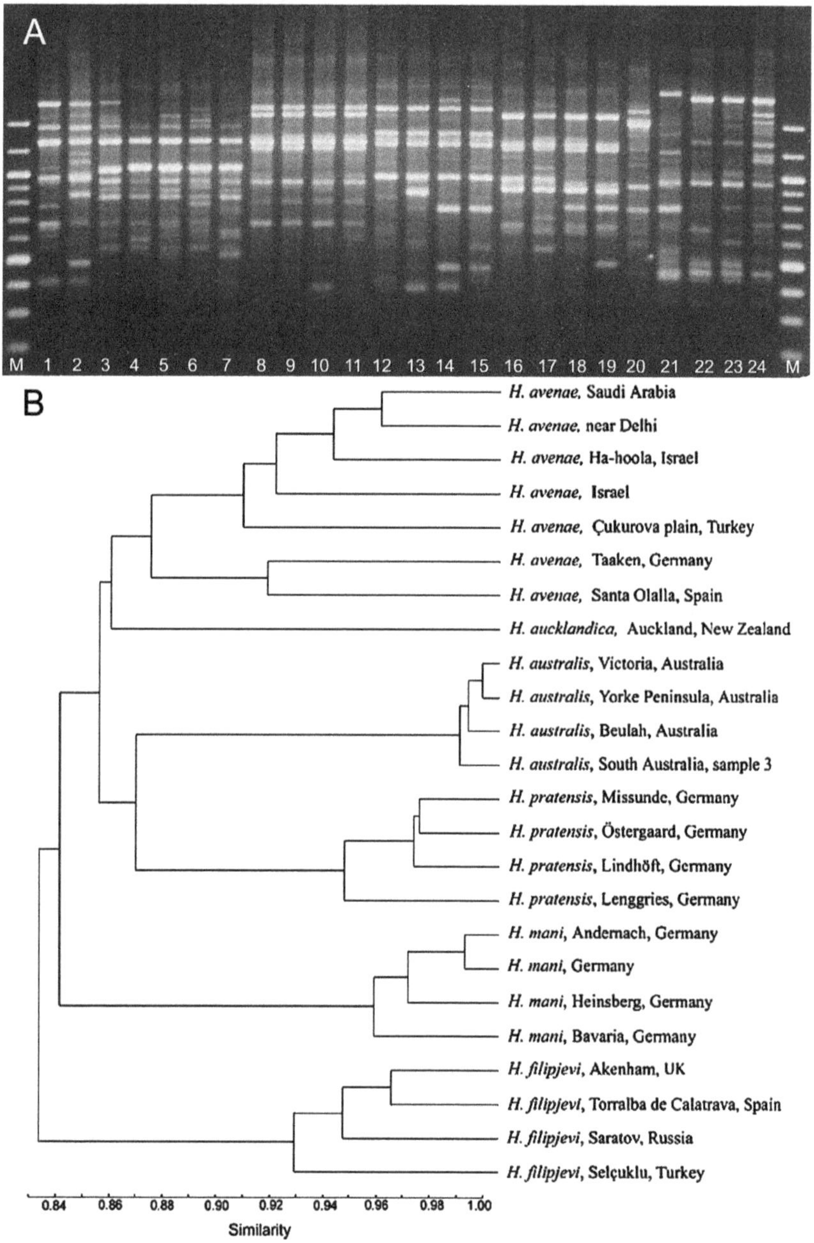
A
M 1 2 3 4 5 6 7 8 9 10 11 12 13 14 15 16 17 18 19 20 21 22 23 24 M
B
H. avenae, Saudi Arabia
H. avenae, near Delhi
H. avenae, Ha-hoola, Israel
H. avenae, Israel
H. avenae, Çukurova plain, Turkey
H. avenae, Taaken, Germany
H. avenae, Santa Olalla, Spain
H. aucklandica, Auckland, New Zealand
H. australis, Victoria, Australia
H. australis, Yorke Peninsula, Australia
H. australis, Beulah, Australia
H. australis, South Australia, sample 3
H. pratensis, Missunde, Germany
H. pratensis, Östergaard, Germany
H. pratensis, Lindhöft, Germany
H. pratensis, Lenggries, Germany
H. mani, Andernach, Germany
H. mani, Germany
H. mani, Heinsberg, Germany
H. mani, Bavaria, Germany
H. filipjevi, Akenham, UK
H. filipjevi, Torralba de Calatrava, Spain
H. filipjevi, Saratov, Russia
H. filipjevi, Selçuklu, Turkey
0.84 0.86 0.88 0.90 0.92 0.94 0.96 0.98 1.00
Similarity

amplicon by restriction enzymes. PCR-RFLP techniques have been used for diagnostics of many cyst nematode species (Table 12).

Szalanski *et al.* (1997) successfully applied PCR-RFLP of the ITS1 to differentiate *H. cruciferae*, *H. goettingiana*, *H. glycines*, *H. trifolii*, *H. schachtii* and *H. zeae*. Ferris *et al.* (1993, 1994) and Blok *et al.* (1998), all studying cyst nematodes, reported more variation in the ITS1 than in the ITS2 region; however, the use of both of these regions appear to be promising, because, after digestion, the larger size of the PCR amplified fragment yields more informative patterns. Moreover, some restriction sites useful for identification of several agriculturally important species are situated in the ITS2 region.

A comprehensive analysis of PCR-RFLP of the ITS1-5.8-ITS2-rRNA gene of cyst nematodes has been conducted by Subbotin and co-authors and published in a series of articles (Subbotin *et al.*, 1997, 1999a, b, 2000a, 2003; Eroshenko *et al.*, 2001; Amiri *et al.*, 2002; Tanha Maafi *et al.*, 2003; Madani *et al.*, 2004). In these studies, the universal TW81 and AB28 primers generating amplicons from approximately 1005-1060 bp in length (depending on species) were used for diagnostics. Twenty-six restriction enzymes were tested to analyse ITS polymorphism among cyst nematode species. Application of the RFLP technique to separate cyst nematode species revealed several restriction enzymes, *Alu*I, *Bsu*RI, *Bsh*1236I, *Cfo*I and *Scr*FI, with greater discriminatory

Fig. 36. *A: RAPD patterns of 26 populations of the* Heterodera avenae *complex. Primer G-10: 5′-AGGGCCGTCT-3′. 1:* H. avenae *(Taaken, Germany); 2:* H. avenae *(Santa Olalla, Spain); 3:* H. avenae *(Çukurova plain, Turkey); 4:* H. avenae *(Saudi Arabia); 5:* H. avenae *(Ha-hoola, Israel); 6:* H. avenae *(Israel); 7:* H. avenae *(near Delhi); 8:* H. australis *(South Australia, sample 3); 9:* H. australis *(Beulah, Australia); 10:* H. australis *(Victoria, Australia); 11:* H. australis *(Yorke Peninsula, Australia); 12:* H. mani *(Bavaria, Germany); 13:* H. mani *(Heinsberg, Germany); 14:* H. mani *(Andernach, Germany); 15:* H. mani *(Germany); 16:* H. pratensis *(Missunde, Germany); 17:* H. pratensis *(Östergaard, Germany); 18:* H. pratensis *(Lindhöft, Germany); 19:* H. pratensis *(Lenggries, Germany); 20:* H. aucklandica *(Auckland, New Zealand); 21:* H. filipjevi *(Saratov, Russia); 22:* H. filipjevi *(Akenham, UK); 23:* H. filipjevi *(Torralba de Calatrava, Spain); 24:* H. filipjevi *(Selçuklu, Turkey). M = 100 bp DNA ladder (Biolab). After Subbotin* et al. *(2003); B: UPGMA cluster diagram based on Simple Matching similarity measure estimated from 237 RAPD markers for 24 populations of the* Heterodera avenae *complex. After Subbotin* et al. *(2003).*

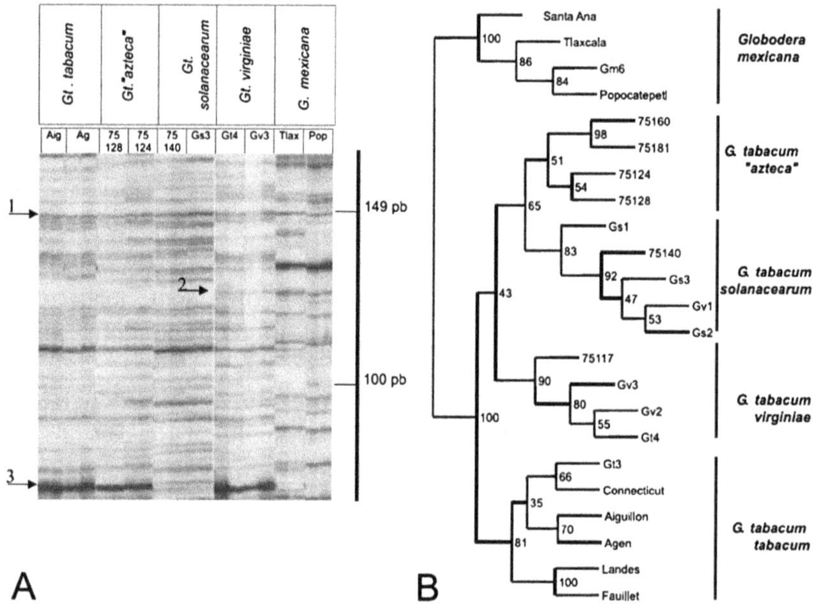

Fig. 37. *AFLP analysis of* Globodera tabacum *populations. A: AFLP fingerprint obtained with E63-M34 primers showing two replicates per population. Markers are indicated with an arrow: 1, the marker is present for all presented populations; 2, marker present for Gt4, Gv3, Tlaxcala (Tlax) and Popocateptl (Pop) populations; 3, marker present for Aiguillon (Aig), Agen (Ag), 75128, 75124, Gt4, and Gv3 populations; B: UPGMA tree derived from AFLP data as the consensus tree obtained after 1000 bootstraps. The bootstrap values are given in percentages. After Marché* et al. *(2001) with modifications.*

Table 12. *References for diagnostics of cyst nematodes using PCR-ITS-RFLP.*

Species	References
Cactodera cacti	Tanha Maafi *et al.* (2003)
Globodera artemisiae	Uehara *et al.* (2005)
G. mexicana **sp. n.**	Thiéry & Mugniéry (1996); Grenier *et al.* (2001)
G. pallida	Thiéry & Mugniéry (1996); Fleming & Powers (1998); Subbotin *et al.* (1999b, 2000b); Fleming *et al.* (2000); Grenier *et al.* (2001); Sirca *et al.* (2003); Reid & Pickup (2005); Skantar *et al.* (2007); Hlaoua *et al.* (2008); Madani *et al.* (2010)

Table 12. *(Continued).*

Species	References
G. rostochiensis	Thiéry & Mugniéry (1996); Orui (1997); Fleming & Powers (1998); Subbotin *et al.* (1999b, 2000b); Fleming *et al.* (2000); Grenier *et al.* (2001); Sirca *et al.* (2003); Sirca & Urek (2004); Reid & Pickup (2005); Uehara *et al.* (2005); Knoetze *et al.* (2006); Skantar *et al.* (2007); Hlaoua *et al.* (2008); Madani *et al.* (2010)
G. tabacum	Thiéry & Mugniéry (1996); Fleming *et al.* (2000); Grenier *et al.* (2001); Sirca *et al.* (2003); Uehara *et al.* (2005); Skantar *et al.* (2007)
G. millefolii	Sirca & Urek (2004)
Heterodera arenaria	Subbotin *et al.* (1999a, 2000a, 2003); Rivoal *et al.* (2003)
H. avenae	Bekal *et al.* (1997); Subbotin *et al.* (1999a, 2000a, 2002, 2003); Gäbler *et al.* (2000); Zheng *et al.* (2000); Rivoal *et al.* (2003); Tanha Maafi *et al.* (2003)
H. aucklandica	Subbotin *et al.* (1999a, 2000a, 2003); Gäbler *et al.* (2000)
H. australis	Subbotin *et al.* (2002, 2003)
H. betae	Wouts *et al.* (2001); Amiri *et al.* (2002)
H. cajani	Subbotin *et al.* (2000a)
H. cardiolata	Subbotin *et al.* (2000a)
H. carotae	Fleming *et al.* (2000); Subbotin *et al.* (2000a); Madani *et al.* (2004)
H. ciceri	Subbotin *et al.* (2000); Rivoal *et al.* (2003); Madani *et al.* (2004)
H. cruciferae	Szalanski *et al.* (1997); Subbotin *et al.* (2000a)
H. cyperi	Subbotin *et al.* (2000a)
H. elachista	Orui (1997); Tanha Maafi *et al.* (2003)
H. fici	Subbotin *et al.* (1997, 2000a); Tanha Maafi *et al.* (2003); Madani *et al.* (2004)
H. filipjevi	Bekal *et al.* (1997); Subbotin *et al.* (1999a, 2000a, 2003); Rivoal *et al.* (2003); Tanha Maafi *et al.* (2003); Madani *et al.* (2004); Umarao & Vangapandu (2008)
H. galeopsidis	Wouts *et al.* (2001)
H. glycines	Orui (1997); Szalanski *et al.* (1997); Fleming *et al.* (2000); Subbotin *et al.* (2000a); Zheng *et al.* (2000); Tanha Maafi *et al.* (2003)
H. goettingiana	Szalanski *et al.* (1997); Fleming *et al.* (2000); Subbotin *et al.* (2000a); Tanha Maafi *et al.* (2003); Madani *et al.* (2004)

Table 12. *(Continued).*

Species	References
H. goldeni	Tanha Maafi *et al.* (2003)
H. hordecalis	Subbotin *et al.* (1999a, 2000); Rivoal *et al.* (2003); Tanha Maafi *et al.* (2003); Madani *et al.* (2004)
H. humuli	Subbotin *et al.* (1997, 2000a); Eroshenko *et al.* (2001); Tanha Maafi *et al.* (2003); Madani *et al.* (2004)
H. koreana	Mundo-Ocampo *et al.* (2008)
H. latipons	Bekal *et al.* (1997); Subbotin *et al.* (1999a, 2000a); Rivoal *et al.* (2003); Tanha Maafi *et al.* (2003); Madani *et al.* (2004)
H. litoralis	Subbotin *et al.* (1999, 2000a)
H. mani	Bekal *et al.* (1997); Fleming *et al.* (2000); Rivoal *et al.* (2003); Subbotin *et al.* (2003)
H. medicaginis	Subbotin *et al.* (2000); Amiri *et al.* (2002)
H. mediterranea	Madani *et al.* (2004)
H. mothi	Tanha Maafi *et al.* (2003)
H. orientalis	Mundo-Ocampo *et al.* (2008)
H. oryzicola	Subbotin *et al.* (2000a)
H. persica	Tanha Maafi *et al.* (2003)
H. pratensis	Subbotin *et al.* (1999a, 2003); Gäbler *et al.* (2000)
H. ripae	Subbotin *et al.* (1997); Eroshenko *et al.* (2001); Madani *et al.* (2004)
H. sacchari	Rivoal *et al.* (2003)
H. salixophila	Subbotin *et al.* (2000a)
H. schachtii	Bekal *et al.* (1997); Szalanski *et al.* (1997); Subbotin *et al.* (1999a, 2000a); Fleming *et al.* (2000); Wouts *et al.* (2001); Amiri *et al.* (2002); Rivoal *et al.* (2003); Tanha Maafi *et al.* (2003)
H. trifolii	Orui (1997); Szalanski *et al.* (1997); Subbotin *et al.* (2000a); Wouts *et al.* (2001); Amiri *et al.* (2002); Tanha Maafi *et al.* (2003)
H. turcomanica	Tanha Maafi *et al.* (2003)
H. urticae	Subbotin *et al.* (2000a)
H. ustinovi	Subbotin *et al.* (1999a, 2000a, 2003)
H. vallicola	Eroshenko *et al.* (2001)
H. zeae	Szalanski *et al.* (1997); Umarao *et al.* (2008)
Punctodera punctata	Fleming *et al.* (2000)

powers and generating more polymorphic profiles than other enzymes. This enabled more species of cyst nematode to be characterised. Digestion of PCR product by one of seven restriction enzymes (*Alu*I, *Ava*I, *Bsu*RI, *Bsh*1236I, *Cfo*I, *Mva*I and *Rsa*I) distinguishes most cyst nematode species from each other (see: Vol. 2, Table 4; Fig. 1). For differentiation of species belonging to cyst nematode groups with similar ITS sequences, additional sets of restriction enzymes were proposed: for the *H. avenae* species complex – *Hinf*I, *Ita*I, *Pst*I, *Taq*I, *Tru9*I (see: Vol. 2, Fig. 3); for the *Schachtii* group – *BseN*I, *Eco72*I, *Scr*FI (see: Vol. 2, Fig. 2); and for the *Humuli* group – *Alw21*I, A*lw26*I, *Dde*I, *Ita*I, *Pvu*II, *Tru9*I (see: Vol. 2, Fig. 5). In several cases, however, the PCR-ITS-RFLP technique is not able to discriminate species, for example: *H. avenae* (type A) from *H. arenaria*; *H. cruciferae* from *H. carotae*; *H. trifolii*, *H. daverti* and *H. ciceri* from one another.

Although rRNA genes are present in many copies in nematode genomes, their sequences are almost identical in that they undergo a homogenisation process known as concerted evolution. However, sometimes this process can be disrupted and several different copies of rRNA genes are present in a single genome. Heterogeneity of the ITS of rRNA is revealed by RFLP analysis in many cyst nematode species (Szalanski *et al.*, 1997; Subbotin *et al.*, 2000a). On an RFLP gel this may be expressed by the presence of additional, sometimes poorly visualised, bands. Such diversity of paralogous ITS-rRNA can complicate species diagnostics based on RFLPs because, under standard PCR conditions, the low-stabile and high-stabile paralogues have different rates of amplification. Although the proportion of different ITS copies might vary among individuals and populations of a given species, for some other species it could be relatively constant and useful for diagnostics. Amiri *et al.* (2002) showed that digestion of ITS-PCR product of *H. schachtii* by *Mva*I produced rather complex RFLP profiles containing several additional fragments with different intensities on gels, but their number was constant and not subject to intra- and inter-population variation under PCR conditions. Presently, *Mva*I is considered as the most reliable enzyme for identification of the sugar beet cyst nematode using PCR-ITS-RFLP (see: Vol. 2, Fig. 2A).

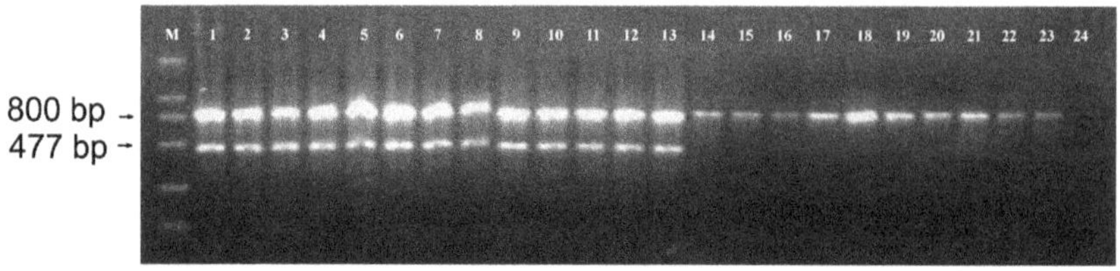

Fig. 38. *PCR products amplified by multiplex PCR with species specific SCAR primers for* Heterodera glycines *(477 bp) and control universal primers D2A and D3B (800 bp). M = DNA marker, lanes 1-13 =* H. glycines *samples from different populations, lanes 14-23 = other nematode samples, 24 = negative control. After Ou* et al. *(2008).*

PCR WITH SPECIFIC PRIMER

This type of PCR enables the detection of species in a nematode mixture by a single PCR test. Oligonucleotide primers for this PCR are designed to be bound to regions of the gene that are conserved over the particular taxa or a group of taxa, so that it may be species-specific or it may target a group of species that differ from other such groups. Detection of a specific size amplicon in a gel indicates the presence of a certain species within a sample. This method excludes an enzyme digestion step, which is necessary for PCR-RFLP. This PCR type constitutes a major development in DNA diagnostics. It enables the detection not only of a single species but also of several species in a nematode mixture by a single PCR test, provided several sets of specific primers for different species are mixed. This approach can significantly decrease diagnostic time and costs. PCR with specific primers has been developed for identification of *G. rostochiensis*, *G. pallida*, *H. glycines* (Fig. 38), *H. schachtii* (Table 13) and the *Schachtii* group species (Amiri *et al.*, 2001).

REAL TIME PCR

This PCR technique also permits estimating of the number of nematodes in a sample. This technique measures the nematode number indirectly by assuming that the number of target DNA copies in the sample is proportional to the number of targeted individuals. The real-time technique allows continuous monitoring of the DNA amplification during PCR using hybridisation probes such as TaqMan or dyes such as SYBR Green I. Real time PCR has been developed for detection and

Table 13. *Species-specific primers for identification of cyst nematodes using conventional PCR.*

Species	Primer code and sequences (5′-3′)	Amplicon size in bp	Amplified gene	References
Globodera rostochiensis	PITSr3 AGCGCAGACATGCCGCAA ITS5 GGAAGTAAAAGTCGTAACAAGG	434	ITS rRNA	Bulman & Marshall (1997)
G. rostochiensis	GCAAGCCCAGCGTCAGCAAC GAACATCAACCTCCTATCGG	315	SCAR	Fullaondo *et al.* (1999)
G. rostochiensis	GGTGACTCG ACGATTGCTGT GCAGTTGGCTAGCGATCTTC	391	ITS rRNA	Mulholland *et al.* (1996)
G. pallida	PITSp4 ACAACAGCAATCGTCGAG ITS5 GGAAGTAAAAGTCGTAACAAGG	265	ITS rRNA	Bulman & Marshall (1997)
G. pallida	TGTCCATTCCTCTCCACCAG CCGCTTCCCCATTGCTTTCG	768	SCAR	Fullaondo *et al.* (1999)
G. pallida	GGTGACTCGACGATTGCTGT GCAGTTGGCTAGCGATCTTC	238	ITS rRNA	Mulholland *et al.* (1996)
Heterodera schachtii	SHF6 GTTCTTACGTTACTTCCA TW81 GTTTCCGTAGGTGAACCTGC	200	ITS rRNA	Amiri *et al.* (2002)
H. glycines	SCNFI GGACCCTGACCAAAAAGTTTCCGC SCNRI GGACCCTGACGAGTTATGGGCCCG	477	SCAR	Ou *et al.* (2008)
H. glycines	GlyF1 TTACGGACCGTAACTCAA 26S TTTCACTCGCCGTTACTAAGG	181	ITS rRNA	Subbotin *et al.* (2001b)

quantification of *G. pallida*, *G. rostochiensis*, *G. tabacum*, *G. artemisiae* and *H. schachtii* (Bates *et al.*, 2002; Madani *et al.*, 2005, 2008; Bacic *et al.*, 2008; Nowaczyk *et al.*, 2008; Toyota *et al.*, 2008; Quader *et al.*, 2008).

Problems with PCR-based methods

The application over many years of PCR-RFLP and PCR with specific primers for diagnostic purposes has revealed several limitations associated with these techniques. First, so-called universal primers, which are designed based on existing knowledge of DNA sequences for target species, might not work for some other species, which have different sequences in this target DNA region. Second, universal primers presently available for application in diagnostics can also amplify other organisms than nematodes, such as fungi and plants. However, the size of amplified ITS-rRNA product for other organisms is often smaller than for those of cyst nematodes. Third, the presently generated PCR-ITS-RFLP profiles do not allow the differentiating of some species of cyst nematode from others.

It is important to note that an RFLP profile that does not fit known restriction patterns can be explained as the result of a certain mutation in restriction sites for some geographical isolates of the same species. The presence of additional bands in RFLP profiles could be the result of a higher level of heterogeneity for some isolates or the presence of a natural hybrid of two species within a single sample. For example, Reid and Pickup (2005) observed cysts of *G. pallida* showing similar restriction patterns for *G. rostochiensis* for some enzymes and combined patterns for both species for others. The complex RFLP patterns was also shown for natural hybrids of these two species of potato cyst nematode (Thiéry *et al.*, 1996).

Application of PCR with a specific primer could face similar problems. One limitation is the potential for obtaining a false positive reaction. As a specific primer design is always based on analysis of DNA sequences for a target species and closely related nematodes (see Skantar *et al.*, 2007), there is a possibility that similar fragments can be obtained for another, previously non-investigated, nematode species. A second limitation is the possibility of a false negative reaction. Although a region with a conserved sequence should be used for primer design, the possibility cannot be excluded that some mutations might have occurred in this region in some specimens or populations of the target taxa. As a result, such specimens or populations might not be detectable by the PCR test. A third limitation is the possibility of sample contamination, which might occur during sample preparation in a laboratory. This might give a false positive reaction due to the great sensitivity of the PCR method with a specific primer. Following strict rules to prevent contamination during preparation of the PCR mixture is imperative for all diagnostics tests.

DNA SEQUENCING

Comparison of the ITS rRNA gene sequences has emerged as a preferred technique for identification of cyst nematodes. Nucleic acid sequencing methods have undergone tremendous advances over the past decade. Now the ITS rRNA gene sequences have been determined for a large number of cyst nematode species and they are deposited in the GenBank database (http://www.ncbi.nlm.nih.gov/). In general, the comparison of the ITS rRNA genes with reference data using sequence and phylogenetic analysis allows classification of nematode samples and establishing identification. Determinations of clades to which samples belong and level of the interspecific variation are two approaches used together for molecular identification. The intraspecific variation for the ITS-rRNA sequence gene varies between nematode genera. For example, it typically does not exceed 1.4% for *Heterodera* spp. (Tanha Maafi *et al.*, 2003; Madani *et al.*, 2004) or 1.8% for *Globodera* spp. (Subbotin *et al.*, 2000b). Observed differences greater than these values between a sample and a standard indicate absence of co-specificity. Recent detailed sequence and phylogenetic studies indicated the presence of several sibling species that probably exist within currently defined species of plant-parasitic nematodes, such as *G. pallida* (Subbotin *et al.*, 2000a, b; Madani *et al.*, 2010), *H. avenae* (Subbotin *et al.*, 2003) and several others. However, in some cases, the ITS-rRNA does show differences between sequences and does not distinguish among some closely related and recently diverged species such as *H. avenae* and *H. arenaria*, and *H. daverti* and *H. trifolii* (Subbotin *et al.*, 2003; Subbotin, unpubl.).

These above mentioned limitations indicate that the PCR and sequence techniques should be used intelligently and the researcher or advisor must be aware of the peculiarities of each method. Where there is doubt it is always necessary to confirm identification by several methods, including the use of biochemical methods and traditional morphological features.

Nematology Monographs & Perspectives, 2010, Vol. 8A, 101-106

Chapter 5

Systematics of the Heteroderinae

Taxonomic history

Schmidt (1871) proposed *Heterodera* Schmidt, 1871, as the type genus of the cyst nematodes. Filipjev and Schuurmans Stekhoven (1941) established the subfamily Heteroderinae that included, in addition to *Heterodera*, *Tylenchulus* and *Paratylenchus* Micoletzky, 1922. Later, in the classification by Thorne (1949), only two genera, *Heterodera* and *Meloidogyne*, were recognised within the subfamily. Skarbilovich (1947) recognised the subfamily at family rank, the Heteroderidae, a family that, in addition to cyst nematodes, also included the genus *Tylenchulus*. Subsequently, different authors proposed various compositions for this family. Thus, the family Heteroderidae included subfamilies Heteroderinae, Hoplolaiminae and Nacobbinae (Chitwood & Chitwood, 1950); Heteroderinae, Tylenchulinae, Sphaeronematinae, Nacobbinae and Meloidogyninae (Skarbilovich, 1959); Heteroderinae, Meloidoderinae and Meloidogyninae (Golden, 1971; Andrassy, 1976) and Heteroderinae, Meloidoderinae and Ataloderinae (Wouts, 1973). Wouts's classification of Heteroderidae (Wouts, 1973) was accepted by Stone (1977), Hopper (1978), Robbins (1978) and Coomans (1979). Husain (1976) added the subfamily Sarisoderinae. Fotedar and Handoo (1978) recognised three subfamilies under Heteroderidae: Heteroderinae, Meloidoderinae and Ataloderinae. Krall and Krall (1978) recognised three subfamilies: Heteroderinae with two newly erected genera, *Bidera* and *Cactodera*; Sarisoderinae; and Punctoderinae. Another perspective on classification of Heteroderidae was developed by Luc and co-authors (Luc *et al.*, 1978), in which division into subfamilies was rejected. However, Luc and co-authors (1988) subsequently accepted three subfamilies: Heteroderinae, Meloidogyninae and Nacobboderinae within the family. In a phylogenetic classification Wouts (1985) recognised six subfamilies under Heteroderidae: Verutinae, Meloidoderinae, Cryphoderinae, Heteroderinae, Ataloderinae and Punctoderinae and did not consider cyst nematodes as a monophyletic group. Baldwin and Schouest (1990), based on phylogenetic analysis of morphological characters, gave a classification of Heteroderinae with several tribes: Heteroderini, Verutini, Meloidoderini, Cryphoderini, Sarisoderini and Ataloderini. Siddiqi (2000) considered several classifications and accepted the division of the family Heteroderidae into subfamilies as proposed by Wouts (1973). This classification is herein accepted.

The taxonomic history of Heteroderinae is discussed in greater detail by Stone (1977, 1986), Krall and Krall (1978), Wouts (1985), Luc *et al.* (1988) and Siddiqi (2000).

Classification (according to Siddiqi, 2000)

Order Tylenchida Thorne, 1949
  Suborder Hoplolaimina Chizhov & Berezina, 1988
    Superfamily Hoplolaimoidea (Filipjev, 1934) Paramonov, 1967
      Family Heteroderidae Filipjev & Schuurmans Stekhoven, 1941 (Skarbilovich, 1947)
        Subfamily Heteroderinae Filipjev & Schuurmans Stekhoven, 1941
          = Heteroderini Filipjev & Schuurmans Stekhoven, 1941 (Coomans, 1979)
          = Punctoderinae Krall & Krall, 1978
          = Sarisoderinae Husain, 1976

DIAGNOSIS OF THE SUBFAMILY HETERODERINAE

After Siddiqi (2000) with modifications.

Heteroderidae. *Mature female* spherical, pear- or lemon-shaped with a short neck, turning into a tough, hard-walled, yellowish, light to dark brown or black cyst containing eggs and juveniles, in some species eggs may be laid in a gelatinous matrix. Cuticle surface with a zigzag or lace-like pattern, annulation absent on mature female and cyst. Vulva and anus usually close together, almost terminal. Clear translucent vulval fenestra usually present (absent in species of *Afenestrata* group of *Heterodera*), an anal fenestra present only in *Punctodera*. *Male* developed through metamorphosis, labial region annulated, lateral field with four or three incisures, tail short, hemispherical, without bursa. J2 with four or three incisures in lateral field.

TYPE GENUS

Heterodera Schmidt, 1871

OTHER GENERA

Betulodera Sturhan, 2002
Cactodera Krall & Krall, 1978
Dolichodera Mulvey & Ebsary, 1980
Globodera Skarbilovich, 1959
Paradolichodera Sturhan, Wouts & Subbotin, 2007
Punctodera Mulvey & Stone, 1976

Identification key to genera of Heteroderinae

1 – Cysts circumfenestrate 2
– Cysts ambifenestrate or bifenestrate, or without fenestration .... *Heterodera*

2 – Cysts with terminal cone 3
– Cysts without terminal cone 4

3 – Vulval slit in cysts 12-28 μm long, J2 with four lateral incisures... .... *Cactodera*
– Vulval slit in cysts 5-8 μm long, J2 with three lateral incisures .... *Betulodera*

4 – Anal region with fenestration........................ *Punctodera*
– Anal region without fenestration 5

5 – Mature female and cyst spheroidal, perineal tubercles usually present.. *Globodera*
– Mature female and cyst elongate-oval shape, perineal tubercles absent 6

6 – Bullae absent in cysts, J2 with DGO = 11-15 μm, c = 9-11, stylet < 20 μm.............................. *Paradolichodera*
– Bullae present in cysts, J2 with DGO = 7-8 μm, c = 5-6, stylet = 22-24 μm .... *Dolichodera*

NOTE

In 2006, the Missouri Extension Nematology Laboratory discovered a new species of corn cyst nematode from a sample collected in a corn field of north-western Tennessee, Obion County, USA. This nematode is considered as a representative of a new genus of the subfamily Heteroderinae. It resembles representatives of the genera *Cactodera* and *Betulodera* in having lemon-shaped cysts and a circumfenestrate vulval area. Cysts lack bullae, an underbridge and internal denticles. Similar cyst specimens had previously been collected in 1978 from Lauderdale County, Tennessee, on goosegrass, *Eleusine indica*. J2 are 320-400 μm long with a stylet 14-16 μm long and the tail has a bluntly rounded terminus (Bernard *et al.*, 2007). Phylogenetic analysis of the D2-D3 expansion regions of 28S rRNA gene sequences revealed that this nematode shares a unique clade with *Betulodera* (Fig. 26; E. Bernard, pers. comm.; T. Powers, pers. comm.). This nematode reproduced well on all tested hybrid corn cultivars. Of the monocotyledons tested, only corn and goose grass were good hosts. No dicotyledonous hosts have been found. In a glasshouse the nematode reproduced best in hosts planted in sandy soil, then clay soil, and least well in loamy soil (Heinz *et al.*, 2007).

Species delimitation

The first cyst nematode was recorded on sugar beet by Schacht (1859) and was later described as *Heterodera schachtii* Schmidt, 1871. Within the next two decades it became recognised as an important pest in the beet-growing regions in Europe. It also became apparent that cyst nematodes induce a similar damage on other crops and it was initially thought that *H. schachtii* was responsible for yield losses. For many years, host range was considered as the main guide to the recognition and designation of cyst nematode species, which were later separated morphologically (Franklin, 1951; Wouts & Baldwin, 1998). Morphological characters for species delimitation and identification started to be more widely used after 1960, when the taxonomic importance of structures in the vulval area and characters of the J2 become evident (Wouts & Baldwin, 1998). The recognition of new morphological taxonomic characters led to a rapid increase in the number of nominal cyst nematode species: from ten in 1951 (Franklin, 1951), 15 in 1959 (Hesling, 1959), to 20 in 1964, over 40 in 1970, 73 in 1985 (Wouts, 1985), 89 in 1998 (Wouts & Baldwin, 1998) and 110 valid species recognised here. During recent years, there has been a new step in the development of a species concept for this group, a step characterised by moving from morphology-based to integrative taxonomy.

Defining a species has been a topic of much debate that has matured with growing understanding of phylogenetics. Most of these discussions, which have also occurred in nematology (Adams, 1998), centre on the consideration of species as being on a distinct evolutionary trajectory, although there is often a gap between such a philosophical view and what is operationally practical when hypothesising and testing species status. The challenges implicit in approaches to species definition have been specifically considered for cyst nematodes (Sturhan, 1983; Stone, 1986). Historically, separating species of cyst nematodes has relied mainly on phenotypic characters, although morphological distinctiveness is widely understood to be a fallible criterion for this purpose. The biological species concept (Mayr, 1942), that species are groups of interbreeding natural populations that are reproductively isolated from other such groups, has also been applied to clarify species bondaries for cyst nematodes (Miller, 1986). However, application of the biological species concept is restricted to sexual out-crossing populations over a short period of evolutionary time, and excludes parthenogenetic organisms. Furthermore, the diagnosis of biological species among cyst nematodes, including evaluating their propensity to interbreed and produce fertile offspring, can rarely be accomplished in practice. Introducing genetic data has further contributed to a process of delimiting species in cyst nematodes. That is, the degree of genetic distance among populations estimated from allozyme frequency, nucleotide divergence, AFLP, RAPD, microsatellites or other markers can be used to argue whether or not species status should be assigned. These approaches to delimiting species are based on an indirect inference of the presence or absence of gene flow. In practice, however, all methods occasionally fail to be definitive. Different measures of speciation may, in some cases, imply different conclusions. In some cases, results may not be definitive because the process of speciation is incomplete, as reflected by diffuse boundaries, or a hybrid zone, between diverging lineages. Groups of populations, which are not fully reproductively isolated from each other, are sometimes named subspecies or semispecies.

Since methods and measures of divergence differ in sensitivity it is argued that a particularly effective approach to delimiting species is to employ integrative or polyphasic taxonomy, using multiple datasets for inferring species. An integrative approach is necessary because the complexity of species biology requires that species boundaries be studied from multiple, complementary, perspectives. The use of integrative taxonomy to infer classifications is based on a consensus of all available data and information (phenotypic, genotypic and phylogenetic) used for delimiting taxa at all levels. This approach integrates any significant information on the organisms and these data are used to develop a consensus type of classification to satisfy most, but not necessarily all, users of taxonomic results. This taxonomy is also an attempt to synthesise the real data, a step towards synthetic taxonomy, which is made possible through the development of new informative strategies. For example, an integrated approach clearly indicates that worldwide diversity in cyst nematode groups remains seriously underestimated. Some species, for example, the cereal cyst nematode (*H. avenae sensu lato*), formerly regarded as widespread, actually now comprises several species: *H. avenae sensu stricto*, *H. arenaria*, *H. filipjevi*, *H. aucklandica*, *H. pratensis*, *H. riparia*, *H. mani* and *H. australis*, each with more restricted geographic and host ranges (Subbotin *et al.*, 2003). Differentiation of some species from this complex based on morphological and morphometric characters is rather difficult and sometimes not even possible. So, *H. avenae* can be clearly distinguished from *H. australis* using biochemical and molecular methods, but not by morphological or morphometric characters. It is also remarkable that the status of several populations of the cereal cyst nematode, *H. avenae sensu lato*, still remains controversial and undefined, and further resolution may ultimately result in descriptions of additional new species. Serious problems of differentiation using phenotype characters exist for many other cyst nematodes consisting of species complex (*e.g.*, *H. cruciferae* and *H. carotae*; *G. pallida* and *G. mexicana* **sp. n.**; *G. artemisiae* and *G. millefolii* and others). However, because of a lack of comprehensive comparative studies dealing with these nematodes, these problems are not yet clearly recognised.

Thus, we emphasise that integrating morphological data with biological (host plants) and biochemical (IEF of proteins and isozymes), together with molecular (PCR-AFLP, PCR-RAPD, rRNA, mtDNA and protein-coding gene sequences) datasets are operational criteria for defining the species boundary within cyst nematodes. Ultimately, all of these approaches test the hypothesis of a given putative species as an entity evolving independently from other such entities.

Nematology Monographs & Perspectives, 2010, Vol. 8A, 107-177

Chapter 6

Systematics of the genus *Globodera*

History

Round cyst nematodes lacking a terminal cone were originally included in the genus *Heterodera*. To accommodate the potato cyst nematodes and related species having round cysts, Skarbilovich (1959) erected the subgenus *Globodera* (Latin *globus* = globe, and Greek ($\delta\varepsilon\rho o\varsigma$) *deros* = skin), leaving the lemon-shaped species in the subgenus *Heterodera*. Behrens (1975) elevated the subgenus *Globodera* to generic status. Almost simultaneously, after careful consideration based on significant morphological and biological differences of the female, cyst and male, Mulvey and Stone (1976) also proposed recognising *Globodera* at genus level. Baldwin and Mundo-Ocampo (1991) provided a review of *Globodera* species. They emphasised the importance of species causing economic losses on potatoes around the world but also included other species of lesser economic impact. Mulvey and Stone (1976) and Siddiqi (2000) gave references and a review of the authorship for the genera and species. Identification keys for *Globodera* were given by Mulvey (1974), Wouts (1984), Golden (1986) and Wouts and Baldwin (1998).

Genus *Globodera* Skarbilovich, 1959
= *Heterodera* (*Globodera* Skarbilovich, 1959)

DIAGNOSIS (after Baldwin and Mundo-Ocampo, 1991 and Siddiqi, 2000)

Mature female and cyst spheroid, lacking terminal cone. Vulval area circumfenestrate. Vulva located in a cavity beneath outline of body, vulval slit < 15 μm. No anal fenestra. Vaginal remnants, underbridge and bullae rarely present. Cuticle with distinct D-layer. All eggs retained in body, egg sac absent. Egg surface smooth. *Male* lateral field with four lines, spicules > 30 μm, distally pointed. *J2* with four incisures in lateral field. Tail conical, pointed, phasmids punctiform. *En face* pattern typically with six separate lips, sometimes with fusion of adjacent submedial lips.

TYPE SPECIES

1. *Globodera rostochiensis* (Wollenweber, 1923) Skarbilovich, 1959
   = *Heterodera schachtii rostochiensis* Wollenweber, 1923
   = *Heterodera rostochiensis* Wollenweber, 1923
   = *Heterodera (Globodera) rostochiensis* (Wollenweber, 1923) Skarbilovich, 1959
   = *Heterodera schachtii solani* Zimmermann, 1927
   = *Heterodera solani* Zimmermann, 1927
   = *Heterodera pseudorostochiensis* Kirjanova, 1963
   = *Globodera pseudorostochiensis* (Kirjanova, 1963) Mulvey & Stone, 1976
   = *Globodera arenaria* Chizhov, Udalova & Nasonova, 2008 **syn. n.**

OTHER SPECIES

2. *G. artemisiae* (Eroshenko & Kazachenko, 1972) Behrens, 1975
   = *Heterodera artemisiae* Eroshenko & Kazachenko, 1972
   = *Globodera hypolysi* Ogawa, Ohshima & Ichinohe, 1983 **syn. n.**
3. *G. bravoae* Franco, Cid del Prado & Lamothe-Argumedo, 2000
4. *G. leptonepia* (Cobb & Taylor, 1953) Skarbilovich, 1959
   = *Heterodera leptonepia* Cobb & Taylor, 1953
   = *Heterodera (Globodera) leptonepia* (Cobb & Taylor, 1953) Skarbilovich, 1959
5. *G. mali* (Kirjanova & Borisenko, 1975) Behrens, 1975
   = *Heterodera mali* Kirjanova & Borisenko, 1975
   = *Globodera mali* (Kirjanova & Borisenko, 1975) Mulvey & Stone, 1976
6. *G. mexicana* **sp. n.**
   = *Heterodera mexicana* Campos-Vela, 1967 **syn. n.** (= *nomen nudum*)
7. *G. millefolii* (Kirjanova & Krall, 1965) Behrens, 1975
   = *Heterodera millefolii* Kirjanova & Krall, 1965
   = *Heterodera (Globodera) millefolii* (Kirjanova & Krall, 1965) Mulvey, 1973

= *Globodera achilleae* (Golden & Klindić, 1973) Behrens, 1975 **syn. n.**
= *Heterodera achilleae* Golden & Klindić, 1973

8. *G. pallida* Stone, 1973a
   = *Heterodera pallida* Stone, 1973a
   = *Heterodera* (*Globodera*) *pallida* Stone, 1973a
9. *G. tabacum tabacum* (Lownsbery & Lownsbery, 1954) Skarbilovich, 1959
   = *Heterodera tabacum* Lownsbery & Lownsbery, 1954
   = *Globodera tabacum* (Lownsbery & Lownsbery, 1954) Behrens, 1975
   G. tabacum solanacearum (Miller & Gray, 1972) Behrens, 1975
   = *Heterodera solanacearum* Miller & Gray, 1972
   = *Heterodera tabacum solanacearum* (Miller & Gray, 1972) Stone, 1983
   = *Globodera solanacearum* (Miller & Gray, 1972) Behrens, 1975
   G. tabacum virginiae (Miller & Gray, 1968) Stone, 1973
   = *Heterodera virginiae* Miller & Gray, 1968
   = *Heterodera tabacum virginiae* (Miller & Gray, 1968) Stone, 1983
   = *Globodera virginiae* (Miller & Gray, 1968) Behrens, 1975

10. *G. zelandica* Wouts, 1984

MORPHOLOGICAL IDENTIFICATION

The diagnostics of *Globodera* species is mainly based on several morphometric and morphological characters, including the number of cuticular ridges between anus and fenestra, Granek's ratio, stylet, hyaline region, shape of stylet knobs, *etc.* Many diagnostically useful morphometric characters used in the key below, and given in Table 14, have relatively stable means for *Globodera* species and may be useful for identification, although in some populations the range may be extended by a few individuals, something that should be taken into account.

Key to species of the genus *Globodera*

1 – Cuticle of cyst thin, transparent . . . . *G. mali*
– Cuticle of cyst thick, dark in colour 2

2 – Mean length of J2 stylet $\leqslant 26\ \mu m$ 3

Table 14. *Morphological and morphometric characters useful for identification of* Globodera *species (for measurements the range of means are given in μm).*

Stage	Cyst					J2				
Character / Species	Length	Width	Number of cuticular ridges	Granek's ratio	Anus-fenestral edge distance	Body length	Stylet	Shape of stylet knob	Hyaline region	Tail
G. artemisiae	452-493	356-461	8	1.0-3.2	26-56	413-524	21-24	Rounded	23-26	40-55
G. bravoae	739	579	11	1.8	48	485	23	Rounded	32	56
G. leptonepia	500	300	–	–	–	567	18	Anteriorly projected	–	–
G. mali	545	404	7	1.3	–	–	22	Rounded	17	41
G. mexicana **sp. n.**	815	635	–	2.8	59	469	23-24	Anteriorly projected	25-26	52-54
G. millefolii	480-639	380-479	6	1.2-1.6	24-29	483-516	25	Rounded	23-30	49-55
G. pallida	510-675	451-648	12	2.1-2.5	48-54	452-486	23-24	Anteriorly projected	26-27	50-53
G. rostochiensis	445-690	382-559	17-20	3.0-4.5	51-70	392-468	20-22	Rounded	20-27	44-51
G. tabacum	536-767	459-558	7-9	1.6-2.8	41-52	476-556	21-25	Rounded to slightly anteriorly projected	21-28	50-56
G. zelandica	425*	190*	–	–	25*	484	28	Slightly anteriorly projected	37	62

* – Measurements are given for females.

– Mean length of J2 stylet > 26 μm . . . . *G. zelandica*

3 – Mean length of J2 stylet < 19 μm . . . . *G. leptonepia*

– Mean length of J2 stylet ⩾ 19 μm 4

4 – Mean Granek's ratio > 2.9, number of cuticular ridges between anus and fenestra > 14 . . . . *G. rostochiensis*

– Mean Granek's ratio ⩽ 2.9, number of cuticular ridges between anus and fenestra ⩽ 14 5

5 – Hyaline region of J2 > 31 μm . . . . *G. bravoae*

– Hyaline region of J2 ⩽ 31 μm 6

6 – Mean Granek's ratio usually > 2, parasites of Solanaceae 7

– Mean Granek's ratio ⩽ 2, parasites of Asteraceae 9

7 – Mean DGO ⩾ 5.5 μm, stylet knobs rounded or slightly anteriorly projected in J2 . . . . *G. tabacum*

– Mean DGO < 5.5 μm, stylet knobs distinctly anteriorly projected in J2 8

8 – Mean Granek's ratio = 2.1-2.5 . . . . *G. pallida*

– Mean Granek's ratio = 2.8 . . . . *G. mexicana* **sp. n.**

9 – Mean stylet ⩾ 25 μm in J2, gubernaculum = 11.2-12.9 μm. . . . . *G. millefolii*

– Mean stylet < 25 μm in J2, gubernaculum = 6.0-9.9 μm. . . . . *G. artemisiae*

MOLECULAR IDENTIFICATION

Applications of PCR-ITS-RFLP, PCR with species-specific primers and sequencing of the ITS-rRNA gene are the main methods for molecular diagnostics of *Globodera* species. Although currently PCR-ITS-RFLP profiles have been generated and verified for only a few species (Table 15; Fig. 39), it could be considered as a reliable method for identification of *Globodera* species. Restriction of PCR-ITS-rRNA by five restriction enzymes clearly distinguishes several species of *Globodera* from each other. Conventional PCR with species-specific ITS-rRNA primers have been developed for *G. pallida*, *G. rostochiensis* and *G. tabacum* and described by Bulman and Marshall (1997), Fullaondo *et al.* (1999) and Skantar *et al.* (2007) (Table 13).

Table 15. *Approximate sizes (in bp)*[*] *of restriction fragments generated by five diagnostic enzymes after digestion of PCR products of the ITS-rRNA regions amplified by TW81 and AB28 primers for* Globodera *spp.*

Species	Unrestricted	*Alu*I	*Bsh*1236I	*Hinf*I	*Pst*I	*Rsa*I	References
G. rostochiensis	**976**	**380, 362, 148, 85, 4**	**826, 130, 20**	**902, 74**	**642, 334**	**584, 221, 162, 9**	Subbotin *et al.* (1999b); Skantar *et al.* (2007); Nakhla (2008b); Madani (unpubl.); this study
G. pallida type A (European type)	**979**	**511, 382, 82, 4**	***(484)*, 352, 342, 135, 130, 20**	**753, 152, 74**	***(643)*, 504, 336, 139**	**586, 384, 9**	Subbotin *et al.* (1999b); Skantar *et al.* (2007); Nakhla (2008b); Madani (pers. comm.); this study
G. pallida type B (South American type)	**979**	511, 382, 82, 4	829, 130, 20	753, 152, 74	**640, 339**	586, 222, 162, 9	This study
G. tabacum	**980**	**511, 383, 82, 4**	**431, 342, 130, 57, 20**	**906, 74**	**643, 337**	**587, 222, 162, 9**	Skantar *et al.* (2007); Nakhla (2008b); Madani (unpubl.), this study
G. millefolii	990	981, 5, 4	581, 258, 131, 20	990	649, 331	595, 274, 112, 5, 4	This study
G. artemisiae	**989**	903, 77, 5, 4	**839, 130, 20**	**989**	**649, 330**	**594, 274, 115, 5, 4**	This study
G. zelandica	**1013**	923, 90, 4	**292, 282, 160, 110, 113, 25, 20, 11**	**543, 470**	**664, 349**	**605, 168, 116, 115, 5, 4**	This study
G. mexicana **sp. n.**	**980**	511, 383, 82, 4	**488, 342, 130, 20**	**754, 152, 74**	**644, 336**	**587, 384, 9**	This study
Globodera sp. Argentina	977	509, 382, 82, 4	*900*, *500*, 486, 341, 130, 20	*900*, 749, *620*, 151, 73	641, 336	586, 221, 161, 9	This study; Doucet (pers. comm.).
Globodera sp. Portugal	986	900, 82, 4	836, 130, 20	986	646, 340	593, 384, 5, 4	This study

[*] – Fragments obtained in results of virtual sequence digestion using Webcutter 2; **Bold** font – verified by PCR-RFLP study; Normal font – without RFLP verification; *Italic* font – additional fragments, sometimes poorly visualised on a gel.; () – additional fragments for some samples only.

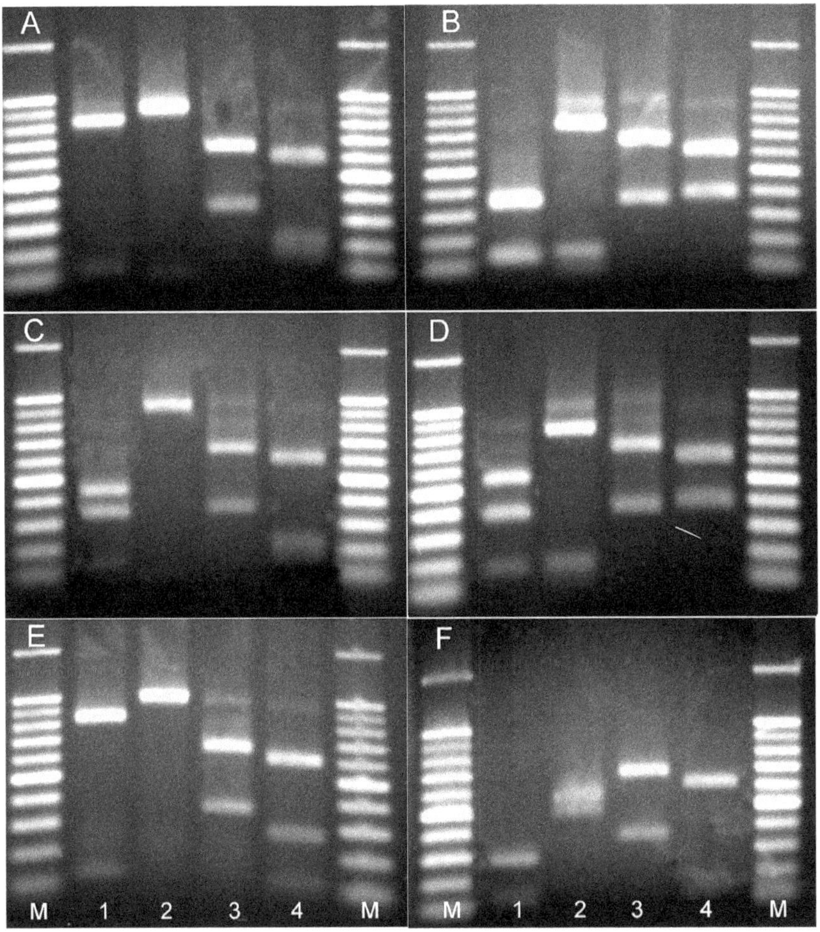

Fig. 39. *PCR-ITS-RFLP profiles for* Globodera *species. A:* G. rostochiensis*; B:* G. pallida*; C:* G. tabacum*; D:* G. mexicana *sp. n.; E:* G. artemisiae*; F:* G. zelandica*; M = 100 bp DNA ladder (Promega); 1 =* Bsh1236*I; 2 =* Hinf*I; 3 =* Pst*I; 4 =* Rsa*I. (Original.)*

SPECIES DESCRIPTIONS

1. *Globodera rostochiensis* (Wollenweber, 1923) Skarbilovich, 1959
(Figs 7E; 40, 41A, B; 42)

Common name: yellow or golden potato cyst nematode.

The golden cyst nematode associated with potato plants was first reported by Kühn (1881) and was considered to be *H. schachtii*, this be-

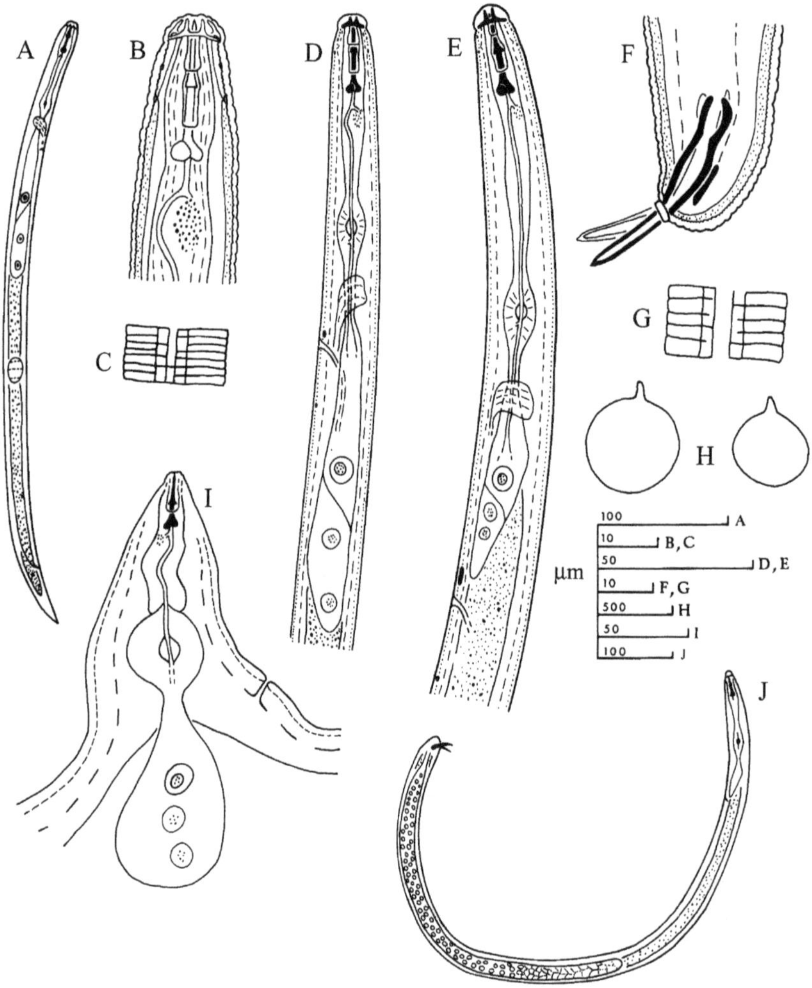

Fig. 40. Globodera rostochiensis. *A-D: J2. A: Entire body; B: Lip region; C: Lateral field; D: Pharyngeal region; E-G: Male. E: Pharyngeal region; F: Tail; G: Lateral field; H: Entire cysts; I: Anterior region of female; J: Entire male. After Stone (1973).*

ing the only known species of cyst nematode at that time. This was, however, an isolated observation and later, during the early 1900s, the potato cyst nematode became more widely known throughout Europe. Many concluded that at least one form of *H. schachtii* was specialised on solanaceous plants, causing "soil sickness of potatoes" (Spears, 1968).

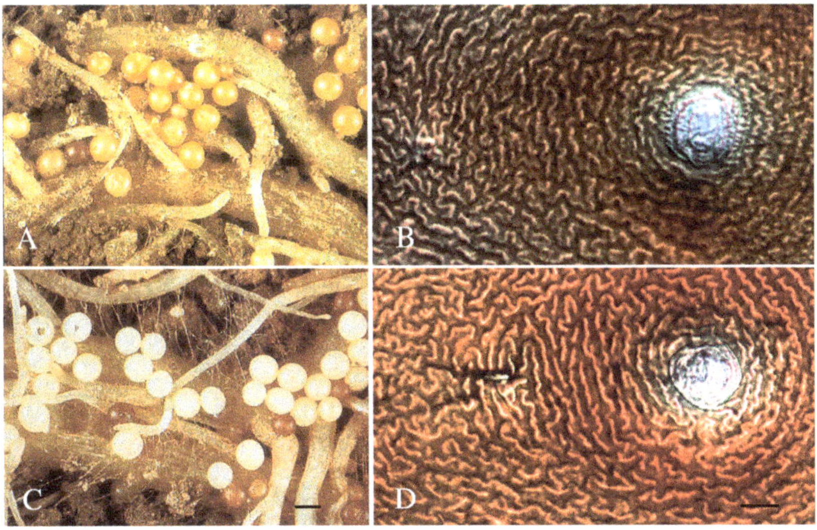

Fig. 41. *Potato cyst nematodes. A, B:* Globodera rostochiensis*; C, D:* G. pallida. *A, C: Females and cysts on roots; B, D: Vulval plates. (A, C courtesy of I. Cid del Prado, Mexico; B, D courtesy of V.N. Chizhov, Russia.) (Scale bars: A, C = 500 μm; B, D = 10 μm.)*

Wollenweber (1923) recognised morphological differences between nematodes parasitising potato and sugar beet and proposed to distinguish two separate species, *H. schachtii* on sugar beet and *H. rostochiensis* on potato. However, some researchers continued to consider potato cyst nematode as a subspecies of *H. schachtii* until 1940 when Franklin published a more complete morphological description and diagnosis of *H. rostochiensis* (Franklin, 1951). Golden and Ellington (1972) designated the neotype and gave a redescription of this species. Morphometrics and descriptions of different stages of *G. rostochiensis* have been provided and reviewed by many authors (Filipjev & Schuurmans Stekhoven, 1941; Franklin, 1951; Trudgill *et al.*, 1970; Webley, 1970; Green, 1971; Behrens, 1972; Øydvin, 1973; Stone, 1973a, b; Evans & Franco, 1977; Baldwin & Mundo-Ocampo, 1991; Gonzales *et al.*, 1992; Manduric & Andersson, 2004; Manduric *et al.*, 2004; Sirca & Urek, 2004; Knoetze *et al.*, 2006; Bacic *et al.*, 2008). Today, *G. rostochiensis* has been reported from many countries and is probably the most economically important nematode pest of potato.

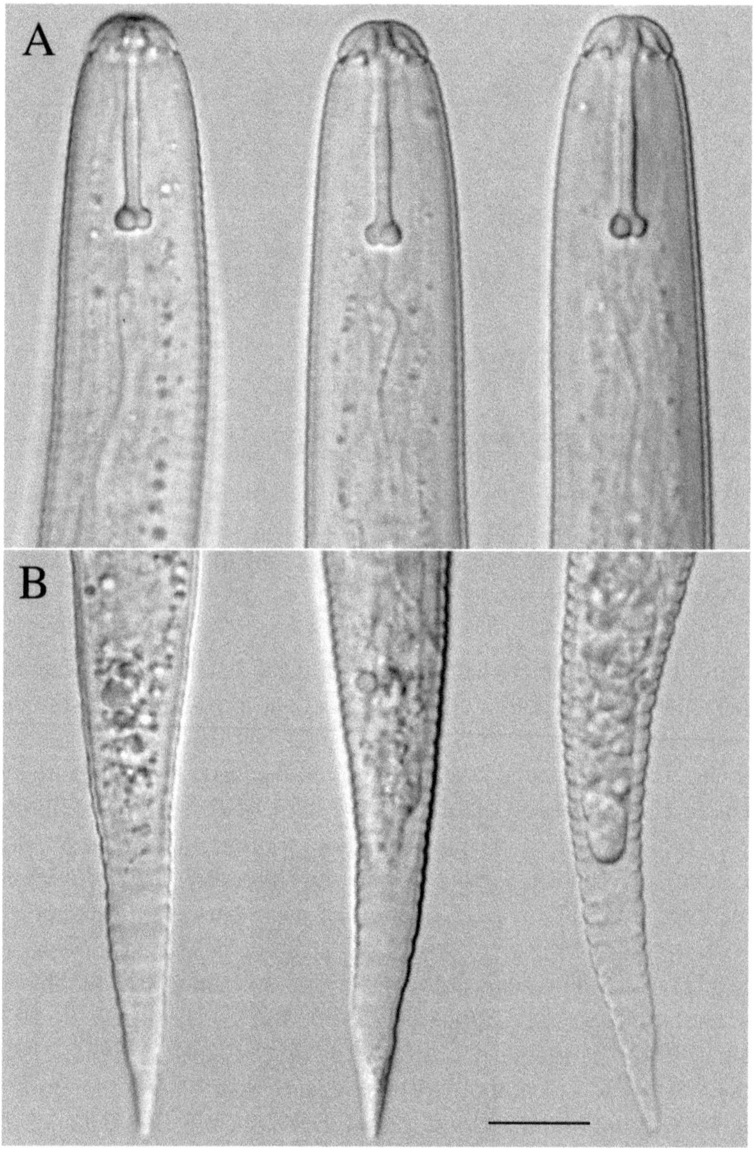

Fig. 42. Globodera rostochiensis. *J2. A: Anterior region; B: Tail. (Original.) (Scale bar = 10 μm.)*

MEASUREMENTS

After Golden and Ellington (1972).

– *Neotype female*: Length = 470 μm; W = 320 μm; L/W ratio = 1.4; stylet = 23.8 μm; DGO = 6.4 μm; anterior end to excretory pore = 130 μm; vulval slit = 10 μm; vulva-anal distance = 41 μm.

– *Female* (n = 50): L (including neck) = 0.52 (0.42-0.64) mm; W = 0.34 (0.27-0.43) mm; L/W ratio = 1.5 (1.2-2.0); stylet = 23 (22-24) μm; DGO = 6.2 (5.8-7.0) μm; anterior end to excretory pore = 131 (105-175) μm; vulval slit = 9 (6-11) μm; fenestral length = 15-30 μm; fenestral width = 18-20 μm; distance from anus to vulva edge = 47 (39-80) μm.

– *Cyst*: see Table 16.

– *Male* (n = 50): L = 1.08 (0.89-1.27) mm; W = 39 (31-46) μm; a = 27 (22-36); b = 5.9 (4.9-7.3); c = 267 (161-664); stylet = 26 (25-27) μm; DGO = 6.4 (5.3-7.0) μm; spicules = 35 (32-39) μm; gubernaculum = 12 (10-14) μm; tail = 4.4 (1.7-6.7) μm.

– *J2*: see Table 16.

– *Egg* (n = 50): L = 105 (95-115) μm; W = 45 (42-48) μm; L/W ratio = 2.3 (2.0-2.6).

After Stone (1973b).

– *Female* (n = 25): Stylet = 22.9 ± 1.2 μm; DGO = 5.7 ± 0.9 μm; labial region diam. = 5.2 ± 0.7 μm; anterior end to: median bulb valve = 73.2 ± 14.6 μm, excretory pore = 145.3 ± 17.4 μm; fenestral diam. = 22.4 ± 2.8 μm; vulval slit = 9.7 ± 1.9 μm; anus to vulval basin = 60.0 ± 10.1 μm; number of cuticular ridges on anal vulval axis = 21.6 ± 3.5.

– *Male* (n = 50): L = 1197 ± 100 μm; labial region diam. = 11.8 ± 0.6 μm; labial region height = 6.8 ± 0.3 μm; stylet = 25.8 ± 0.9 μm; DGO = 5.3 ± 0.9 μm; anterior end to: median bulb valve = 98.5 ± 7.4 μm, excretory pore = 172.3 ± 12.1μm; tail = 5.4 ± 1.1 μm; body diam. at anus level = 13.5 ± 0.4 μm; spicules = 35.5 ± 2.8 μm; gubernaculum = 10.3 ± 1.5 μm.

Onions (1953) reported giant J2 of *G. rostochiensis*: L = 783-817 μm; stylet = 33 μm.

Table 16. *Morphometrics of cysts and J2 of* Globodera rostochiensis *(all measurements in μm).*

Population Characters	Germany, Golden & Ellington (1972)	Germany, Rostock, Stone (1973b)	UK, Feltwell, Evans & Franco (1977)	Russia, Kazachenko (1993)	Pakistan, Shahina & Maqbool (1995)	Slovenia, Libelice, Sirca & Urek (2004)	Serbia, Bata, Bacic *et al.* (2008)
Cyst (n)	50	25	–	25	25	10	10
Length (excl. neck)	680 (450-990)	445 ± 50	–	589 (467-725)	533 (480-665)	–	–
Width	540 (250-810)	382 ± 61	–	559 (430-713)	502 (387-654)	–	–
Length/width	1.27 (1.0-1.8)	–	–	1.04 (0.9-1.3)	–	–	–
Fenestral diam.	15.0 (8-20)	18.8 ± 2.2	20.7	14.1 (10.4-19.5)	17.2 (16-19.5)	17.9 ± 1.7 (14.6-19.4)	18.4 ± 2.1 (14.6-20.7)
Number of ridges		–	17.2	–	–	18.3 ± 1.6 (16-22)	20-24
Anus-fenestra distance	68 (29-116)	66.5 ± 10.3	70.2	50.7 (33.8-78)	61 (53.5-72.3)	64.8 ± 4.7 (58.1-73.3)	55.9 ± 8.3 (44.9-72.9)
Granek's ratio	4.5 (2.0-7.0)	3.6 ± 0.8	3.4	3.7 (2.1-5.4)	–	3.6 ± 0.4 (3.0-4.3)	3.0 ± 0.8 (2.3-4.6)
J2 (n)	50	50	–	25	20	30	10
Length	430 (370-470)	468 ± 20	457.3	392 (366-406)	465 (439-502)	434.4 ± 16.7 (402.7-468.0)	426.1 ± 32.7
a	19 (16-23)	–	–	21.3 (18.6-25.4)	23 (19-29.5)	21.4 ± 1.1 (18.9-23.0)	–
b	2.3 (2.2-2.5)	–	–	3 (2.7-3.4)	–	–	–
c	8 (7-9)	–	–	9 (8.1-11)	12.5 (10.0-15.5)	9.0 ± 0.6 (7.8-11.0)	–

Table 16. *(Continued).*

Population Characters	Germany, Golden & Ellington (1972)	Germany, Rostock, Stone (1973b)	UK, Feltwell, Evans & Franco (1977)	Russia, Kazachenko (1993)	Pakistan, Shahina & Maqbool (1995)	Slovenia, Libelice, Sirca & Urek (2004)	Serbia, Bata, Bacic *et al.* (2008)
Stylet	22 (21-23)	21.8 ± 0.7	21.0	19.9 (19.5-20.8)	22.5 (20.5-23.5)	20.7 ± 0.6 (18.9-22.4)	21.2 ± 0.7 (19.8-21.7)
Labial region height	–	4.6 ± 0.6	–	–	–	–	–
Labial region diam.	–	9.9 ± 0.4	–	–	–	–	–
DGO	5.5 (5.0-6.7)	2.6 ± 0.6	–	5.0 (3.9-5.2)	4.5-5.5	3.8 ± 0.8 (2.4-5.0)	–
Anterior end to excretory pore	–	100.5 ± 2.4	95.5	84.7 (75.6-91)	80-90	–	–
Anterior end to median bulb valve	68 (64-76)	69.2 ± 1.9	64.5	61 (54.6-66.3)	–	65.0 ± 6.9 (55.3-77.6)	–
Body diam. at mid-body	23 (19-26)	–	–	–	–	20.1 ± 0.6 (18.9-22.4)	–
Body diam. at anus level	–	11.4 ± 0.6	–	10.9 (10.4-11.7)	–	12.7 ± 0.7 (11.3-14.4)	–
Hyaline region	24 (18-30)	26.5 ± 1.8	–	19.9 (19.5-20.8)	25.0 (22.2-28.4)	25.1 ± 1.6 (22.5-29.9)	22.4 ± 2.0 (19.4-24.3)
Tail	51 (44-57)	43.9 ± 11.6	46.4	43.5 (36.4-46.8)	48.5 (41.0-51.4)	48.5 ± 2.7 (42.6-53.8)	44.9 ± 4.8 (36.5-53.5)

Description

After Stone (1973b).

Female

Pearly white, subspherical to ovate, with elongate, protruding neck. Colour changing from white to yellow to light golden as female matures to cyst stage. Cuticle thick, with superficial, rugose, lace-like pattern, D-layer present, punctations resolved near or beneath surface. Labial region slightly offset, bearing two annuli. Labial framework weakly developed. Stylet fairly strong, straight to slightly curved, with well developed rounded basal knobs, sloping posteriorly. Median bulb large, nearly spherical, with well developed valve. Pharyngeal glands often obscured but appearing clustered. Excretory pore conspicuous, always at or near base of neck. Vulva terminal, slit of medium length. Vulval area circumfenestrate. No anal fenestration, but anus and vulva both lying in a 'vulval basin', anal area not encircled by cuticular rings. Often beneath vulva, generally in a cluster, are vulval bodies of highly variable size and shape, large superficial tubercles clumped near vulva. Vulva ellipsoid in shape, anus shorter than vulva. All eggs retained in body, no egg mass.

Cyst

Brown, ovate to spherical in shape with protruding neck, circumfenestrate, abullate, without distinct 'vulval bodies' commonly seen in white females. Fenestra circular, anus conspicuous at apex of a V-shaped subsurface cuticular mark. Cyst wall pattern basically as in female but often more prominent, especially near mid-body, tending to form wavy lines going around body. Subcrystalline layer absent. Punctations generally present but variable in intensity and arrangement. Each cyst containing 200-1000 eggs.

Male

Body vermiform, slightly tapering at both anterior and posterior regions. Cuticle with prominent annulation. Labial region slightly offset, hemispherical, with six annuli. Labial framework heavily sclerotised. Stylet strong, with prominent knobs. Anterior and posterior cephalids present. Lateral fields with four equally spaced lines. Median bulb ellipsoidal. Excretory pore *ca* two annuli posterior to often distinct hemizonid. One testis. Spicules slightly arcuate, tips rounded, not notched. Tail short, variable in length and shape.

J2

Body tapering at both extremities but more at posterior end. Cuticular annulation well defined. Lateral fields with four lines extending for most of body length, outer two lines crenate but without areolation. Labial region slightly offset, bearing 4-6 annuli, considerably wider at base than anteriorly, presenting a rounded, though rather anteriorly flattened, appearance. Labial framework heavily sclerotised. Stylet well developed, with prominent rounded knobs as viewed laterally. Anterior and posterior cephalids present. Valve of median bulb prominent, ellipsoidal. Isthmus and pharyngeal glands typical for the genus. Excretory pore almost adjacent yet slightly posterior to hemizonid. Genital primordium slightly posterior to mid-body, with four cells commonly resolved. Tail tapering to small, rounded terminus. Phasmids generally difficult to see, when visible, located about halfway along tail.

TYPE HOST AND LOCALITY

Roots of potato, *Solanum tuberosum* (Solanaceae, Solanales) from Rostock, Germany.

OTHER HOSTS

Tomato, *Solanum lycopersicum* L. and eggplant, *S. melongena* L., are agronomic crops other than potato that are attacked by *G. rostochiensis*. Other hosts include many *Solanum* spp., *Datura* spp. *Hyoscyamus niger* L., *Nicotiana acuminata* Hook., *Physalis* spp. *Physochlaina orientalis* G. Don, *Salpiglossis* spp., *Capsicum annuum* L., *Saracha jaltomata* Schlecht (Mai & Lownsbery, 1948; Mai, 1952; Ellenby, 1954; Stelter & Engel, 1973; Stone, 1973b; Roberts & Stone, 1981; Brodie, 1984; Stelter, 1987; CAB International, 2005a; Sullivan *et al.*, 2007 and others) (Table 17). *Chenopodium album* var. *centrorubrum* Makino (CAC), *Polygonum nepalense* Meisn. and *P. convolvus* L. are poor hosts for this species (Inagaki, 1984).

DISTRIBUTION

The centre of origin of potato cyst nematodes is the Andes Mountains in South America from where they were introduced to Europe with potatoes, probably around 1850 (Evans *et al.*, 1975). From here they were spread with seed potatoes to other areas. The present distribution covers temperate zones down to sea level and in the tropics at higher

Table 17. *Differential host range of* Globodera *parasitising solanaceous plants (after Campos-Vela, 1967; Stelter, 1987 and Mugniéry and Phillips, 2007 with modification).*

	Solanum tuberosum (potato)	*Solanum lycopersicum* (tomato)	*Nicotiana tabacum* (tobacco)	*Solanum nigrum* (nightshade)	*Solanum dulcamara* (bittersweet)
G. rostochiensis	+	+	−	−	+
G. pallida	+	+	−	−	+
G. tabacum	−	+	+	+	+
G. mexicana **sp. n.**	−	+	−/+	+	+

altitudes. In these areas, the distribution is linked with that of the potato crop.

Europe: Albania, Austria, Belarus, Belgium, Bulgaria, Croatia, Cyprus, Czech Republic, Denmark, Estonia, Finland, France, Germany, Greece, Hungary, Iceland, Ireland, Latvia, Lithuania, Luxembourg, Malta, The Netherlands, Norway, Poland, Portugal, Romania, Russia, Serbia, Slovakia, Slovenia, Spain, Sweden, Switzerland, UK, Ukraine, Yugoslavia (former); *Asia*: Armenia, Georgia, India, Indonesia, Iran, Japan, Lebanon, Oman, Pakistan, Philippines, Sri Lanka, Tajikistan, Turkey; *Africa*: Algeria, Egypt, Libya, Morocco, Sierra Leone, South Africa, Tunisia; *North and Central America*: Canada, Costa Rica, Mexico, Panama, USA (New York; eradicated in Delaware); *South America*: Argentina (eradicated), Bolivia, Brazil, Chile, Colombia, Ecuador, Guatemala, Peru, Venezuela; *Oceania*: Australia, New Zealand (EPPO Quarantine Pest).

Note

Globodera arenaria, recently described by Chizhov *et al.* (2008), is synonymised here with *G. rostochiensis*, because it does not differ morphologically and molecularly from the golden potato cyst nematode (Subbotin, unpubl.).

Relationships

Globodera rostochiensis is morphologically similar to *G. pallida* and *G. tabacum*. It differs from *G. pallida* by yellow or gold *vs* cream coloured maturing females, higher number of ridges between the vulva and anus (16-31 *vs* 8-20), larger mean for Granek's ratio (3.0-4.5 *vs* 2.1-2.5), stylet knob shape (rounded *vs* anteriorly projected), shorter average stylet length (20-22 *vs* 23-24 μm) and rounded *vs* more pointed J2 tail terminus. It differs from *G. tabacum* by shorter mean J2 body length (392-468 *vs* 476-556 μm), shorter mean stylet (20-22 *vs* 21-25 μm) and in the cyst by higher mean number of cuticular ridges (17-20 *vs* 7-9) (Table 14).

PCR-RAPD (Thiéry *et al.*, 1997a) and ITS-rRNA analysis (Thiéry & Mugniéry, 1996; Subbotin *et al.*, 2000b) have revealed that *G. rostochiensis* is more closely related to *G. tabacum* than to other species (Fig. 27).

BIOLOGY

In temperate regions, *G. rostochiensis* usually completes only one generation (Fig. 2) (Morris, 1971), although a second generation may be initiated but not completed (Jones, 1950; Evans, 1969). In subtropical regions two generations might occur (Philis, 1980; Greco *et al.*, 1988). Development of one generation requires 6-10 weeks. The J2 can go into diapause and remain viable for many years, hatching continuing for 25 or more years.

PATHOGENICITY

Heavily infected plants become yellow and stunted. Infected plants have reduced root systems which are abnormally branched and are brownish in colour. Symptoms in the field first appear in small patches. At low nematode densities tuber sizes are reduced, whereas at higher densities both number and size of tubers can be reduced. At eight and 64 eggs/g of soil, yield losses of about 20 and 70%, respectively, can be expected (Greco, 1988). Seinhorst (1965, 1982), Brown (1969) and Elston *et al.* (1991) proposed a model that described the relationships of potato cyst nematode population densities before planting with potato yield and post-harvest nematode populations. The damaging effect of potato cyst nematodes is not only determined by nematode density, but also by such factors as cultivar, crop husbandry and environmental conditions (Mulder *et al.*, 1997). Potato cyst nematodes are responsible for annual potato tuber losses of up to 9% in Europe (Evans & Stone, 1977).

BIOCHEMICAL AND MOLECULAR CHARACTERISATION

Application of the IEF of proteins and PCR-ITS-RFLP for differentiation of *G. rostochiensis* from other *Globodera* species has been tested by several authors (Tables 9, 12). The enzymes *Alu*I generated a unique profile for *G. rostochiensis* after restriction of the PCR-ITS product. RFLP profile and approximate sizes of restriction fragments generated by diagnostic enzymes for this species are given in Figure 39 and Table 15, respectively. Species specific PCR primers for identification of this species are given in Table 13. IEF profiling is a reliable tool to distinguish the two species of potato cyst nematodes from each other (Fig. 33).

2. *Globodera artemisiae* (Eroshenko & Kazachenko, 1972) Behrens, 1975

(Figs 43, 44)

Common name: wormwood cyst nematode.

The wormwood cyst nematode was isolated from roots of *Artemisia rubripes* Nakai in the Primorsky Territory of the Russian Far East and was described by Eroshenko and Kazachenko (1972) as *H. artemisiae*. In 1976, similar cyst nematodes were found in roots of *A. princeps* Pamp. in the Shimabara Peninsula, Nagasaki Prefecture, Japan, and described by Ogawa *et al.* (1983) as *G. hypolysi*. Ogawa *et al.* (1983) reported that there were differences between *G. hypolysi* and *G. artemisiae* with respect to the distance from the anus to the fenestra in the cysts and the female and in body length and shape of stylet knobs of the J2. However, comparison of morphometrics of different stages of different populations provided by Chen *et al.* (1994), Brzeski (1998), Choi and Kim (2001), Manduric and Andersson (2004) and Dobosz *et al.* (2006) did not support this claim. Comparison of the ITS-rRNA gene sequences for both species did not reveal any significant differences and thus this data supported the synonym of *G. hypolysi* with *G. artemisiae* (Subbotin *et al.*, unpubl.).

MEASUREMENTS

After Eroshenko and Kazachenko (1972).

– *Paratype females* (n = 25): L = 562 (504-650) μm; W = 345 (275-388) μm; neck = 105 (10-110) μm; stylet = 23.2 (22.4-23.8) μm;

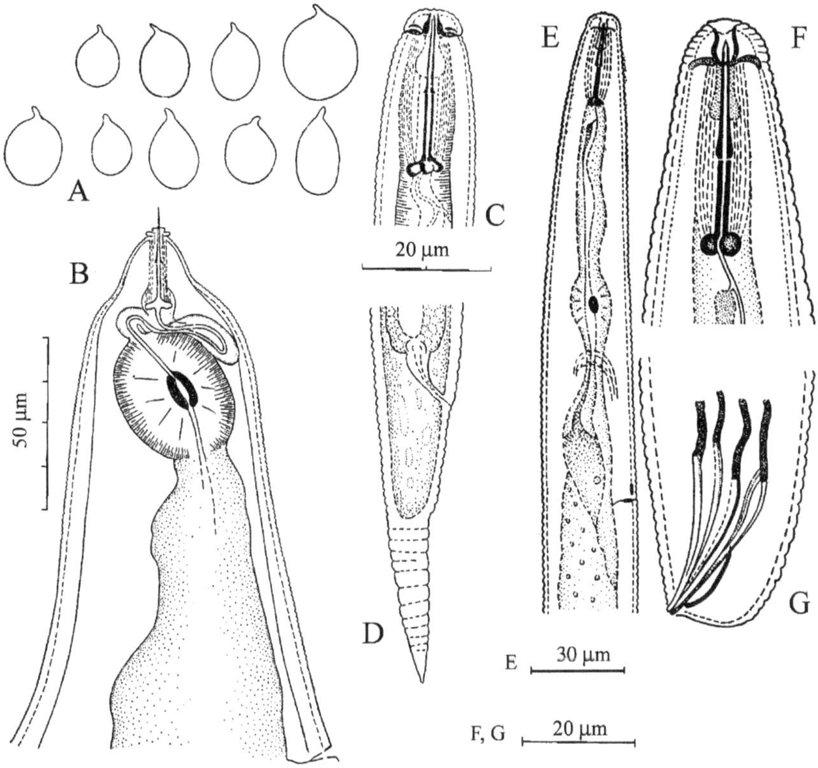

Fig. 43. Globodera artemisiae. *A: Cysts; B: Anterior region of female; C: Anterior region of J2; D: Tail of J2; E: Anterior end of male; F: Head of male; G: Tail of male. After Eroshenko and Kazachenko (1972, 1983).*

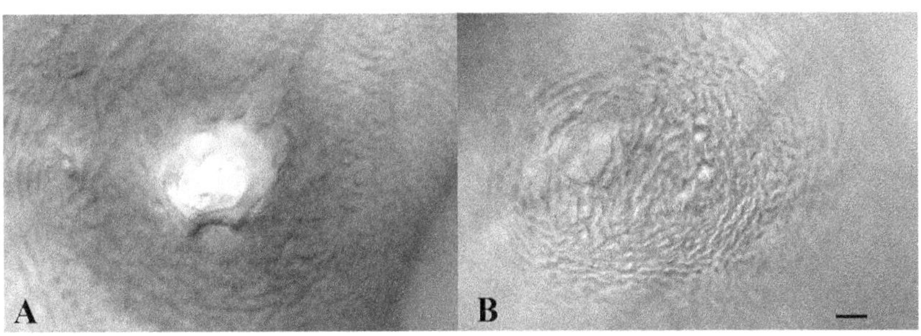

Fig. 44. Globodera artemisiae. *A, B: Vulval plates. (Original.) (Scale bar = 10 μm.)*

DGO = 5 μm; anterior end to excretory pore = 103 (85-128) μm; vulval slit = 4.3 (3-6.3) μm; fenestral diam. = 13.3 (10-16.8) μm; distance from anus to fenestral edge = 31.8 (25.2-42) μm.

– *Paratype cysts*: see Table 18.

– *Paratype J2*: see Table 18.

– *Paratype eggs*: L = 98.7 (69-114) μm; W = 44.7 (33-60) μm; L/W ratio = 2.0.

After Ogawa *et al.* (1983).

– *Female* (n = 85): L = 415 (320-540) μm; W = 317.6 (210-470) μm; L/W ratio = 1.3 (1.0-1.7); neck = 96.7 (70-210) μm; stylet = 24.4 (22-26) μm; DGO = 4.3 (2-5) μm; anterior end to excretory pore = 130.7 (99-178) μm; vulval slit = 5.1 (4-6) μm; vulva to anus distance = 47.5 (32-57) μm.

– *Male*: see Table 19.

– *Egg* (n = 100): L = 100.9 (85-120) μm; W = 42.6 (38-50) μm; L/W ratio = 2.4 (1.7-3.1).

After Choi and Kim (2001).

– *Female* (n = 34): L = 382.3 $\pm$ 30.9 (326.4-438.4) μm; W = 305.3 $\pm$ 38.3 (220.8-380.8) μm; L/W ratio = 1.3 $\pm$ 0.1 (1.1-1.5); neck = 77.1$\pm$17.3 (41.6-112.0) μm; stylet = 25.5$\pm$2.2 (18.0-28.5) μm.

DESCRIPTION

Female

Body with rounded posterior end, egg-shaped to almost spherical with elongated, protruding neck. Labial region offset from neck, with two cuticular annuli. Stylet strong with well developed knobs sloping posteriorly. Procorpus short. Median bulb large, oval. Cuticle white, maturing to yellow-brown. Vulval lips thick, hemispherical. Vulval slit short. Anus small, pore-like.

Cyst

More rounded than female, light to dark brown in colour. Generally circumfenestrate. Bullae-like structures forming compact circles around fenestra. Cyst wall pattern basically as in female but often more prominent, especially near mid-body, tending to form wavy lines extending around body.

Table 18. *Morphometrics of cysts and J2 of* Globodera artemisiae *(all measurements in μm).*

Population / Characters	Paratypes Russia, Eroshenko & Kazachenko (1972)	Japan, Ogawa *et al.* (1983)	China, Chen *et al.* (1994)	Poland, Brzeski (1998)	Korea, Choi & Kim (2001)	Japan, Yamada & Takakura (2003)	Germany, Sturhan, unpubl.; Manduric & Andersson (2004)	Sweden, Manduric & Andersson (2004)
Cyst (n)	25	50	–	–	73	50	–	20-100
Length (excl. neck)	493 (414-587)	475.6 (370-600)	–	410-590	452 ± 55.2 (338-590)	519 ± 77.8 (375-750)	–	
Width	356 (299-575)	448.1 (320-620)	–	300-570	377.8 ± 84.0 (216-576)	461.6 ± 84.4 (350-625)	–	
Length/width	–	1.15 (0.9-1.4)	–	–	1.2 ± 0.2 (0.9-1.6)	1.1 ± 0.1 (1.0-1.4)	–	
Fenestral length	24.9 (19.8-32.4)	20.9 (16-27)	25	18 (12-30)	21.7 ± 2.9 (16.2-28.8)	23.4 ± 3.8 (18-29)	–	22.9 ± 1.8 (17.6-26.4)
Fenestral width	23 (14.4-27.6)	18.7 (15-25)	–	–	18.6 ± 3.0 (10.8-25.2)	–	–	
Number of ridges	–	–	–	–	–	–	–	7.8 ± 2.0 (5-16)
Anus-fenestral edge distance	26 (19.2-46.8)	43.6 (25-80)	33.5	56 (26-116)	35 ± 8.9 (23.3-54)	38.6 ± 9.3 (23-52)		28.8 ± 4.1 (24.2-37.4)
Granek's ratio	1.0 (0.8-1.7)	2.1 (1.2-3.5)	1.3	3.2 (1.6-8.3)	2.0 ± 0.6 (0.7-3.1)	1.7 ± 0.4 (0.9-2.2)	–	1.3 ± 0.2 (1.0-1.7)
J2 (n)	25	50	–	–	22	30	–	20-100
Length	413 (357-490)	460.6 (410-510)	476.5	420 (390-450)	457 ± 18.5 (421.2-486.0)	524.2 ± 28.3 (460-590)	458.2 ± 21.2 (397-495)	461.8 ± 23.5 (417.0-520.1)
a	23.3 (21.2-26.8)	26.8 (24.4-31.4)	23.7	22.4 (20-24)	26.2 ± 2.2 (23.6-33.3)	28.7 ± 1.6 (26.0-32.6)	–	27.8 ± 1.2 (24.9-30.7)

Table 18. *(Continued).*

Population / Characters	Paratypes Russia, Eroshenko & Kazachenko (1972)	Japan, Ogawa *et al.* (1983)	China, Chen *et al.* (1994)	Poland, Brzeski (1998)	Korea, Choi & Kim (2001)	Japan, Yamada & Takakura (2003)	Germany, Sturhan, unpubl.; Manduric & Andersson (2004)	Sweden, Manduric & Andersson (2004)
b	3.3 (2.1-4.9)	4.8 (4.3-5.2)	–	4.3 (3.9-4.9)	–	5.2 ± 0.2 (4.7-5.5)	–	–
c	9.6 (6.1-14)	9.8 (7.4-11.0)	–	8.9 (7.2-9.4)	9.4 ± 1 (7.0-11.8)	9.6 ± 0.6 (8.0-10.9)	–	9.9 ± 0.8 (7.4-12.0)
Stylet	22.6 (18-29)	24.6 (24-25)	23.9	21 (19-22.5)	24.4 ± 3.1 (21.2-35.8)	24.1 ± 0.4 (23.0-24.5)	23.7 ± 0.7 (21.8-24.6)	22.3 (20.4-23.6)
Labial region height	–	4.1 (4-5)	–	4-5	–	–	–	–
Labial region diam.	–	9.8 (9-10)	–	9-10	–	–	–	–
DGO	–	–	–	2.5-9	–	5.2 ± 0.6 (4.0-6.5)	–	–
Anterior end to excretory pore	88	–	–	93 (81-101)	–	–	–	–
Anterior end to median bulb valve (MB)	–	68.5 (65-72)	–	64 (54-69)	–	–	–	–
Body diam. at mid-body	–	–	–	–	–	18.3 ± 1.0 (16-20)	–	–
Body diam. at anus level	–	12.1 (11-13)	–	–	–	–	–	–
Hyaline region	23 (18-28)	23.6 (21-27)	25.4	24 (17-28)	24.8 ± 4.5 (19.0-41.6)	25.3 ± 3.0 (17.8-36.0)	25.7 ± 2.3 (21.5-29.5)	23.1 ± 1.5 (20-26)
Tail	40 (33-64)	47.2 (42-61)	47.2	47 (44-57)	49.2 ± 5.5 (40.2-69.4)	54.9 ± 3.8 (47.0-65.0)	49.5 ± 3.2 (43.1-53.4)	46.7 ± 3.6 (41.2-62.4)

Table 19. *Morphometrics of males of* Globodera artemisiae *(all measurements in μm).*

Characters	Armenia, Poghossian & Karapetyan (1975)	Paratypes, Russia, Eroshenko & Kazachenko (1983)	Japan, Ogawa *et al.* (1983)	Korea, Choi & Kim (2001)
n	15	25	50	23
L	1115 (975-1400)	1021 (816-1211)	982.8 (715-1200)	893.3 ± 83.8 (838.8-1087.2)
a	42.5 (33.6-56)	38.7 (34.8-45.6)	42.2 (35.2-53.2)	37.6 ± 3.9 (35.8-43.8)
b	6.5 (5.7-8.7)	5.8 (4.6-6.7)	7.3 (6.3-8.0)	7.3 ± 0.8 (0-9.3)
b′	–	–	5.0 (3.7-6.4)	4.6 ± 0.4 (4.0-5.3)
c	103.3 (80.8-169.3)	263.9 (157-378)	–	–
Stylet	26.8 (22.5-30)	26.5 (24-28.2)	27.4 (24-30)	27.4 ± 1.1 (26.3-29.9)
DGO	3.1 (2.2-3.3)	3.5 (3-5.4)	3.6 (3-5)	–
Anterior end to				
junction of pharynx and intestine	–	156-192	132.5 (119-153)	–
median bulb valve	–	–	86.4 (75-98)	84.3 ± 11.2 (67.9-93.4)
pharyngeal gland base	–	–	198.5 (154-238)	–
excretory pore	–	124-148	145.9 (104-169)	145.4 ± 13.1 (111.0-171.6)
Spicules	28.6-35.2	27.9 (20.8-32.5)	29.2 (24-33)	27.6 ± 2.6 (23.4-32.9)
Gubernaculum	9.0 (7.7-9.9)	8.3 (7.2-9.1)	7.2 (6-8)	–

Male

Body slender, vermiform with short bluntly rounded tail of variable shape. Labial region offset, semispherical with 5-7 annuli. Labial framework heavily sclerotised. Stylet well developed, with rounded basal knobs. Hemizonid two annuli long, situated four or five annuli anterior to excretory pore. Testis single, *ca* 50% of body length. Spicules slightly arcuate, tips rounded, not notched.

J2

Body vermiform, slightly curved ventrally when killed by gentle heat. Labial region slightly offset, with five or six annuli. Stylet well developed, with rounded or anteriorly concave basal knobs. Lateral field with four incisures. Valve of median bulb prominent. Hemizonid two annuli long, situated one annulus anterior to excretory pore. Tail tapering to small, sharply pointed, terminus. Phasmid minute.

TYPE HOST AND LOCALITY

Wormwood, *Artemisia rubripes* (Asteraceae, Asterales) Nakai, at 224 km on Far East Railway, Khasan district, Primosky Territory, Russia.

OTHER HOSTS

Artemisia princeps Pamp., *A. vulgaris* L. and *A. montana* (Nakai) Pamp. (Ogawa *et al.*, 1983; Manduric & Andersson, 2004). Experiments with several species of the family Asteraceae, including hosts of *G. millefolii*, revealed that the Russian Far East population of *G. artemisiae* failed to develop on them. Of many wormwood species, native to the Russian Far East, none has been recorded as a host of *G. artemisiae* under natural conditions (Krall, 1977).

DISTRIBUTION

Asia: Armenia (Poghossian & Karapetyan, 1976), China (Chen *et al.*, 1994), Japan (Ogawa *et al.*, 1983), Kazakhstan (Kirjanova *et al.*, 1976), Russian Far East (Eroshenko & Kazachenko, 1972; Krall, 1977); *Europe*: Germany (Sturhan, 1988), Poland (Brzeski, 1998; Dobosz *et al.*, 2006), Sweden (Manduric & Andersson, 2004).

BIOLOGY

Observations of the life cycle in pots showed that the species developed only one generation per season (Dobosz, 2007).

RELATIONSHIPS

Globodera artemisiae is closely related to *G. millefolii* from which it is distinguished by smaller mean J2 stylet (21-24 *vs* 25 μm), smaller gubernaculum (6.0-9.9 *vs* 11.2-12.9 μm) and a distinctive host range (Table 14).

BIOCHEMICAL AND MOLECULAR CHARACTERISATION

Uehara *et al.* (2005, 2006) provided a diagnostic PCR-ITS-RFLP profile for *G. artemisiae*. Restriction of the PCR-ITS product by enzymes *Alu*I generated a unique profile for this species. A combination of several enzymes distinguishes this species from other *Globodera* (Fig. 39; Table 15). Dobosz *et al.* (2006) developed primers for identification of this species.

3. *Globodera bravoae* Franco, Cid del Prado & Lamothe-Argumedo, 2000
(Figs 45, 46)

In 1998, a *Globodera* population was detected at 'La Cañada de Contreras' in Magdalena Contreras Delegation, Mexico City, parasitising *Jaltomata procumbens* (Cav.) J.L. Centy. Detailed examination of this species revealed significant differences between it and other species and it was described as a new species by Franco *et al.* (2000).

MEASUREMENTS

After Franco *et al.* (2000).

– *Holotype female*: L (including neck) = 650 μm; W = 420 μm; L/W ratio = 1.5; neck = 156 μm; anterior end to: median bulb valve = 81.6 μm, excretory pore = 132 μm; labial region height = 2.0 μm; labial region diam. = 6.8 μm; stylet = 24.8 μm; DGO = 7.6 μm; vulval slit = 12.4 μm; fenestral length = 20.8 μm; fenestral width = 21.2 μm; vulva to anus distance = 44 μm; fenestral diam. (length plus width/2) =

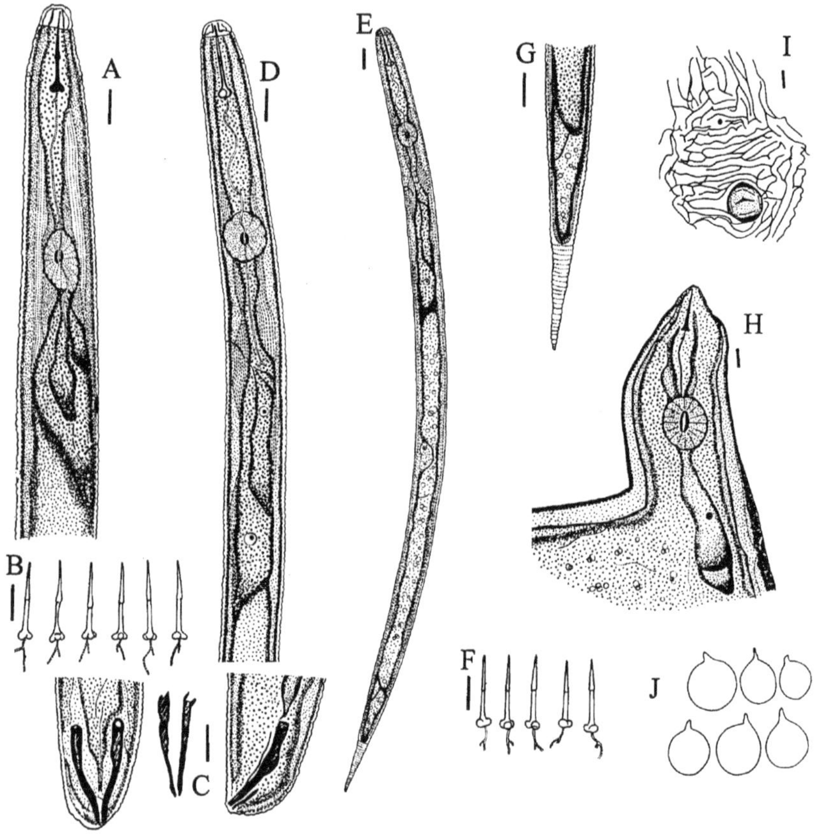

Fig. 45. Globodera bravoae. *A-C: Male. A: Anterior region; B: Stylet; C: Tail and spicules; D-G: J2; E: Entire body; G: Tail; F: Stylet; H: Anterior region of female; I: Fenestral pattern; J: Cysts. After Franco* et al. *(2000). (Scale bars: A-H = 10 μm; I = 20 μm.)*

21 μm; distance anus to vulva edge = 34 μm; number of ridges between vulva and anus: 10; Granek's ratio = 1.62.

– *Paratype females* (n = 50): L (including neck) = 534 ± 26.7 (152-836) μm; W = 487.5 ± 17.7 (368-620) μm; L/W ratio = 1.4 ± 0.1 (0.4-1.7); neck = 36.2 ± 7.9 (76-196) μm; anterior end to: median bulb valve = 87.4 ± 3.0 (73.2-106.4) μm, excretory pore = 144.2 ± 3.7 (129.2-169.2) μm; median bulb length = 34.6 ± 1.9 (24.8-52.4) μm; median bulb diam. = 30.1 ± 1.1 (24.8-36.4) μm; labial region height = 2.7 (1.6-4.0) μm; labial region diam. = 6.8 ± 0.4 (4.8-9.6) μm; stylet = 24.5 ± 0.7 (21.6-28.0) μm; DGO = 7.8 ± 0.3 (6.0-9.2) μm; vulval

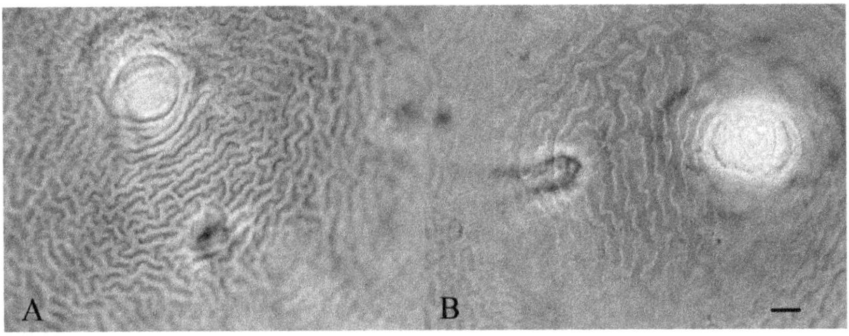

Fig. 46. Globodera bravoae. *Vulval plate. (Courtesy of Cid del Prado, Mexico.) (Scale bar = 10 μm.)*

slit length = 12.6 ± 0.4 (9.6-14.8) μm; fenestral length = 21.6 ± 0.8 (15.2-28.4) μm; fenestral width = 22.8 ± 0.8 (16.8-28.4) μm; fenestral diam. (length plus width/2) = 22.2 ± 0.7 (16.4-28.4) μm; vulva to anus distance = 49.3 ± 2.4 (36-72) μm; distance from vulva to anus border = 39.5 ± 2.2 (28.0-60.0) μm; number of ridges between vulva and anus = 11.9 ± 0.5 (8.0-15.0).

– *Paratype cysts* (n = 40): L (including neck) = 739 + 35.6 (459-936) μm; W = 579 ± 40.1 (296-749) μm; L/W ratio = 1.3 ± 0.1 (1.1-1.6); neck = 123.6 ± 5.8 (100-192) μm; vulval slit = 13.6 ± 0.5 (12.0-14.8) μm; fenestral length = 27.8 ± 1.1 (22-35) μm; fenestral width = 26.1 ± 1.3 (20.0-35.6) μm; fenestral length plus width/2 = 27.0 ± 1.0 (22.0-35.4) μm; distance from anus to external edge of fenestra = 47.5 ± 4.6 (34-80) μm; number of ridges between vulva and anus = 10.7 (8-15); Granek's ratio = 1.8 (1.3-2.7).

– *Paratype males* (n = 40): L = 1017 ± 26.6 (819-1186) μm; W = 32.1 ± 1.6 (24.4-44.0) μm; a = 32.5 ± 1.9 (22.9-41.3); c = 268.6 ± 6.2 (134.9-459.7); c′ = 0.3 ± 0.01 (0.2-0.5); anterior end to: median bulb valve = 83.3±2.7 (63.6-103.2) μm, excretory pore = 141.4±5.0 (96.4-169.9) μm, hemizonid = 140.9 ± 7.8 (112.4-153.6) μm; labial region height = 6.2 ± 0.4 (4.4-8.4) μm; labial region diam. = 11.3 ± 0.3 (8.4-12.4) μm; stylet = 24.2 ± 0.7 (18.8-28.0) μm; DGO = 4.7 ± 0.2 (4.0-6.4) μm; testis = 460.6±43.7 (206.0-751.6) μm; T = 44.9±11.9 (22.5-66.4); spicules = 31.1 ± 1.2 (24-42) μm; gubernaculum = 11.1 ± 0.5 (8.8-12.0) μm; tail = 4.1 ± 0.4 (2.4-7.6) μm; diam. at anus level = 13.9 ± 0.8 (8.4-19.6) μm.

– *Paratype J2* (n = 59): L = 484.6 ± 6.9 (442-553) μm; W = 19.2 ± 0.6 (16-24) μm; a = 25.5 ± 0.6 (20.4-31.1); b = 4.4 ± 0.15

(4.25-4.78); b′ = 2.6 ± 0.1 (2.3-3.1); c = 8.6 ± 0.1 (7.5-10.2); c′ = 4.9 ± 0.1 (4.2-6.2); stylet = 22.6 ± 0.5 (18.4-26.8) μm; stylet knob length = 2.1 ± 0.1 (1.6-3.2) μm; stylet knobs width = 4.3 ± 0.1 (3.6-6.0) μm; labial region height = 4.3 ± 0.1 (3.2-5.6) μm; labial region diam. = 9.2 (8.0-11.6) μm; DGO = 6.8 ± 0.2 (5.6-8.0) μm; anterior end to: median bulb valve = 67.6 ± 1.2 (54-83) μm, excretory pore = 101.5 ± 2.2 (87-140) μm, genital primordium = 270 ± 5.6 (236-305) μm; tail = 56.2 ± 1.1 (45-66) μm; anal body diam. = 11.5 ± 0.3 (8.8-13.6) μm; hyaline region = 32.2 ± 1.0 (24.8-41.2) μm.

– *Paratype eggs* (n = 67): L = 113.5 ± 2.3 (94-133) μm; W = 44.2 ± 0.8 (39-50) μm; L/W ratio = 2.6 ± 0.1 (2.0-3.1).

DESCRIPTION

Female

Body spherical to oval, white in colour, progressively changing to yellow, pale brown and dark brown as female reaches cyst stage. Labial region with two annuli, labial framework weak, labial disc rectangular, with four small lobes, well developed stylet with a strong base and prominent round knobs posteriorly directed. Pharyngeal glands forming a small gland bulb. Neck with transverse ridges at base changing posteriorly to longitudinal ridges beginning about mid-body, excretory pore ovoid, prominent, located at base of neck. Circular fenestra with abundant small tubercles dorsal to vulval slit, but decreasing on ventral side. Anus close to vulva, parallel ridges between vulva and anus with few anastomoses, cuticular overgrowths in form of small vesicles found close to the fenestra.

Cyst

Spherical, large, with long and robust neck, pale to dark brown in colour. Cuticular ridges longitudinally arranged at mid-body but in perineal region concentric and encircling fenestra. Circular fenestra with abundant small tubercles dorsally but fewer ventral to vulval slit. Anus conspicuous, delimited by small parallel ridges between vulva and anus, ridges parallel with few anastomoses. Abundant punctations, following annular pattern of cuticle.

Male

Body vermiform. Labial region with five annuli, labial disc oval, submedial, subdorsal and subventral lips usually separate, but rarely fused. Labial framework heavily sclerotised, anterior cephalids on third annulus and posterior cephalids between ninth and eleventh annuli. Lateral fields with four incisures, reaching end of body, outer lines with regular areolation. Stylet knobs rounded, conspicuous, directed posteriorly. Pharyngeal gland short, overlapping intestine ventrally. Excretory pore at level of posterior end of pharyngeal gland. Testis extending for half of body length. Tail short, rounded.

J2

Body elongate with cephalic region broad and rounded, body annuli conspicuous, labial disc octagonal, rounded dorsally and ventrally, labial framework heavily sclerotised. Lateral fields with four incisures, outer bands irregularly areolated, width of lateral fields at mid-body = 4.0 (3.6-4.8) μm, occupying 20% of body diam. Anterior cephalids on third body annulus, posterior cephalids on sixth to eighth annuli, median bulb rounded, pharyngeal glands elongate, filling body cavity. Stylet strong with prominent rounded knobs, dorsal knob rounded, projecting anteriorly. Tail terminus conical, acute, hyaline region with 25 annuli, phasmid openings pore-like, 8-9 annuli posterior to anus.

Type host and locality

Roots and rhizosphere of *Jaltomata procumbens* (Cav.) J.L. Gentry (Solanaceae, Solanales), Gentry at 'La Cañada de Contreras', Delegación Magdalena Contreras, Mexico City, Mexico.

Distribution

North America: Mexico.

Relationships

The morphology of *G. bravoae* differs from other members of the genus by longer mean cyst length (739 *vs* <690 μm, except for *G. tabacum*, where the upper range extends to 767 μm), longer mean DGO in J2 (6.8 *vs* 4.5-6.1 μm, except for *G. leptonepia* with DGO = 12 μm) (Table 14).

Biochemical and molecular characterisation

No available data.

4. *Globodera leptonepia* (Cobb & Taylor, 1953) Skarbilovich, 1959 (Figs 47, 48)

This species was found in a sample of cysts submitted for identification by Mr P.F. Frink of the Bureau of Entomology and Plant Quarantine, USDA, USA. The material was collected at the Oakland, California, USA, port of entry. The cysts had been washed from soil from small potatoes taken from the ship's stores at Callao, Peru, and presumably grown in that country. Of the 15 cysts in this sample, 12 were identified as the golden potato nematode and three belonged to a new species described as *H. leptonepia* by Cobb and Taylor (1953). Two of these cysts contained eggs with J2 and the strikingly slender appearance of these J2 indicated that they were different from any known species of the genus.

MEASUREMENTS

After Cobb and Taylor (1953).

- *Paratype cysts* (n = 3): L = 500 μm; W = 300 μm.
- *Paratype J2* (n = 20): L = 567 (520-600) μm; a = 39; b = 2.7; c = 9.1; stylet = 18 μm; DGO = 12 μm.
- *Paratype eggs*: L = 95 μm; W = 45 μm.

DESCRIPTION

Cyst

Light brown, more or less ovate, with distinct neck, smoothly rounded posteriorly. Fenestra round, much larger than minute, pore-like anus. Outer layer of cyst wall with rugose pattern of striae extending from neck to near vulva, immediately around vulva, striae interrupted, forming an irregular pattern, lower layer of cyst wall distinctly punctate, with minute dots arranged in closely spaced parallel rows at right angles to cyst axis.

J2

Much more slender than in other known species of *Globodera*. Labial region offset by constriction, no annuli seen. Lateral fields with four faint lines or grooves, apparently ending slightly posterior to anus. Phasmid not observed. Stylet slender, knobs anteriorly convex. Median bulb ellipsoidal. Dorsal pharyngeal gland forming a more or less distinct lobe on dorsal side of body, two subventral glands forming a much elongated

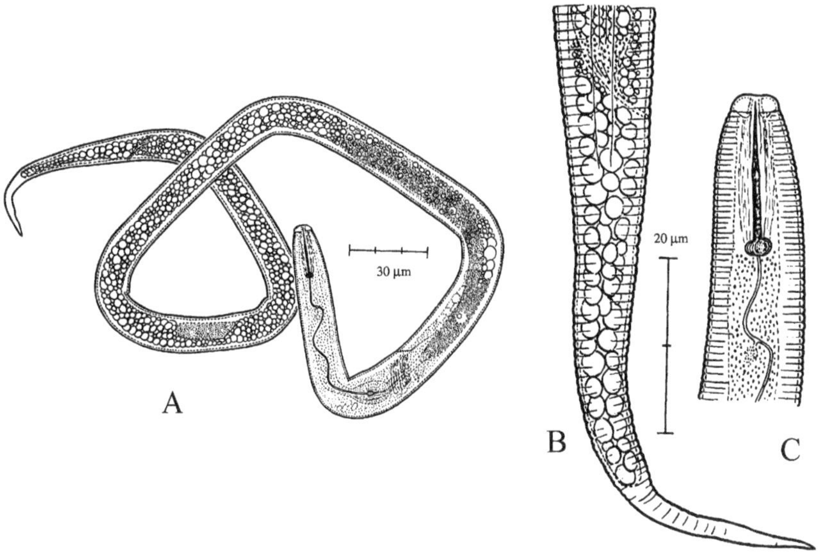

Fig. 47. Globodera leptonepia. *A: Entire body of J2; B: Head of juvenile; C: Tail of J2. After Cobb and Taylor (1953).*

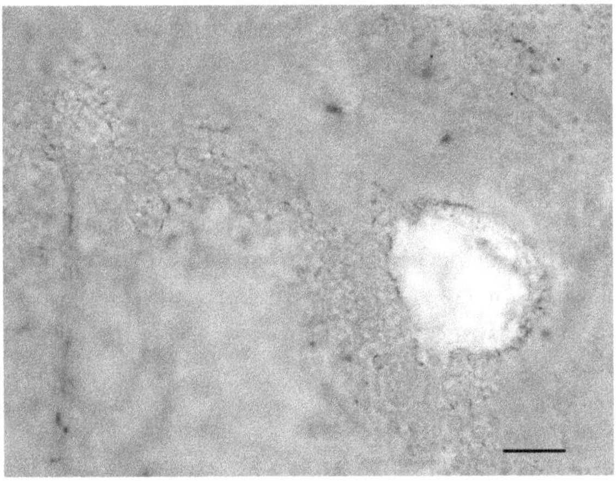

Fig. 48. Globodera leptonepia. *Vulval cones. (Original.) (Scale bar = 10 μm.)*

lobe ventral to intestine. Excretory pore more or less opposite nerve ring. Genital primordia visible as group of cells *ca* two-thirds of body length from anterior end. Hyaline region longer than stylet, a little less than one-third of tail. Terminus very sharply pointed.

Egg

Shell without visible markings.

TYPE HOST

Potato, *Solanum tuberosum* L. (Solanaceae, Solanales).

DISTRIBUTION

South America: possibly Chile and Peru.

RELATIONSHIPS

This species differs from other species by shorter stylet (<19 *vs* >19 μm), longer DGO (12 *vs* <7 μm) and more slender body (a = 39 *vs* <30) in the J2 (Table 14).

BIOCHEMICAL AND MOLECULAR CHARACTERISATION

No available data.

5. *Globodera mali* (Kirjanova & Borisenko, 1975) Behrens, 1975
(Fig. 49)

Common name: apple cyst nematode.

In 1973, an unknown species of cyst nematode was found by A.V. Borisenko on roots of apple in the Palvodar region of Kazakhstan. This species was described by Kirjanova and Borisenko (1975) based on several cysts. Sturhan (2010) provided additional information on unhatched and partly squashed J2 from a paratype egg slide.

MEASUREMENTS

After Kirjanova and Borisenko (1975).

– *Holotype cyst*: L = 460 μm; W = 345 μm; neck = 50 μm; vulva to anus distance = 58 μm; Granek's ratio = 1.3; number of ridges = 10; vulval slit = 11 μm.

– *Paratype cysts* (n = 25): L (including neck) = 545 (420-676) μm; W = 404 (272-496) μm; neck 68 (38-93) μm; vulva to anus distance = 42.8 (22-58) μm; Granek's ratio = 1.3 (1.0-1.8), number of ridges = 7 (4-11); vulval slit = 9-11 μm.

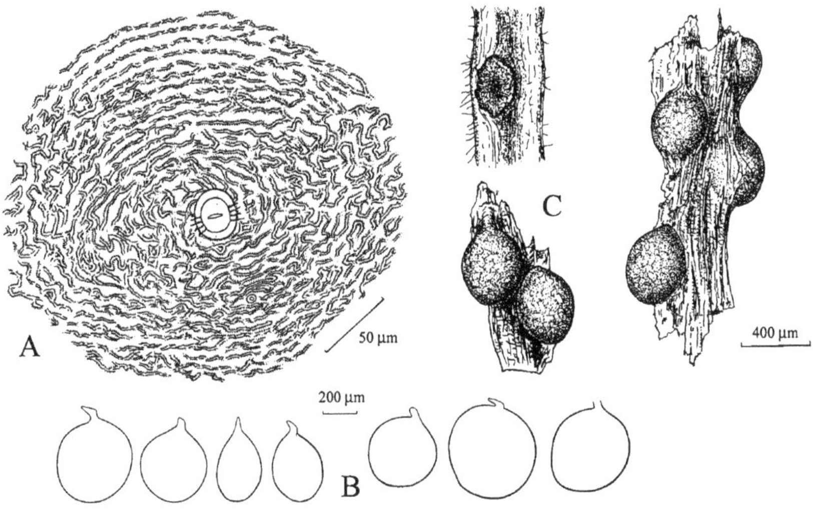

Fig. 49. Globodera mali. *A: Vulval plate; B: Cysts; C: Cysts on roots. After Kirjanova and Borisenko (1975).*

– *Paratype eggs* (n = 30): L = 101 (88-112) μm; W = 46 (42-60) μm.

After Sturhan (2010).

– *Paratype J2 from squashed eggs* (n = 20): Stylet = 22 (21.5-22.5) μm; tail (n = 5) = 41 (39-43) μm; hyaline region = 17 (13-20) μm.

DESCRIPTION

Cyst

Small. Round or pear-shaped, circumfenestrate, without vulval cone. Cuticle exceptionally transparent, mostly light brown.

J2 (after Sturhan, 2010)

Stylet knobs rounded and slightly inclined posteriad, 4 μm in diam., labial region diam. = 8.2-8.5 μm; labial region height = 3.6-3.8, with four annuli, tail end rather blunt, annulus width at mid-body *ca* 1.5 μm.

TYPE HOST AND LOCALITY

Apple, *Malus domestica* L. (Rosaceae, Rosales), Michurin state farm, Pavlodar region, Kazakhstan.

DISTRIBUTION

Asia: Kazakhstan.

RELATIONSHIPS

This species can be distinguished from other *Globodera* by the thin and transparent cuticle. Krall (1977) suggested that *G. mali* was closely related to *G. chaubattia* Gupta & Edward, 1973 and might be synonymous. *Globodera chaubattia* was found in Asia, also around the roots of apple, but without definitive indication of the host. Siddiqi (2000) considered *G. mali* as a synonym of *G. chaubattia*. However, Sturhan (unpubl.) noticed morphological differences between these species and suggested that this synonymisation was incorrect. Hesling (1978) and Sharma and Swarup (1984) proposed that *G. chaubattia* should be considered as a *species inquirenda* under the original name *Heterodera chaubattia* and we agree with that opinion herein. Re-examination of both species is evidently needed (Krall, 1977).

BIOCHEMICAL AND MOLECULAR CHARACTERISATION

No available data.

6. *Globodera mexicana* sp. n.
(Figs 50, 51)

Common name: Mexican cyst nematode.

In 1963, during nematological surveys of wild potatoes and solanaceous plants in the Central Highlands of Mexico's Toluca Valley in the state of Mexico and Huamanta Valley in the state of Tlaxcala, many cysts closely resembling potato cyst nematode were found from *Solanum rostratum*. Detailed biological and morphological study made by Campos-Vela (1967) revealed that this nematode was a new species. The data from this work, together with a description of a species named as *H. mexicana*, was presented in Campos-Vela's Ph.D. thesis. Reportedly, the host plants for this species also included *Solanum lycopersicum* L. and *S. carolinense* L. This cyst nematode does not complete its life cycle on potato, *S. tuberosum* L., or *Nicotiana* species, including *Nicotiana tabacum*. Several years later, however, Golden and Ellington (1972) considered *H. mexicana* as a *nomen nudum* because the description of this

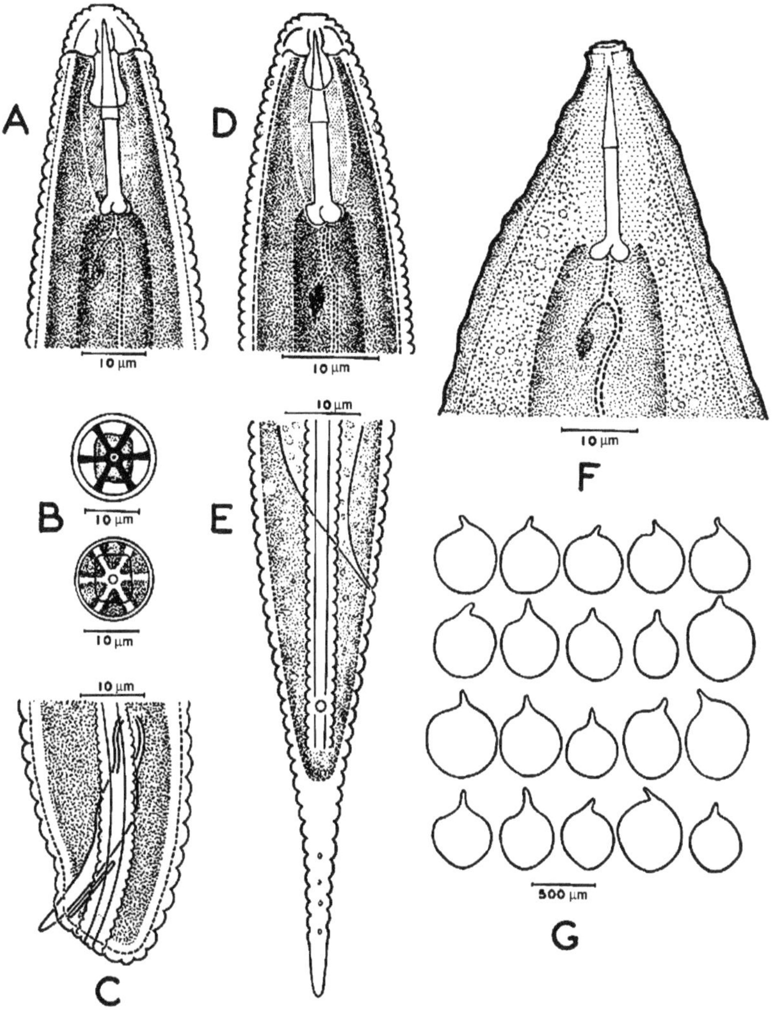

Fig. 50. Globodera mexicana ***sp. n.*** *A: Male anterior region; B: Male face views; C: Male tail; D: J2 anterior region; E: J2 tail; F: Mature female anterior region; G: Cysts. After Campos-Vela (1967).*

nematode was presented in a Ph.D. thesis which could not be considered as a valid publication for the purposes of species nomenclature. Moreover, after examination of the material of the Mexican cyst nematode, Golden and Ellington (1972) concluded it to be conspecific with *H. tabacum virginiae*. Greet (1972) also agreed with this synonymisation and showed that protein pattern profiles obtained from electrophoresis

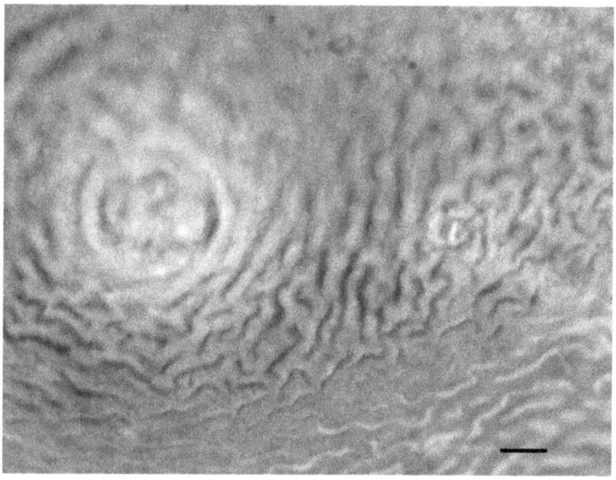

Fig. 51. Globodera mexicana ***sp. n.*** *Vulval plate. (Courtesy of Cid del Prado, Mexico.) (Scale bar = 10 μm.)*

were almost identical between *H. tabacum* subspecies and the Mexican cyst nematode. Stone (1983) suggested that the Mexican cyst nematode should not be given any nomenclatural status, especially in view of the results of comparative analysis of morphometrics and morphology of *G. tabacum* and several Mexican populations, where these species were indistinguishable. A contrasting point of view on relationships of the Mexican cyst nematode, named as *G. mexicana* by Quiñones and Sosa Moss (1977), with other *Globodera* was presented by French nematologists in several publications (Mugniéry *et al.*, 1992; Bossis & Mugniéry, 1993; Thiéry & Mugniéry, 1996, 2000; Thiéry *et al.*, 1997a, b; Marché *et al.*, 2001). Results of two dimensional gel electrophoresis of total proteins (Bossis & Mugniéry, 1993), PCR-ITS-RFLP (Thiéry *et al.*, 1997a; Grenier *et al.*, 2001) and PCR-RAPD (Thiéry *et al.*, 1997a), satellite DNA sequences (Grenier *et al.*, 2001) showed that the Mexican cyst nematode was clearly different from all other *Globodera*. These biochemical and molecular studies showed that *Globodera* '*mexicana*' was more closely related to *G. pallida* than to *G. tabacum*. Moreover, experimental biological studies revealed that *G.* '*mexicana*' did not hybridise with *G. tabacum* (Mugniéry *et al.*, 1992), but its males produced viable hybrids when mated with *G. pallida* females (Thiéry *et al.*, 1997b). Marché *et al.* (2001) also found differences in AFLP profiles between *G. tabacum* and *G.* '*mexicana*' and showed that the mean of the genetic distance between *G.* '*mexicana*' and *G. tabacum* was greater than the mean genetic distance between the *G. tabacum* subspecies. Such a contradiction in the results given by Golden and Ellington (1972), Greet (1972) and Stone (1983) on the one hand, and from results obtained by French nematologists on the other, might be explained by the study of two different species, *viz.*, *G. tabacum* and 'true' Mexican cyst nematode, respectively, in these studies, although the material for both came from the same source. As publications providing lists of species of cyst nematode genera either do not mention the Mexican cyst nematode (Hesling, 1978; Wouts, 1985; Golden, 1986; Baldwin & Mundo-Ocampo, 1991; Wouts & Baldwin, 1998) or list it as a *nomen nudum* (Siddiqi, 2000), and, considering the facts indicating that the Mexican cyst nematode is biochemically and molecularly different from other *Globodera*, we propose herein to re-establish this species by accepting it as a valid species and naming it as *G. mexicana* **sp. n.** A modified description of this species based on that of Campos-Vela (1967) is provided.

MEASUREMENTS

After Campos-Vela (1967).

– *Paratype females* (n = 33): L (including neck) = 695.5 (540-980) μm; W = 483.6 (350.4-670.0) μm; neck = 139.6 (85.7-220.0) μm; stylet = 24.7 (23.2-28.8) μm; DGO = 6.02 (3.0-10.5) μm.

– *Paratype cysts* (n = 100): L (including neck) = 815 (671-999) μm; W = 635 (429-800) μm; neck = 131.1 (71.4-228.5) μm; fenestral diam. = 20.85 (15.2-28.5) μm; anus to nearest fenestral margin distance = 58.6 (34.2-110.0) μm; Granek's ratio = 2.8 (1.7-5.3).

– *Paratype males* (n = 33): L = 1150 (800-1300) μm; a = 35.9 (28.5-45.1) μm; b = 6.13 (4.50-7.85); c = 216 (130-547); stylet = 26.7 (22-29.5) μm; DGO = 3.1 (1.4-4.8) μm; anterior end to pharyngeal gland tip = 189.2 (147.0-231.0) μm; distance from excretory pore to hemizonid = 5.6 (4.2-6.4) μm; distance from hemizonion to hemizonid = 17.3 (14.0-19.2) μm; T = 66.4 (51.2-68.3); spicules = 33.9 (27.8-36.3) μm; gubernaculum = 9.2 (8.2-11.0) μm; tail = 5.8 (2.0-8.8) μm.

– *Paratype J2* (n = 125): L = 468.5 (333-587) μm; a = 23.6 (18.0-26.0); b = 2.7 (2.1-3.7); c = 8.6 (5.7-11.6); stylet = 23.3 (20.0-27.0) μm; DGO = 4.5 (2.8-6.0) μm; anterior end to: median bulb valve = 63.1 (55.0-71.0) μm, pharyngeal gland tip = 171.1 (140.0-231.0) μm, excretory pore = 100.4 (88.0-110.0) μm, genital primordia = 175.7 (140-200) μm; tail terminus to phasmids = 35.6 (31.0-41.7) μm; hyaline region = 24.5 (19.8-28.8) μm; tail = 54.0 (44.2-74.2) μm.

– *Paratype eggs* (n = 100): L = 104.7 (92-107) μm; W = 43.0 (41.0-49.2) μm.

After Mugniéry *et al.* (1992).

– *J2*: Stylet = 23.5 $\pm$ 0.2 μm; DGO = 5.2 $\pm$ 0.2 μm; hyaline region = 25.8 $\pm$ 0.5 μm; tail = 51.7 $\pm$ 0.7 μm.

Description

Mature female

Body ovate to spherical with elongate neck tapering to a narrow labial region with weak labial framework. Labial region variable, without well defined limits, apparently having only one prominent annulus and a labial disc. Stylet well developed with massive asymmetrical knobs the dorsal knob resembling that of a male. Median bulb large, nearly spherical. Pharyngeal glands indistinguishable, apparently contained in a single lobe. Two ovaries filling body cavity at maturity. Egg number very variable.

Cyst

Light to medium brown in colour. Ovate to spherical with round posterior end. Perineal pattern comprising wavy lines and rows of subsurface dots. Anus minute, pore-like.

Male

Body slender, vermiform, cuticle marked by transverse striae. Lateral field with three incisures 5-7 annuli posterior to anterior end, increasing to four just beyond stylet knobs and continuing throughout length of body and around tail. Incisures occupying one-third of body diam. at mid-body. Outer incisures crenate, central incisures plain. Labial region hemispherical, with 5-7 annuli, anterior-most forming a labial disc, face view typical for genus with six sectors and pore like amphidial openings at lateral borders of labial disc and adjacent lips, labial framework strongly sclerotised, stylet with strong, well defined, symmetrical knobs sloping posteriorly. Anterior cephalids usually three, sometimes two, annuli posterior to labial region constriction, posterior cephalids 7 (5-8) annuli posterior to labial region constriction. Annuli 5-7 posterior to labial region constriction distinctly smaller than cephalic lips. Labial region distinctly offset from body. Median bulb ellipsoid with well developed valve, two subventral pharyngeal glands and dorsal pharyngeal gland highly variable in length and size. Nerve ring conspicuous, encircling pharynx just anterior to junction with intestine; excretory pore = 13.1 (10.8-14.6)% of body length from anterior end. Hemizonion posterior to excretory pore. Testis with reflexed anterior end, spicules slightly arcuate with tips finely rounded. Gubernaculum slender. Tail with a posterior protuberance, phasmid openings near terminus, inconspicuous.

J2

Cuticle marked by transverse striae, lateral fields with four incisures, extending from labial region to slightly posterior to phasmids, labial framework heavily sclerotised, labial region bearing four annuli. Stylet robust with well developed, concave knobs. Anterior cephalids located usually three, sometimes two, annuli from head constriction, posterior cephalids situated seven annuli posterior to head constriction. Median bulb of pharynx ellipsoid with sclerotised valve. Hemizonid located anterior to excretory pore, hemizonion posterior to excretory pore. Tail terminus finely rounded, with transverse rows of subcuticular dots visible in hyaline region, number and size of dots very variable.

Type host and locality

Buffalo burr, *Solanum rostratum* Dunal. (Solanaceae, Solanales), El Sauce ranch, 6.4 km north east of Huamantla, Tlaxcala State, Mexico.

Type material

Holotype and allotype slides deposited in the Collection of the Plant Pathology Department, University of Wisconsin, Madison, Wisconsin, USA (Campos-Vela, 1967).

Other hosts

Solanum carolinense L. (horsenettle) and *S. lycopersicum* L. (tomato). *Globodera mexicana* **sp. n.** does not complete its life cycle on *Solanum tuberosum*, *S. nigrum*, *Nicotiana tabacum* or 11 other tested species of

Nicotiana (Campos-Vela, 1967). However, Thiéry *et al.* (1997a) reported development of cysts on *N. tabacum* and *N. acuminata* with some populations of *G. mexicana* **sp. n.** (Table 17).

Distribution

North America: Mexico in Central Highlands of Mexico, Toluca Valley in Mexico State and Huamantla Valle, Tlaxcala State (Campos-Vela, 1967) and Municipality of Amecameca in Mexico State (Cid del Prado, pers. comm.).

Relationships

Morphologically *G. mexicana* **sp. n.** closely resembles *G. tabacum*, *G. rostochiensis* and *G. pallida*. It differs from *G. rostochiensis* by longer mean stylet (23-24 *vs* 19-22 μm) in J2, shorter DGO (3.1 *vs* 5.3-5.7 μm) in male and smaller Granek's ratio (2.8 *vs* 3.0-4.5); and from *G. pallida* by larger Granek's ratio (2.8 *vs* 2.1-2.5). It differs from *G. tabacum* by a mean Granek's ratio of 2.8 in *G. mexicana* **sp. n.**, with only 4% of the cysts having a ratio of 2.0 or less, *vs* 1.5 in *G. tabacum* with 93% of the cysts having a ratio of 2.0 or less (Campos-Vela, 1967) (Table 14). Analyses of two-dimensional gel electrophoresis profiles of total proteins (Bossis & Mugniéry, 1993), PCR-RAPD (Thiéry *et al.*, 1997a) and ITS-rRNA gene sequences (Subbotin *et al.*, unpubl.) show that *G. mexicana* **sp. n.** is closely related to *G. pallida*.

Biology

J2 penetration into plant host roots occurs at 16-24°C and at 28-32°C penetration of roots is completely inhibited. The life cycle is completed in 23-24 days (Campos-Vela, 1967).

Pathogenicity

Under glasshouse conditions, *G. mexicana* **sp. n.** significantly reduced fresh weight of tomato roots when plants were inoculated with 4500 and 9000 cysts (Campos-Vela, 1967).

Biochemical and molecular characterisation

Globodera mexicana **sp. n.** has been characterised by analyses of two dimensional electrophoresis of proteins (Bossis & Mugniéry,

1993), RAPD-PCR (Thiéry *et al.*, 1997a), PCR-ITS-RFLP (Thiéry & Mugniéry, 1996; Grenier *et al.*, 2001), AFLP (Marché *et al.*, 2001), ITS-rRNA sequences (Picard *et al.*, unpubl.; Subbotin *et al.*, unpubl.) and *cytb* sequences (Picard *et al.*, 2007). The ITS-rRNA profile for *G. mexicana* **sp. n.** is given in Figure 39 and Table 15.

7. *Globodera millefolii* (Kirjanova & Krall, 1965) Behrens, 1975
(Figs 52, 53)

Common name: yarrow cyst nematode.

This species was based on a single female collected in 1956 from roots of *Achillea millefolium* L. in Tallinn, Estonia, the description being published in 1965 (Kirjanova & Krall, 1965). Such a poorly described species may be cited as the reason why some authors (Wouts, 1984; Stone, 1986) consider it as *species inquirenda.* Subsequently, re-examination of the holotype by Krall (1977) revealed an error in the description of the vulval slit shape and length as the strongly developed fenestral edge was erroneously thought to be the vulval slit. Krall (1977) concluded that *G. millefolii* came very close to *G. achilleae* which had been described from the former Yugoslavia, and from the same host, by Golden and Klindić (1973). He noticed that some morphological features indicated that *G. millefolii* might be different from *G. achilleae.* However, in this book we consider *G. achilleae* as a synonym of *G. millefolii.* Our comparisons of the ITS-rRNA sequences of an Estonian population of *G. millefolii* collected close to the type locality with those identified as *G. achilleae* did not reveal any differences (Subbotin *et al.*, unpubl.). Nor were significant differences in morphology and morphometrics found between these two nominal species.

MEASUREMENTS

After Kirjanova and Krall (1965).

– *Holotype female*: L = 642.6 μm; W = 428.4 μm; stylet = 18 μm; fenestral diam. = 16 μm.

– *Paratype eggs*: L = 131.7 (122.5-140) μm; W = 48.9 (45-52) μm.

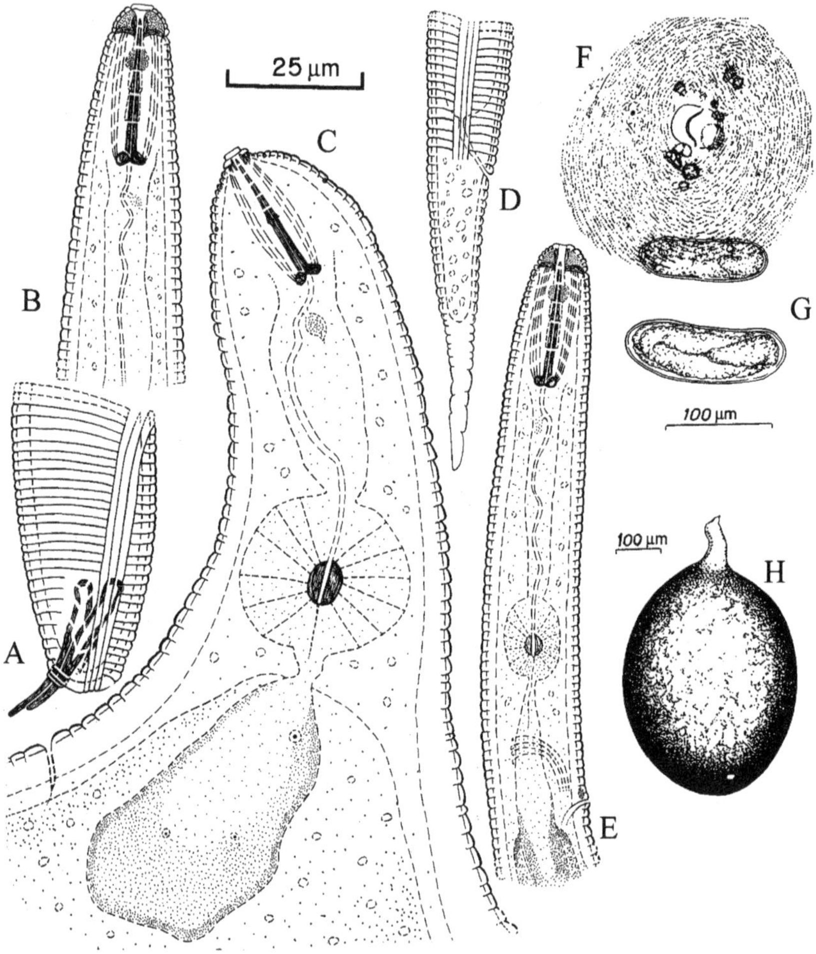

Fig. 52. Globodera millefolii. *A: Tail of male; B: Anterior region of male; C: Anterior region of female; D: Tail of J2; E: Anterior region of J2; F: Vulval plate; G: Egg; H: Cyst. A-E after Golden and Klindić (1973), F-H after Kirjanova and Krall (1965).*

After Mulvey (1972).

– *Cyst* (n = 5): L = 400-500 μm; W = 320-420 μm; fenestral length = 20-28 μm; fenestral width = 15-28 μm; vulval slit = 14 μm; Granek's ratio = 2.1 (1.8-2.2).

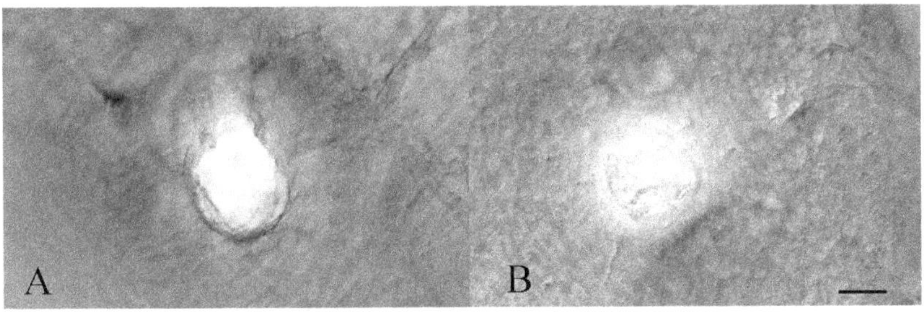

Fig. 53. Globodera millefolii. *A, B: Vulval plates. (Original.) (Scale bar = 10 μm.)*

After Golden and Klindić (1973).

– *Female*: (n = 25): L (including neck) = 564 (434-938) μm; W = 321 (230-454) μm; L/W ratio = 1.8 (1.2- 2.9); neck = 126 (83-188) μm; stylet = 26 (23-28) μm; DGO = 5.6 (3.4-6.1) μm; anterior end to excretory pore = 126 (112-154) μm; fenestral diam. = 15.8 (11.2-18.5); vulval slit = 5 (2.8-5.6) μm; anus to nearest edge of fenestra = 29 (20-39) μm.

– *Cyst*: see Table 20.

– *Male*: (n = 25): L = 1145 (1048-1219) μm; W = 37 (34-39) μm; a = 31 (28-34); b = 5.9 (4.8-7.0); c = 895 (0-2176); stylet = 28 (27-29) μm; DGO = 5.7 (5.6-6.2) μm; anterior end to median bulb valve = 98 (80-108) μm; spicules = 33 (30-37) μm; gubernaculum = 11.8 (11.2-12.9) μm.

– *J2*: see Table 1.

– *Egg* (n = 25): L = 110 (102-119) μm; W = 50 (45-56) μm; L/W ratio = 2.2 (1.9-2.5).

Description

After Golden and Klindić (1973).

Female

Body pearly white, ovate to subspherical, posteriorly rounded, neck elongate, protruding. Cuticle thick. Cuticular pattern rugose in posterior portion, with lines extending anteriorly on to neck. Punctations just beneath surface, often appearing as latitudinal lines as in cysts. Labial region slightly offset, variable in shape, bearing two annuli. Labial framework distinct but weakly developed. Stylet strong, slightly dorsally curved, with well developed knobs sloping posteriorly. Cephalids

Table 20. *Morphometrics of cysts and J2 of* Globodera millefolii *(all measurements in* μm*).*

Population Characters	Bosnia, Golden & Klindić (1973)	Poland, Brzeski (1998)	Zadraga, Slovenia, Sirca & Urek (2004)	Trbonje, Slovenia, Sirca & Urek (2004)
Cyst (n)	25	–	10	10
Length	639 (459-852)	480 (380-600)	–	–
Width	479 (316-602)	430 (310-570)	–	–
Length/width	1.3 (1.2-1.5)	1.1 (1.0-1.5)	–	–
Fenestral diam.	16 (12-18)	25 (12-44)	20.1 ± 1.6 (18.1-22.1)	22.3 ± 3.8 (18.7-32.3)
Number of ridges			5.6 ± 0.8 (5-7)	5.7 ± 1.2 (4-7)
Anus-fenestral edge distance	27 (22-34)	24 (12-42)	28.6 ± 3.7 (23.8-34.1)	29.0 ± 5.7 (18.1-34.9)
Granek's ratio	1.6 (1.3-1.9)	1.2 (0.3-1.9)	1.4 ± 0.1 (1.3-1.5)	1.3 ± 0.3 (0.9-1.7)
J2 (n)	25	–	15	40
Length	492 (472-515)	490 (460-520)	483.5 ± 24.7 (455.5-535.6)	516.2 ± 39.9 (423.8-585.0)
a	21 (18-22)	22 (18-26)	22.5 ± 1.9 (18.7-26.1)	24.2 ± 1.5 (20.7-27.3)
b	3.0 (2.5-3.7)	4.1 (3.6-4.4)	–	–
c	9.0 (7.9-10.4)	9.7 (7.9-11.6)	9.3 ± 0.4 (9.3-10.5)	9.7 ± 0.6 (8.5-11.1)
Stylet	25 (24-25)	25 (24-26.5)	24.9 ± 0.8 (23.6-26.4)	25.1 ± 0.8 (23.6-26.6)
Labial region height	3.4 (3.3-3.9)	–	–	–
Labial region diam.	9.0 (8.4-9.5)	8.5-10	–	–
DGO	5.7 (5.6-6.1)	5-8	5.6 ± 1.4 (3.6-8.7)	4.9 ± 1.1 (2.9-7.3)
Anterior end to excretory pore	–	106 (97-118)	–	–
Anterior end to median bulb valve (MB)	73 (68-77)	75 (68-85)	68.2 ± 6.4 (58.5-80.4)	71.0 ± 6.9 (56.1-83.1)
Body diam. at mid-body	24 (21-28)	–	21.6 ± 1.6 (19.3-25.0)	21.4 ± 1.2 (18.7-23.6)
Body diam. at anus level	–	–	12.8 ± 0.8 (11.1-13.8)	13.5 ± 1.0 (10.1-15.5)
Hyaline region	27 (20-32)	23 (15-32)	25.5 ± 3.2 (21.6-32.2)	29.8 ± 3.7 (22.4-39.1)
Tail	55 (47-63)	51 (41-63)	49.0 ± 3.5 (45.3-57.4)	53.2 ± 3.7 (45.0-61.7)

present. Vulval bodies deep underneath vulva, clustered, variable in shape. Anus small but distinct.

Cyst

Circumfenestrate, abullate, light to dark brown, ovate to spherical, with or without distinct vulval bodies as generally seen in female. Cuticular pattern variable. At *ca* 50-100 μm from fenestra, cyst pattern changing from a transverse rugose pattern to a longitudinal pattern of wavy-lines, zone of change between two patterns very distinct. Prominent wavy lines extending to and onto the neck. Subcuticular punctations generally distinct, arranged in close, parallel rows at right angles to longitudinal axis of cyst. Anus small, but distinct.

Male

Body slender, vermiform, tapering slightly at both extremities. Cuticle with prominent annulation, subcuticular annulation less distinct. Lateral fields consisting of four lines with no areolation. Labial region slightly offset, bearing five annuli. Labial framework heavily sclerotised. Stylet strong, slightly dorsally curved, well developed knobs sloping posteriorly. Median bulb ellipsoidal. Excretory pore *ca* two annuli posterior to distinct hemizonid. One testis. Spicules slightly arcuate, with round, not notched tips. Tail very short.

J2

Body vermiform, tapering at both extremities, but much more so posteriorly. Subcuticular annulation less distinct, double in number to that of surface annuli. Lateral field lacking areolation, composed of four lines. Labial region slightly offset, with five annuli. Stylet strong, slightly dorsally curved, with well developed knobs sloping posteriorly. Cephalids present. Pharyngeal region with prominent median bulb. Excretory pore posterior, adjacent to hemizonid. Tail long, becoming slightly reduced in diam. on ventral side immediately posterior to anus, then tapering to a finely rounded terminus. Rectum unusually large, prominent. Phasmids small, located anterior to mid-tail.

TYPE HOST AND LOCALITY

Common yarrow, *Achillea millefolium* L. (Asteraceae, Asterales), Tallinn, Estonia.

DISTRIBUTION

Europe: Bosnia and Herzegovina (Golden & Klindić, 1973), Estonia (Krall, 1977), Germany (Sturhan, 1980, 1984), Latvia (Krall, 1977), Poland (Radziwinowicz, 1977), Slovakia (Sturhan & Liskova, 2004), Slovenia (Sirca & Urek, 2004), UK (Sykes & Webley, 1979).

OTHER HOSTS

Experimental testing revealed several hosts from the Asteraceae: *Achillea*, *Tripleurospermum*, *Matricaria*, *Pyrethrum*, *Leucanthemum*, *Tanacetum*, *Anthemis* and *Chrysanthemum* (Krall, 1977). Brzeski (1998) reported development of this species on *Anthemis cotula* L. and *Matricaria chamomilla* L.

RELATIONSHIPS

Globodera millefolii is similar to *G. artemisiae* from which it is distinguished by longer J2 stylet (mean = 25 *vs* 21-24 μm), longer gubernaculum (11.2-12.9 *vs* 6.0-9.9 μm) and different host ranges (Table 14).

BIOCHEMICAL AND MOLECULAR CHARACTERISATION

The PCR-ITS-rRNA profile for *G. millefolii* is given in Table 15. Enzymes *Alu*I and *Bsh*1236I generated unique profiles after digestion of PCR-ITS-rRNA product.

8. *Globodera pallida* Stone, 1973
(Figs 7F; 12A; 41C, D; 54, 55)

Common name: white or pale potato cyst nematode.

Guile (1966, 1970) was the first to note that females of pathotype A of the golden cyst nematode became golden yellow shortly after they ruptured the root cortex, whereas those of pathotype E remained white until they died and became brown cysts. Webley (1970) found that the J2 of pathotype A differed in measurements from those of pathotypes B and E and that the shape of the stylet knobs also differed among pathotypes. Jones *et al.* (1970) suggested that females with a cream or white colour represented an undescribed species and Trudgill

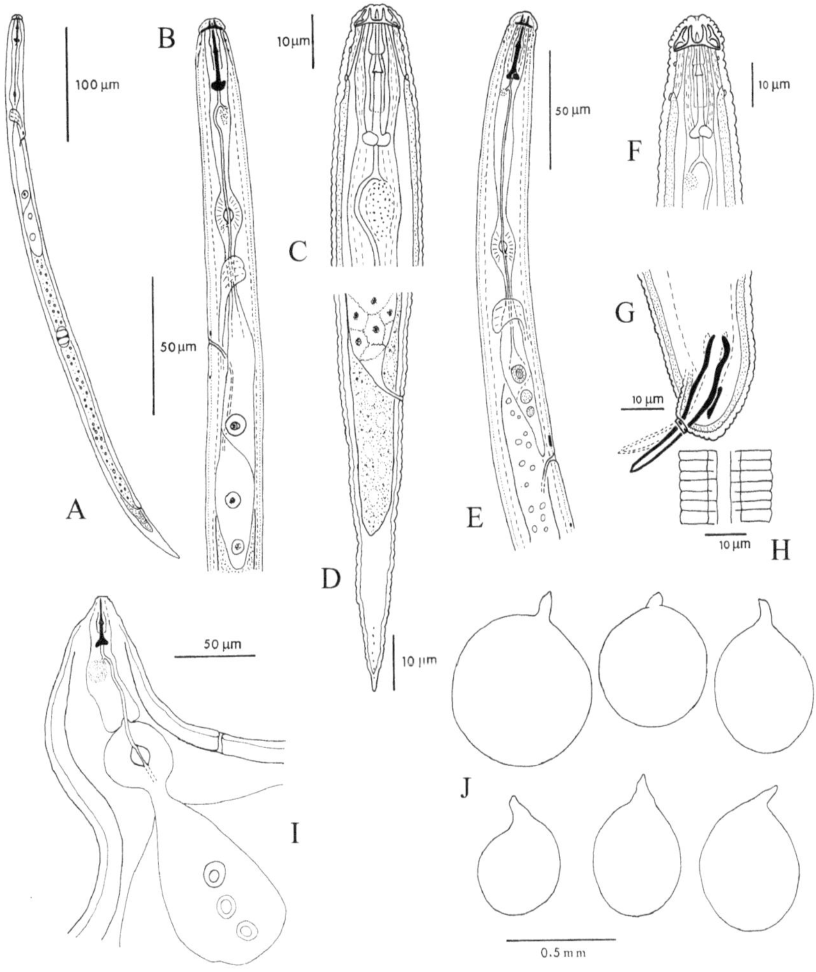

Fig. 54. Globodera pallida. *A-D: J2. A: Entire body; B, C: Anterior region; D: Tail; E-H: Males. E, F: Anterior region; G: Tail; H: Lateral field; I: Anterior region of female; J: Cysts. After Stone (1972).*

and Carpenter (1971) were able to distinguish these females using electrophoresis in polyacrylamide gels. Stone (1973a) described this species as *Heterodera pallida* but considered that it belonged to the subgenus *Globodera*. Measurements and descriptions of different stages were also given by Trudgill *et al.* (1970), Webley (1970), Bowman and Ross (1972), Behrens (1972), Øydvin (1973), Stone (1973c), Hesling and Ellis (1974), Evans and Franco (1977), Hesling (1978), Baldwin

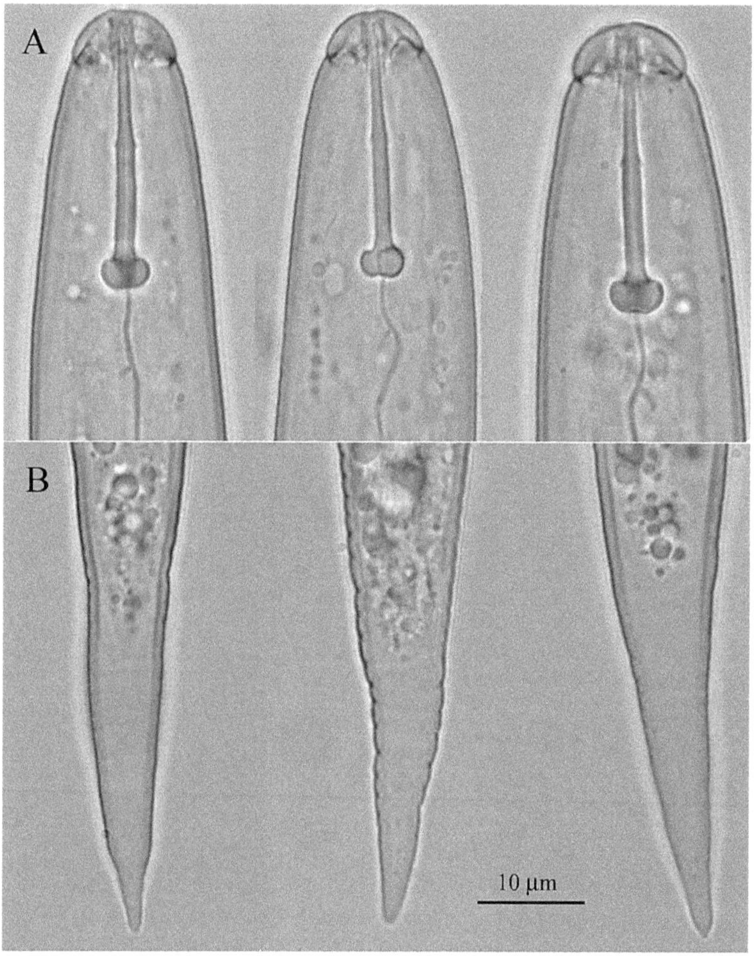

Fig. 55. Globodera pallida. *J2. A: Anterior region; B: Tail. (Courtesy of A.-M. Deeren, ILVO, Belgium.)*

and Mundo-Ocampo (1991), Gonzales *et al.* (1992), Manduric *et al.* (2004), Skantar *et al.* (2007), *etc.* The pale cyst nematode, *G. pallida*, is considered to be a major pest of potato crops in cool temperate climates.

MEASUREMENTS

After Stone (1973a).

– *Paratype females* (n = 25): Stylet = 27.4 ± 1.1 μm; labial region diam. = 5.2 ± 0.5 μm; DGO = 5.4 ± 1.1 μm; anterior end to: median bulb valve = 67.2 ± 18.7 μm, excretory pore = 139.7 ± 15.5 μm; vulval slit = 11.5 ± 1.3 μm; anus to vulval basin = 44.6 ± 10.9 μm; number of cuticular ridges on anal-vulval axis = 12.5 ± 3.1.

– *Paratype males* (n = 50): L = 1198 ± 104 μm; body width at excretory pore = 28.4 ± 1.3 μm; labial region diam. = 12.3 ± 0.5 μm; labial region height = 6.8 ± 0.3 μm; stylet = 27.5 ± 1.0 μm; DGO = 3.4 ± 1.0 μm; anterior end to: median bulb valve = 96.0 ± 7.1 μm, excretory pore = 176.4 ± 14.5 μm; tail = 5.2 ± 1.4 μm; anal body diam. = 13.5 ± 2.1 μm; spicules = 36.3 ± 4.1 μm; gubernaculum = 11.3 ± 1.6 μm.

– *Holotype J2*: L = 510 μm; labial region height = 5 μm; labial region diam. = 9 μm; stylet = 25 μm; DGO = 2 μm; anterior end to: median bulb valve = 74 μm; excretory pore = 106 μm, tail = 53 μm; hyaline region = 31 μm; anal body diam. = 13 μm.

– *Paratype J2*: see Table 21.

DESCRIPTION

After Stone (1973a).

Female

Body subspherical with projecting neck bearing head, pharynx corpus, isthmus and anterior part of pharyngeal glands. White in colour, some populations passing, after 4-6 weeks, through a cream stage, turning glossy brown when dead. Labial region with amalgamated lips and one or two prominent annuli, deep irregular annulations present on neck, changing to reticulate pattern of ridges over most of body surface. Head framework weakly developed, hexaradiate. Stylet knobs sloping backward. Very large median pharyngeal bulb, almost circular with large crescentic valve plates. Pharyngeal gland lobe broad, frequently displaced anteriad, three gland nuclei. Prominent excretory pore situated at base of neck. Internal structures in neck region often obscured by hyaline secretions on cuticle surface. Vulva a transverse slit at posterior end, set in a slight circular depression or vulval basin. Cuticle surface between anus and vulval basin including about 12 parallel ridges with a few cross connections. Subsurface punctations irregularly arranged over much of body surface, may be confused with surface papillae on vulval crescents.

Table 21. *Morphometrics of cysts and J2 of* Globodera pallida *(all measurements in μm).*

Population / Characters	Epworth, UK, Paratypes, Stone (1973a)	Duddingston, UK, Stone (1973a)	Münster, BBA, Germany, Wouts & Weischer (1977)	Pukekole, The Netherlands, Wouts & Weischer (1977)	Pakistan, Shahina & Maqbool (1995)	Idaho, USA, Skantar *et al.* (2007)
Cyst (n)	25	25	–	–	20	80
Length (without neck)	579 ± 70	510 ± 69	–	–	675 (555-748)	574 ± 85 (420-700)
Width	534 ± 66	451 ± 76	–	–	648.5 (540-685.2)	534 ± 80 (400-600)
Length/width	–	–	–	–	–	–
Vulval areas (n)	25	25	–	–	20	80
Fenestral diam.	24.5 ± 5.0	20.7 ± 3.2	–	–	22 (20-23.5)	24.8 ± 5.6 (17.5-45.0)
Number of ridges	12	–	–	–	12	12 ± 2.3 (7-17)
Anus-fenestra distance	49.9 ± 13.4	50.2 ± 9.8	–		48 (40-53)	53.8 ± 11.8 (30-80)
Granek's ratio	2.1 ± 0.9	2.5 ± 0.8	–	–	–	2.2 ± 0.5 (1.2-3.6)
J2 (n)	50	50	20	20	20	80
Length	486 ± 23	482 ± 18	457 ± 3.5	455 ± 3.6	479 (456-505)	452 ± 36 (380-533)
Stylet	23.8 ± 1.0	23.4 ± 0.6	23.7 ± 0.2	23.3 ± 0.2	23.7 (23.5-23.9)	23.2 ± 0.7 (22.5-25.0)
Labial region height	5.5 ± 0.1	5.2 ± 0.6	4.8 ± 0.1	4.7 ± 0.1	–	–
Labial region diam.	10.6 ± 0.5	10.9 ± 0.4	9.6 ± 0.1	10.0 ± 0.1	–	–
DGO	2.7 ± 0.9	2.9 ± 0.9	–	–	–	–
Anterior end to excretory pore	108.6 ± 4.1	103.6 ± 3.7	–		–	–
Anterior end to median bulb valve	68.7 ± 2.7	66.9 ± 2.6	68.8 ± 0.5	68.9 ± 0.6	–	–
Anal body diam.	12.1 ± 0.4	12.0 ± 0.6	–	–	–	–
Hyaline region	26.6 ± 4.1	26.6 ± 3.4	26.0 ± 0.6	26.4 ± 0.9	25.2 (24.5-27.5)	25.9 ± 2.4 (20-31)
Tail	51.1 ± 2.8	52.6 ± 4.1	50.3 ± 0.7	52.0 ± 0.8	51.8 (51.7-51.9)	49.6 ± 3.0 (40-57)

Cyst

New cysts glossy brown, subspherical with protruding neck. Vulval region intact or fenestrated with single circumfenestrate opening occupying all or part of vulval basin. Abullate, but small darkened or thickened 'vulval bodies' sometimes present in vulval region. Anus visible in most specimens, often at apex of a V-shape mark. Cuticular pattern as in female but more accentuated. Subcrystalline layer absent.

Male

Heat relaxed specimens C- or S-shaped, posterior part twisted 90-180° about longitudinal axis. Cuticle with regular annulations and four incisures in lateral field, terminating on tail. Labial region offset, rounded with large oral disc, six irregular lips, six or seven annuli and heavily sc1erotised hexaradiate framework. Stylet well developed with posteriorly sloping basal knobs and cone forming *ca* 45% of total stylet. Ellipsoid pharyngeal median bulb with strong crescentic valve plates linked by a narrow isthmus encircled by a broad nerve ring, to a narrow, ventrally situated, pharyngeal gland lobe. Hemizonid two annuli long, situated two or three annuli posterior to excretory pore. One testis, commencing with single cap cell 40-65% of body length from head, terminating in a narrow *vas deferens* with glandular walls. Cloaca with small raised circular lip containing two stout arcuate spicules terminating distally in uni-pointed tips. Small dorsal gubernaculum without ornamentation, slightly wider in dorsoventral aspect. Tail short with bluntly rounded terminus of variable shape.

J2

Lateral field with four incisures but with three anteriorly and posteriorly, occasionally completely areolated. Cuticle thickened for first seven or eight body annuli. Labial region rounded, slightly offset with four to six annuli. Oral disc surrounded by two lateral lips bearing amphidial apertures, adjacent dorsal and ventral submedial lips often fused. Contour of lips and oral disc sub-rectangular (Stone, 1973a). Heavily sclerotised hexaradiate head framework, dorsal and ventral radii bifurcate at tips in 60% of specimens. Stylet well developed, basal knobs with distinct anterior projection as viewed laterally. Gland lobe extending posteriorly for *ca* 35% of body length. Excretory pore *ca* 20% of body length from anterior end. Distinct hemizonid two annuli long, located one annulus anterior to excretory pore; hemizonion five or six annuli posterior to excretory pore. Genital primordium at *ca* 60% of body length from anterior end. Tail tapering uniformly with a finely rounded point, hyaline region forming about half of tail region.

TYPE HOST AND LOCALITY

Potato, *Solanum tuberosum* L. (Solanaceae, Solanales), from Epworth, Lincolnshire, UK.

OTHER HOSTS

Eggplant, *Solanum melongena* L., tomato, *S. lycopersicum*, many species of *Solanum*; black henbane, *Hyoscyamus niger* (Stelter & Engel, 1975; Stelter, 1987; CAB International, 2005a; Sullivan *et al.*, 2007 and others) (Table 17). It was experimentally demonstrated that only a few females can develop on NC-95 flue-cured tobacco (Parrot & Miller, 1977).

DISTRIBUTION

It is generally accepted that PCN originated from the Andean region of South America, where these nematodes parasitise wild potatoes and other *Solanum* species (Evans & Stone, 1977). Although a few records of *G. pallida* have been reported from Bolivia (Evans *et al.*, 1975), the main range lies in Peru, Ecuador and Colombia, *i.e.*, to the north of Lake Titicaca (15.6°S), whilst its sibling species, *G. rostochiensis*, is found to the south of this latitude. In South America, this cyst nematode is mainly found between 2000 m and 4000 m a.s.l., with the heaviest infestations between 2900 m and 3800 m a.s.l. (Franco, 1977). From South America PCN has spread, as the result of human activity, into many regions of the world. Although potatoes were introduced into Europe in about 1570, it is thought that PCN probably came with tubers from South America introduced into Europe only in around 1850 (Evans *et al.*, 1975). In Central and North America *G. pallida* has been reported in Panama, USA and Canada, but in the last two countries PCN have a rather restricted distribution with small infested areas, because of rigorous phytosanitary regulations and seed potato certification programmes, compared to the widespread infestations found in European countries. Recently, mtDNA has been used to study genetic relationships among Peruvian populations of *G. pallida*, thus identifying the origin of western

European populations of this species (Picard *et al.*, 2007; Plantard *et al.*, 2008). Using mtDNA gene, cytochrome *b* (*cytb*) sequences and microsatellite loci, Plantard *et al.* (2008) showed that the *G. pallida* presently distributed in Europe derived from a single restricted area in the extreme south of Peru, located between the north shore of Lake Titicaca and Cusco (Fig. 29).

Europe: Austria, Belgium, Cyprus, Denmark, France, Germany, Greece, Iceland, Ireland, Italy, Luxembourg, Malta, The Netherlands, Norway, Poland, Portugal, Slovakia, Spain, Sweden, Switzerland, UK, Ukraine, Yugoslavia (former); *Asia*: India, Pakistan, Turkey; *Africa*: Algeria, Tunisia; *North and Central America*: Canada, Panama, USA; *South America*: Bolivia, Chile, Colombia, Ecuador, Peru, Venezuela; *Oceania*: New Zealand (EPPO Quarantine Pest; CAB International, 2005b).

RELATIONSHIPS

Globodera pallida is most closely related to *G. rostochiensis* and *G. tabacum*. Stone (1973a, c) gave the most complete diagnosis of this species. It differs from *G. rostochiensis* by cream coloured females *vs* yellow or gold, smaller number of ridges between the vulva and anus (mean of 12 *vs* 17-20), smaller mean for Granek's ratio (2.1-2.5 *vs* 3.0-4.5), stylet knob shape (anteriorly projected *vs* rounded), longer stylet length (23-24 *vs* 20-22 μm), tail terminus (more pointed *vs* rounded) and presence of refractive bodies on the hyaline part of the tail (usually 4-7 refractive bodies *vs* absence) in J2; from *G. tabacum* by longer mean body length (452-486 *vs* 476-556 μm) and by J2 stylet knob shape (anteriorly projected *vs* rounded) and higher mean number of cuticular ridges (12 *vs* 7-9) for cysts (Table 14).

BIOLOGY

Globodera pallida develops one generation for a vegetation season. This species is adapted to cool temperatures and is able to hatch earlier in the year and develop at 2°C cooler than *G. rostochiensis* (Mugniéry, 1978; Langeslag *et al.*, 1982).

PATHOGENICITY

The symptoms of attack by *G. pallida* are similar to those for *G. rostochiensis* and the damage threshold is 1-2 eggs/g soil (EPPO Quarantine Pest).

BIOCHEMICAL AND MOLECULAR CHARACTERISATION

Characterisation of *G. pallida* populations using IEF of proteins and PCR-ITS-RFLP have been given by several authors (Tables 9, 12). The IEF profile for *G. pallida* is presented in Figure 33. After studying PCR-ITS-RFLP generated by 16 restriction enzymes for seven European countries and 24 populations from Peru, Ecuador and Colombia, Rumpenhorst and Ayub (2001) revealed three types of RFLP profiles for all studied samples: *i*) type A, or European type; *ii*) type B, or South American type; and *iii*) Peruvian/Ecuadorian type with an RFLP pattern that is a combination of types A and B. Digestion with *Bsh*1236I, *Taq*I, *Rsa*I, *Mva*I, *Pst*I, *Nla*III and *Mse*I can separate the European populations from those from South America. Grenier *et al.* (2001) provided a list of restriction enzymes distinguishing these types of *G. pallida* from each other and from other species parasitising Solanaceae. Differences between European and South American populations of *G. pallida* have been found in PCR-RAPD analyses and IEF of proteins, and Rumpenhorst and Ayub (2001) proposed the presence of a third species among PCN populations that are presently identified as *G. pallida*. The presence of at least two groups within *G. pallida* was also revealed by several other studies using simple repeat primers (Blok & Phillips, 1995), RAPDs (Bendezu *et al.*, 1988; Blok *et al.*, 1997; Grenier *et al.*, 2001), 2-DGE of proteins (Grenier *et al.*, 2001), satellite DNA analysis (Grenier *et al.*, 2001), PCR-ITS-RFLPs and sequences (Blok *et al.*, 1998), and *cytb* sequence analysis (Plantard *et al.*, 2008; Pylypenko *et al.*, 2008; Madani *et al.*, 2010). Madani *et al.* (2010) suggested that clade II and the clade III in the mtDNA tree (Fig. 29) might represent a third and fourth PCN species, respectively. PCR-ITS-rRNA profiles for two types of *G. pallida* are given in Table 15. Enzyme *Pst*I discriminates *G. pallida* type A from *G. mexicana* **sp. n.** Species-specific primers are given in Table 13.

9. *Globodera tabacum* (Lownsbery & Lownsbery, 1954) Skarbilovich, 1959
(Figs 7G; 12B; 56-58)

Common name: tobacco cyst nematode.

In 1951, a cyst nematode, morphologically resembling the golden potato cyst nematode, was found parasitising tobacco on a farm at Hazardville, Connecticut, USA. The occurrence on tobacco suggested that this nematode was not identical to the golden potato cyst nematode. It has been shown in several tests that golden potato cyst nematode does not develop on tobacco cultivars nor on *Nicotiana rustica* L., with the exception of a *G. rostochiensis* population from Chile, which is able to reproduce on *Nicotiana acuminata* (Sullivan *et al.*, 2007). The tobacco

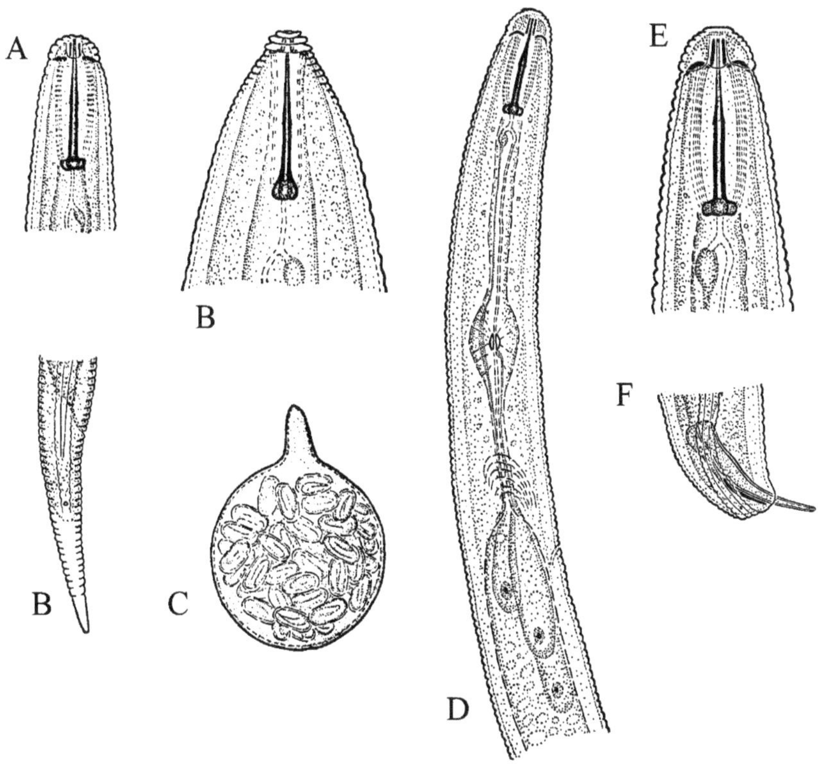

Fig. 56. Globodera tabacum. *A: Anterior region of J2; B: Tail of J2; B: Head of female; C: Cyst; D: Pharyngeal region of male; E: Anterior region of male; F: Tail of male. After Lownsbery and Lownsbery (1954).*

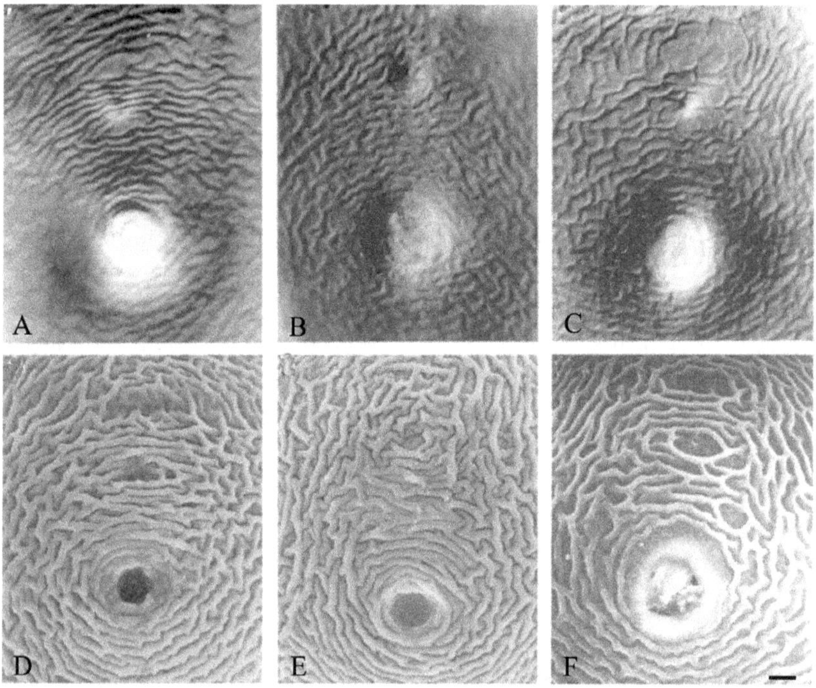

Fig. 57. Globodera tabacum. *Vulval plates. Light microscopy (A-C) and scanning electron microscopy (D-F). A, D:* G. t. tabacum*; B, E:* G. t. virginiae*; C, F:* G. t. solanacearum. *(Scale bar = 10 μm.) After Mota and Eisenback (1993b).*

cyst nematode does not develop on potato, but develops on all varieties of tobacco tested, as well as *N. rustica* L. In addition to these differences in host preference, a study of the tobacco cyst and golden nematodes revealed morphological differences. The tobacco cyst nematode was therefore described and proposed as a new species by Lownsbery and Lownsbery (1954). After the description of *H. tabacum*, two other similar cyst nematodes, colloquially referred to as horsenettle cyst nematode and Osborne's cyst nematode, were later designated by Miller *et al.* (1962) from Virginia, USA. These cyst nematodes were fully described and named as *H. virginiae* and *H. solanacearum* by Miller and Gray (1968) and Miller and Gray (1972), respectively. The type host for these species was *Solanum carolinense* L.; other hosts included different species of *Nicotiana*, *Physalis* and *Solanum*, as well as *Atropa belladonna* L., *Hyoscyamus niger* L., but not *S. tuberosum*. Miller and Gray

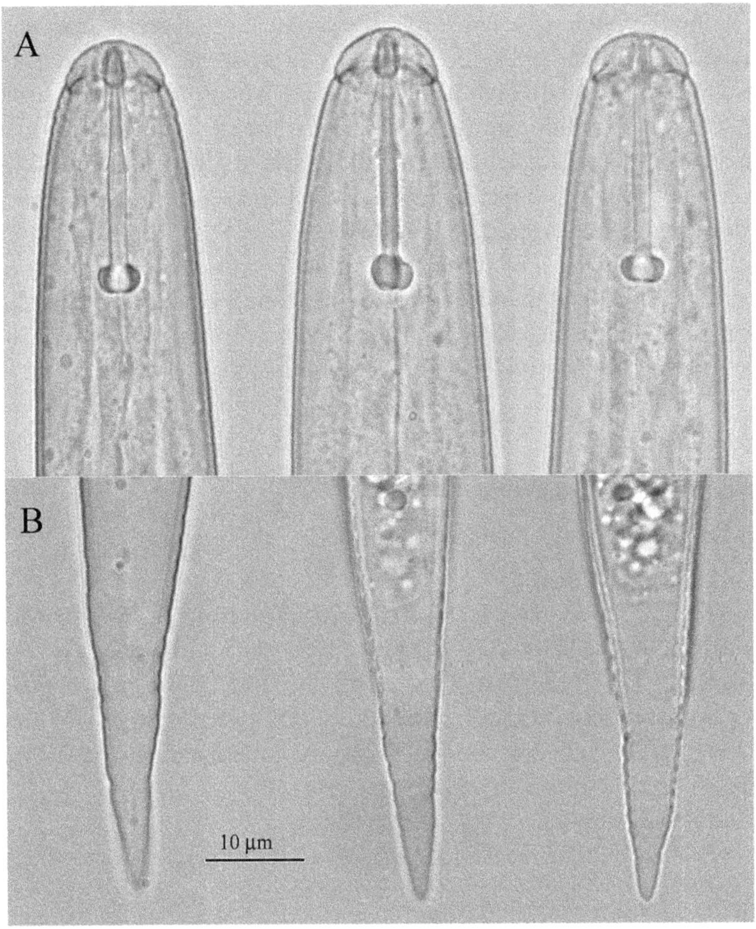

Fig. 58. Globodera tabacum. *J2. A: Anterior region; B: Tail. (Courtesy of A.-M. Deeren, ILVO, Belgium.)*

(1968, 1972) distinguished these nematode species from *H. tabacum* and from each other by minor differences in vulval plate structures and morphometrics. Stone *et al.* (1983), using canonical discriminate analysis of morphometric characters, concluded that *G. tabacum* was a polytypic species containing the following subspecies: *G. tabacum tabacum* (Lownsbery & Lownsbery, 1954); *G. tabacum virginiae* (Miller & Gray, 1968) and *G. tabacum solanacearum* (Miller & Gray, 1972). All three subspecies develop on tobacco and horsenettle, but otherwise differ in host preference. Roberts and Stone (1981) concluded that host range data within *Solanum* spp. failed to differentiate these three subspecies of *G. tabacum*. Miller (1983) and Mugniéry *et al.* (1992) obtained viable hybrids from different combinations of these nematodes that confirmed their subspecies relationships. Mota and Eisenback (1993a, b, c) made comparative morphological and morphometric studies and showed that it is difficult reliably to distinguish these subspecies from each other at any developmental stages. They further considered this complex of three subspecies as having a continuum of values for the majority of the observed characters and therefore confirmed Stone's conclusion regarding their rank as subspecies. *Globodera tabacum* is considered as a serious and important pest of shade and broadleaf tobacco.

MEASUREMENTS

Globodera tabacum tabacum

After Lownsbery and Lownsbery (1954).

– *Cyst*: see Table 22.

– *Female* (n = 50): L = 464 (327-688) μm; W = 310 (206-516) μm; stylet = 22.4 (18.5-24.1) μm; DGO = 5.5 (3.8-8.0) μm.

– *Male*: L = 1123 (710-1355) μm; a = 33 (27-40); b = 7.4 (6.3-8.7); c = 1478 (284-5325); T = 67 (65-69)%; stylet = 26 (24-27) μm; DGO = 3.6 (2.0-5.5); spicules = 26-34 μm; gubernaculum = 9-12 μm; tail = 1.9 (0.2-3.6) μm.

– *J2*: see Table 22.

– *Egg* (n = 20): L = 92-115 μm; W = 43-50 μm.

Globodera tabacum solanacearum

After Miller and Gray (1972).

– *Female* (n = 125): L (including neck) = 632.7 (411-873) μm; neck = 144.4 (58-215) μm; W = 430 (293-614) μm; stylet = 26.2 (21-30) μm; DGO = 6.0 (3-11) μm; anterior end to: excretory pore = 141.7 (103-185) μm, median bulb valve = 26.3 (64-174) μm; median bulb length = 42.7 (22-55) μm; median bulb diam. = 33.5 (17-50) μm; vulval slit = 9.4 (5-13) μm; fenestral length = 17.3 (14-25) μm; fenestral width = 21.2 (16-29) μm; distance of anus to nearest margin of fenestra = 40.6 (25-64) μm.

– *Male* (n = 115): L = 1108 (750-1450) μm; W = 26.0 (22-40) μm; stylet = 26.3 (24-27) μm; DGO = 2.3 (1.5-4) μm; tail = 4.3 (2-8) μm; spicules = 32.9 (29-35) μm.

Table 22. *Morphometrics of cysts and J2 of* Globodera tabacum *(all measurements in μm).*

Population Characters	*G. tabacum tabacum*, Hazardville, Connecticut, USA, Lownsbery & Lownsbery (1954)	*G. tabacum tabacum*, Windsor, Connecticut, USA, Mota & Eisenback (1993c)	*G. tabacum virginiae*, Suffolk, Virginia, USA, Miller & Gray (1968)	*G. tabacum virginiae*, Horton, Suffolk, Virginia, USA, Mota & Eisenback (1993c)	*G. tabacum solanacearum*, Watkins, Virginia, USA, Miller & Gray (1972)	*G. tabacum solanacearum*, Nottoway, Virginia, USA, Mota & Eisenback (1993c)
Cyst (n)	131	20	125	20	125	20
Length (without neck)	536 (337-740)	540 ± 105 (377-754)	767.3 ± 69.2 (537-937)	635.5 ± 84.0 (464.0-817.8)	717.6 (533-907)	579.2 ± 91.1 (446.6-783)
Width	459 (232-645)	536.5 ± 121 (348-812)	528 ± 65.6 (383-660)	557.8 ± 84.2 (400.2-713.4)	496.4 (375-639)	519.5 ± 99.3 (377-730.8)
Length/width	–	1.0 ± 0.1 (0.9-1.3)	–	1.1 ± 0.1 (1.0-1.3)	–	1.1 ± 0.1 (1.0-1.5)
Vulval areas (n)	–	20	125	20	125	20
Fenestral diam.	–	27.6 ± 3.2 (21.0-36.4)	18.7 ± 2.3 (13-25)	23.4 ± 4.6 (16.8-33.6)	22.6 (16-29)	25.3 ± 3.9 (16.8-33.6)
Number of ridges	–	8.8 ± 2.1 (5-15)	–	8.3 ± 1.0 (6-8)	–	7.2 (5-9)
Anus-fenestra distance	–	44.8 ± 7.9 (29.4-67.2)	52.1 ± 10.1 (28-85)	47.8 ± 11.2 (30.8-72.8)	48.0 (31-84)	48.3 ± 10.2 (30.8-79.8)
Granek's ratio	–	1.6 ± 0.3 (1.0-2.4)	2.8 ± 0.5 (1.5-4.2)	2.1 ± 0.4 (1.4-2.9)	–	2.5 ± 0.4 (1.4-2.9)
J2 (n)	40	20	115	20	115	20
Length	476 (410-527)	516.4 ± 32.1 (464-580)	503.7 = 31.7 (419-592)	556.1 ± 44.9 (476-661.2)	497.0 (390-577)	540.7 ± 42.8 (458.9-621.3)
a	23 (22-24)	22.1 ± 1.9 (19.1-26.6)	24.7 ± 1.5 (19.1-27.8)	21.6 ± 1.8 (18.3-25.8)	–	22.8 ± 2.4 (18.9-27.1)

Table 22. *(Continued).*

Population / Characters	*G. tabacum tabacum*, Hazardville, Connecticut, USA, Lownsbery & Lownsbery (1954)	*G. tabacum tabacum*, Windsor, Connecticut, USA, Mota & Eisenback (1993c)	*G. tabacum virginiae*, Suffolk, Virginia, USA, Miller & Gray (1968)	*G. tabacum virginiae*, Horton, Suffolk, Virginia, USA, Mota & Eisenback (1993c)	*G. tabacum solanacearum*, Watkins, Virginia, USA, Miller & Gray (1972)	*G. tabacum solanacearum*, Nottoway, Virginia, USA, Mota & Eisenback (1993c)
b	4.4 (4.1-4.9)	–	–	–	–	–
c	10 (8-11)	–	10.1 ± 0.8 (8.6-14.5)	–	–	–
Stylet	24 (22-26)	21.2 ± 0.8 (19.8-22.5)	23.7 ± 0.9 (19-28)	23.3 ± 1.0 (20.7-24.7)	23.9 (22-26)	24.9 ± 1.0 (22.3-26.6)
Labial region height	–	–	4	–	–	–
Labial region diam.	–	–	9	–	–	–
DGO	5.5 (4.3-6.8)	5.8 ± 0.7 (4.5-7.2)	6.1 ± 0.8 (3-9)	5.9 ± 0.8 (4.3-7.2)	5.5 (5-7)	6.0 ± 0.6 (4.8-7.1)
Anterior end to excretory pore	–	113.9 ± 6.7 (99.0-126.9)	–	109.8 ± 8.7 (96.2-128.7)	–	108.9 ± 7.3 (94.3-122.2)
Anterior end to median bulb valve	–	71.3 ± 3.9 (63.0-79.2)	–	70.4 ± 2.7 (67.5-77.4)	–	71.8 ± 3.3 (65.8-78.4)
Body diam. at mid-body	–	23.5 ± 2.3 (20.3-29.0)	20.5 ± 1.5 (18-24)	26.0 ± 3.4 (21.1-34.2)	21.5 (16-25)	24.0 ± 3.0 (17.1-28.5)
Body diam. at anus level	–	–	–	–	–	–
Hyaline region	28 (22-31)	20.9 ± 3.0 (17.1-27.0)	27.1 ± 3.3 (19-35)	27.8 ± 3.0 (19.9-34.2)	24.1 (16-30)	25.8 ± 2.4 (20.8-29.9)
Tail	52 (46-59)	49.8 ± 3.9 (44.1-60.3)	50.3 ± 3.8 (34-61)	56.4 ± 3.8 (49.5-63.7)	50.3 (33-60)	51.8 ± 5.1 (44.2-61.8)

– *Egg* (n = 125): L = 92.2 (73-113) μm; W = 42.5 (37-52) μm; L/W ratio = 2.2.

Globodera tabacum virginiae

After Miller and Gray (1968).

– *Female* (n = 125): L (including neck) = 688 ± 79 (528-966) μm; neck = 152.1 ± 33 (59-231) μm; W = 450 ± 67.6 (313-614) μm; stylet = 25.0 ± 1.6 (21-29) μm; DGO = 6.1 ± 1.8 (3-11) μm; anterior end to: excretory pore = 153.8 ± 17 (114-207) μm, median bulb valve = 114.8 ± 9.4 (92-138) μm; vulval slit = 9.2 ± 0.8 (8-11) μm; fenestral length = 9.3 ± 2.2 (14-26) μm; fenestral width = 14.1 ± 1.4 (11-18) μm; distance anus to nearest margin of fenestra = 47.2 ± 6.5 (33-64) μm.

– *Male* (n = 115): L = 1164±98.3 (850-1325) μm; W = 27.1±1.7 (23-33) μm; a = 27-52 (43.1 ± 4.0); b′ = 5.2-8.5 (6.5 ± 0.5); c = 536±330 (189-2400); stylet = 26.6±0.9 (24-29) μm; DGO = 3.5±0.8 (2-6) μm; tail = 2.7±1.1 (0.5-5) μm; spicules = 32.8±1.3 (30-35) μm.

Description

After Lownsbery and Lownsbery (1954).

Female

Body ovate to spherical with elongate neck. White, becoming yellow as eggs mature. Cuticle surface marked by a rugose pattern. Inner layer of cuticle marked by minute, often obscure punctations. At mid-body punctations usually arranged in transverse rows, between vulva and anus they either show no alignment or are aligned parallel to vulvar-anal axis. Internal head sclerotisation weak. Labial region variable, without well defined limits, apparently having three prominent annuli and a labial disc (Fig. 6D). Stylet with well developed basal knobs, almost as stout as in male. Median bulb large, nearly spherical. Pharyngeal glands appearing to be contained in a single lobe. Two ovaries filling body cavity at maturity. No markings observed on eggshell. Vulva not protruding from body, situated ventrally from anus. Vulva slit-like, much larger than anal opening.

Male

Subcuticle marked by duplex striations. Lateral fields ranging from one-third body diam. at median bulb to one-fifth body diam. near tail, beginning as three incisures *ca* 15 μm from anterior end and increasing to four incisures just posterior to base of stylet, posteriorly extending around the tail. Outer incisures crenate. Labial region hemispherical with seven annuli, anterior-most annulus forming labial disc. Amphidial apertures minute, located at lateral borders of labial disc. First neck annulus smaller than succeeding annuli and located under head contour, thereby setting off head. Labial framework heavily sclerotised. Stylet robust, with well developed knobs. Median bulb of pharynx ellipsoidal with well developed valve. Two subventral pharyngeal glands and dorsal pharyngeal gland variable in length and size. In some specimens dorsal pharyngeal gland greatly reduced. Nerve ring large, encircling pharynx just anterior to junction with intestine. Excretory pore 10-13% of body length from anterior end, adjacent to subventral glands. Testis occupying one-third body diam. at anterior end, increasing to two-thirds body diam. posteriorly. Just adjacent to cap cell, testis containing double row of spermatocytes, about six rows of spermatozoa anterior to spicules. Spicules lightly arcuate, tips finely rounded, not notched. Tail very short. Phasmid openings near tail terminus.

J2

Cuticle marked by transverse striae averaging 1.8 μm apart at mid-body. Lateral fields with four incisures, extending from head to anterior boundary of hyaline region and slightly posterior to phasmid opening, *ca* one-fifth of body diam. at mid-body. Labial region offset by a slight constriction, marked by four striae. Labial framework heavily sclerotised. Stylet robust, with well developed rounded basal knobs. Median bulb of pharynx ellipsoidal. Pharyngeal glands arranged as in male. Excretory pore 21-24% of body length from anterior end. Genital primordium 53-75% of body length from anterior end. Phasmid openings about one-third distance from anus to tip of tail. Terminus of tail finely rounded.

After Miller and Gray (1972).

Cyst

Body turbinate to subglobose with a protruding neck. Cyst light shiny brown, circumfenestrate, abullate. Ridges and grooves of cyst wall forming circular pattern around fenestra, width of grooves (light discontinuous areas in low focus) between anus and fenestra highly variable with greatest width equal to, usually greater than, width of

Table 23. *Some important characters used for distinguishing* Globodera tabacum *subspecies (after Mota & Eisenback, 1993a, b, c).*

Characters	*G. tabacum tabacum*	*G. tabacum virginiae*	*G. tabacum solanacearum*
Cyst shapes	Round, L/W ratio = 1.0	Ovoid to ellipsoid, L/W ratio = 1.1-1.2	Ovoid to ellipsoid, L/W ratio = 1.1-1.2
Anal-fenestra ridges in cyst	Parallel ridges	Maze-like pattern	More widely separated
Anus in cyst	Distinct	Small and indistinct	Distinct
Perineal tubercles in female	Clumped	Individual	Individual
Average tail length in male (μm)	4.0-4.1	4.8-5.0	4.9-5.6
Average stylet knob width in male (μm)	4.9-5.0	5.2-5.6	5.0

adjacent ridge. Anal opening conspicuous, much smaller than opening formed by disintegrated vulval fenestra.

After careful investigation by Mota and Eisenback (1993a, b, c), the most useful characters to separate the three subspecies are considered to be as follows: female and cyst shape form, stylet knob shape in females, perineal tubercles in females, anal-fenestral ridge patterns, distinctiveness of the anus in the cyst, tail length and stylet knob width in male (Table 23).

TYPE HOST AND LOCALITY OF *G. TABACUM TABACUM*

Tobacco, *Nicotiana tabacum* L. (Solanaceae, Solanales), Hazardville, Connecticut, USA.

TYPE HOST AND LOCALITY OF *G. TABACUM SOLANACEARUM*

Horsenettle, *Solanum carolinense* L. (Solanaceae, Solanales), Watkins Farm, adjacent to Beaver Dam Creek on the north side, west side of Virginia Highway 153 and 0.81 km south of Scott's Fork (Ameali County) Virginia, USA.

TYPE HOST AND LOCALITY OF *G. TABACUM VIRGINIAE*

Horsenettle, *Solanum carolinense* L. (Solanaceae, Solanales), Crutchlow Farm, north side of Virginia Highway 58 and 4.7 km west of the

Thomas Lipton, Inc. tea plant, Suffolk (Nansemond County), Virginia, USA.

DISTRIBUTION

Europe: Bulgaria, France, Greece, Italy, Slovenia, Spain, Yugoslavia (former), *Asia*: China, Japan, Pakistan, South Korea; *Africa*: Madagascar, Morocco; *North America*: Canada (Ontario), Mexico, USA (Connecticut, Maryland, Massachusetts, North Carolina, Virginia); *South America*: Argentina, Colombia (Marché *et al.*, 2001; Esparrago *et al.*, 2002; Anon., 2004; Syracuse *et al.*, 2004; Bélair & Miller, 2006; Tovar Soto *et al.*, 2006).

OTHER HOSTS PLANTS

Solanaceae, especially, tobacco, tomato (Table 17) and species of the genera: *Nicotiana*, *Solanum*, as well as *Atropa belladona* L., *Hyoscyamus niger* L., *Nicandra physalodes* (L.) Gaertn., *Capsicum annuum* L. (Miller & Gray, 1968, 1972; Stelter, 1987; Marché *et al.*, 2000; Suzuki *et al.*, 2003). *Globodera tabacum virginiae* can develop on *Nicotiana* × *sanderae*, 50-B, whereas other subspecies do not. *Globodera tabacum tabacum* is unable to develop on *N. acuminata* 2-G-58, contrary to other subspecies (Bossis & Mugniéry, 1993).

PATHOGENICITY

Infected tobacco plants have small root systems and above-ground symptoms are similar to those associated with severe root-knot and lesion nematode infestations. Nematode infection is often associated with increased damage from bacterial wilt, black shank and *Fusarium* wilt and significantly decreases tobacco leaf quality. Farmers in Virginia, USA, have recorded complete crop failures, but losses generally average 15%. A high density of nematode populations early in the growing season can reduce flue-cured tobacco yield by 25-50%, although tobacco may escape significant losses from moderate populations, especially under favourable growing conditions. In field microplot experiments it has been shown that initial nematode densities of 0.1 to 1097 J2 cm^{-3} of soil was negatively correlated with leaf yield, total shoot weight and normal plant height 5 to 6 weeks after transplanting (LaMondia, 1995). Initial population density above 600 J2 cm^{-3} can reduce shoot weight of shade and broadleaf tobacco by 60 and 40%, respectively (Johnson *et al.*, 2005).

BIOLOGY

Two or more generations usually occur in Virginia and Connecticut, USA (Miller & Gray, 1968, 1972; LaMondia, 1990). In the Mediterranean climate experienced at Camposano, near Naples, Italy, *G. tabacum* completed almost five generations on eggplant (Ambrogioni & D'Errico, 1998).

RELATIONSHIPS

It differs from *G. rostochiensis* by J2 with longer mean values of body length (476-556 *vs* 392-468 μm), mean stylet length (21-25 *vs* 20-22 μm) and by cysts with a smaller mean number of cuticular ridges (7-9 *vs* 17-20). It differs from *G. mexicana* **sp. n.** by J2 with longer mean body length (476-556 *vs* 469 μm) and from *G. pallida* by cysts with a smaller mean number of cuticular ridges (7-9 *vs* 12) and by J2 with longer mean body length (476-556 *vs* 452-486 μm) (Table 14).

BIOCHEMICAL AND MOLECULAR CHARACTERISATION

Bossis and Mugniéry (1993), using two-dimensional electrophoresis of total proteins, showed that the genetic differences observed between *G. tabacum* subspecies corresponded to intraspecific variability. Three subspecies of *G. tabacum* are distinguished by the PCR-RAPD (Thiéry *et al.*, 1997) and AFLP techniques (Marché *et al.*, 2001). Marché *et al.* (2001) also recognised a fourth group from Mexico, named *G. tabacum* '*azteca*' (Fig. 37). The PCR-ITS-RFLP profile for *G. tabacum* has been generated by several authors (Table 12) and is given in Figure 39 and Table 15. The ITS rRNA gene sequences and PCR-ITS-RFLP do not distinguish the *G. tabacum* subspecies.

10. *Globodera zelandica* Wouts, 1984
(Figs 59, 60)

This species was isolated from the New Zealand native tree fuchsia, *Fuchsia excorticate* (Forst. & Forst.f.) L.f. (Onagraceae). It was the second *Globodera* species known to parasitise a woody plant and the

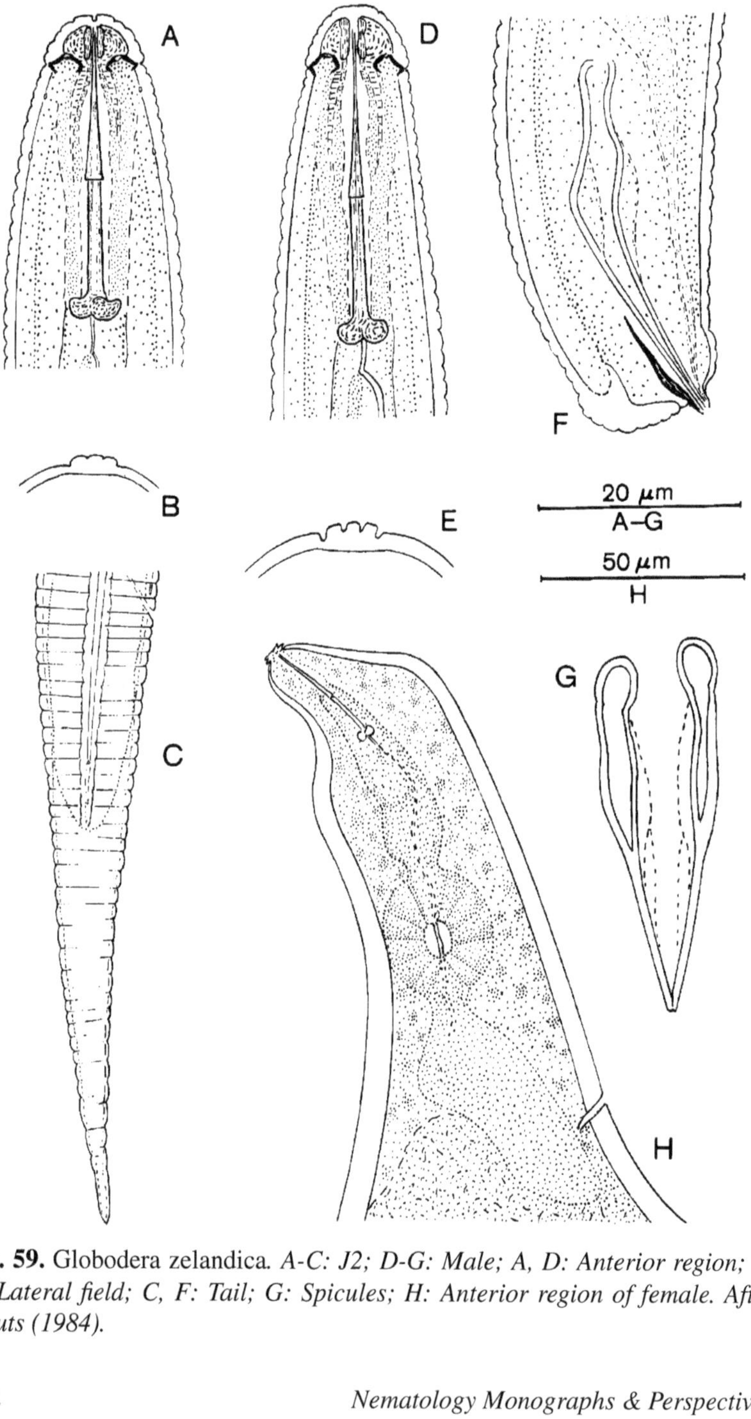

Fig. 59. Globodera zelandica. *A-C: J2; D-G: Male; A, D: Anterior region; B, E: Lateral field; C, F: Tail; G: Spicules; H: Anterior region of female. After Wouts (1984).*

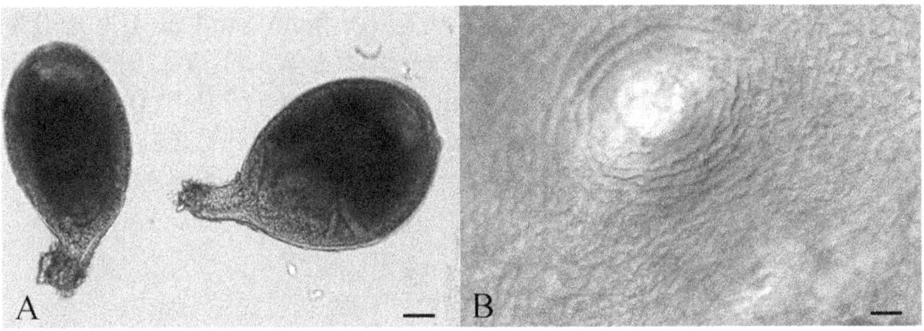

Fig. 60. Globodera zelandica. *A: Cysts; B: Vulval plate. (Original). (Scale bars: A = 100 μm; B = 10 μm.)*

first species of *Globodera* known to parasitise a native plant of New Zealand.

MEASUREMENTS

After Wouts (1984).

– *Holotype female*: L = 420 μm; width = 205 μm; neck = 135 μm; stylet = 28.5 μm; vulva to anus distance = 35 μm.

– *Paratype females* (n = 6): L (including neck) = 425 (400-475) μm; W = 190 (175-215) μm; neck = 124 (100-150) μm; stylet = 28.5 (27.5-30.0) μm; vulva to anus distance = 25 (21-28) μm; anterior end to excretory pore = 120 (100-150) μm; vulval slit = 6.0 μm; fenestral diam. = 20 μm.

– *Paratype male* (n = 10): L = 1125 (942-1180) μm; W = 25 (21-28) μm; a = 44.9 (36.2-53.7); b = 8.6 (8.4-8.8); b′ = 5.7 (5.1-6.1); labial region height = 6.1 (5.0-7.0); labial region diam. = 10.9 (10.0-11.5); stylet = 29.9 (29.0-31.0) μm; stylet knob height = 2.2 (2.0-2.5) μm; stylet knob width = 5.0 (4.5-5.5) μm; DGO = 4.0 (2.5-5.0) μm; anterior end to: median bulb valve = 102 (84-106) μm, hemizonid = 158 (120-172) μm, excretory pore = 165 (126-180) μm; length of genital tract = 612 (525-675) μm; spicules = 37.8 (35.0-39.0) μm; gubernaculum = 11.5 (10.0-12.0) μm.

– *Paratype J2* (n = 30): L = 484 ± 3.5 (429-511) μm; W = 19.6 ± 0.2 (16.0-21.0) μm; a = 24.8 ± 0.3 (22.0-28.0); b′ = 2.6 ± 0.1 (2.3-2.7); c = 7.9 ± 0.1 (6.5-8.8); labial region height = 4.0-5.0 (4.6 ± 0.06) μm; labial region diam. = 9.4 ± 0.2 (9.0-10.5) μm; stylet = 27.9 ± 0.1 (26.5-30.0); stylet knob height = 2.6 ± 0.1 (2.25-3.0); stylet knob width = 5.6 ± 0.1 (5.0-6.5); DGO = 4.5 ± 0.2 (3.0-7.0); anterior end of median bulb valve = 77.7 ± 0.7 (67-86); hemizonid = 106 ± 0.8 (93-112) μm; pharyngeal gland base = 194 ± 2.3 (165-215), genital primordium = 276±2.4 (236-296); tail = 62.0±1.0 (53.0-76.0); hyaline region = 36.6 ± 0.7 (29-44); anal body diam. = 12.7 ± 0.2 (11.0-14.5); body diam. at beginning of hyaline region = 7.7 ± 0.2 (6.5-9.0).

DESCRIPTION

Female

Body elongate, pale brown; subcrystalline layer absent in specimens studied, cuticle with rugose line pattern and distinct subcuticular punctation. Labial region with wide anterior annulus, labial disc distinct. Crescentic area on dorsal and ventral sides of vulva, variable in width, usually only one cuticular fold wide with tubercles in these areas discrete, not densely placed.

Cyst

Generally spherical with short neck; vulval fenestration almost circular, diam. extremely variable (max. measured = 40 μm); distance between fenestra and anus also variable (max. measured = 40 μm); a partial subcrystalline layer observed in one specimen only, cuticle brown with rugose pattern, egg sac not observed.

Male

General morphology typical for genus, stylet knobs anteriorly almost flat, narrower than those of J2, labial region offset with 5-7 annuli and a labial disc. Lateral field with four lines, outer lines slightly crenate, non-areolated. Hemizonid located at anterior end of intestine, several body annuli anterior to excretory pore. Spicules curved.

J2

Body curved ventrally, labial region with four annuli and labial disc distinctly offset from rest of body, labial disc consisting of an oral disc surrounded by two small lateral lips, adjacent submedial lips fused resulting in single elongated dorsal and ventral lips. Lateral fields with four lines, indistinctly crenate anteriorly, crenate posteriorly. Stylet strong, knobs anteriorly concave. Cuticle near labial region thicker than on rest of body, median bulb oval, pharyngeal gland well developed, pharyngeal/intestinal junction not always distinct, hemizonid at level of pharynx base, excretory pore located immediately posterior to hemizonid. Phasmid openings *ca* one anal body diam. posterior to anus, distinct in some specimens, tail terminus sharply pointed.

TYPE HOST AND LOCALITY

Soil and roots of tree fuchsia, *Fuchsia excorticata* (Forst. & Forst.f.) L. f. (Onagraceae, Myrtales) on bank of Riwaka River at Moss Picnic Area no. 2, about 5 km west of Riwaka, at the foot of the Takaka Hill Nelson South Island, New Zealand.

DISTRIBUTION

Oceania: New Zealand.

OTHER HOSTS

Kowhai (*Sophora microphylla* Ait.) (Papilionaceae) is the only other plant native to New Zealand known to be parasitised by *G. zelandica*. Although none of the kowhais in the type locality was found to be infected by the nematode, a stand of kowhais north of Niagara, near Niagara Falls, about 10 km from the southernmost coast of the South Island, was found to be moderately to heavily infected by a nematode population that is morphologically indistinguishable from *G. zelandica*. Whether this population parasitises tree fuchsia was not investigated (Wouts, 1984). Sturhan (unpubl.) found *G. zelandica* from *Plagianthus regius* (Poit) Hochr (Malvaceae, Malvales).

RELATIONSHIPS

Globodera zelandica can be distinguished from other *Globodera* species by the longer mean length of the stylet (28 *vs* 18-25 μm), longer hyaline region (37 *vs* 20-32 μm) and longer tail (62 *vs* 40-56 μm) in the J2 (Table 14).

BIOCHEMICAL AND MOLECULAR CHARACTERISATION

PCR-ITS-RFLPs using one of the following restriction enzymes: *Bsh*1236I, *Hinf*I and *Rsa*I discriminate *G. zelandica* from other *Globodera* species (Table 15).

Undescribed *Globodera* species

Molecular and morphological analyses have revealed several putative new and as yet undescribed species of *Globodera*.

Globodera sp. from Portugal (Reis, 1997)

Cysts of a *Globodera* sp. parasitising *Chamaemelum mixtum* (L.) All. (Asteraceae, Asterales) were found in Portugal and have morphological differences from potato cyst nematodes. Sabo *et al.* (2002) sequenced the ITS-rRNA gene for this isolate. Phylogenetic and sequence analysis revealed that this isolate is different from other *Globodera* species and represent a new species (Sabo *et al.*, 2002; Subbotin *et al.*, unpubl.)

Globodera sp. from South Africa (Knoetze *et al.*, 2006)

This *Globodera* species was found in potato plots infected with *G. rostochiensis* in the Sandvel region, South Africa. This species is morphologically close to *G. millefolii*.

– *Cysts*: Number of cuticular ridges between anus and vulval basin = 11-13; diam. of vulval basin = 14.2-23.6 μm; distance anus-vulva = 17.9-53.8 μm; Granek's ratio = 1.06-2.71. Some cysts contain bullae. Ridges in most of the cysts were arranged in a parallel pattern, more or less separated from one another and with some connections among them. Some individuals exhibited ridges highly anastomosed among one another (maze-like pattern), showing a very irregular pattern.

– *J2*: The anterior face of the basal knobs of the stylet showed different shapes depending on the specimens, *viz.*, rounded, flattened and anteriorly projected.

This species can be distinguished from other *Globodera* species by PCR-ITS-RFLP and ITS-rRNA sequences and from *G. pallida* and *G. rostochiensis* by the absence of a restriction site for *Hinf*I in the ITS1-rRNA region (Knoetze *et al.*, 2006).

Globodera sp. from Argentina (Lax *et al.*, unpubl.)

This *Globodera* identified as *G. pallida* was found in soil samples taken from a potato plot in the locality of Campo Carreras, department of Iruya, province of Salta, Argentina. However, comprehensive phylogenetic analyses showed that this sample might belong to a new and undescribed *Globodera* species (Madani *et al.*, 2010; Subbotin *et al.*, unpubl.).

This species can be distinguished from other *Globodera* species by PCR-ITS-RFLP and ITS-rRNA sequences (Table 15).

Globodera spp. from New Zealand (Sturhan, unpubl.)

Two *Globodera* species were found from samples collected from subalpine vegetation, Desert Road, North Island, and subalpine scrub vegetation Lake Lyndon area, Canterbury, South Island, respectively. Phylogenetic and sequence analyses revealed that these species are different from other *Globodera* (Subbotin *et al.*, unpubl.).

Globodera sp. from Oregon state, USA (Handoo, pers. comm.)

In 2008, during surveys for the federally regulated nematodes *G. pallida* and *G. rostochiensis*, a possible new *Globodera* species parasitising potato was discovered in central Oregon, USA. Molecular analysis revealed that this species differs from other *Globodera* parasitising Solanaceae (Skantar, pers. comm.; Handoo, pers. comm.).

Nematology Monographs & Perspectives, 2010, Vol. 8A, 179-201

Chapter 7

Systematics of the genus *Punctodera*

History

Skarbilovich (1959) placed the spherical and pear-shaped species of *Heterodera* in the subgenus *Globodera* and the lemon-shaped species in the subgenus *Heterodera*. After careful consideration of the morphological and biological differences of the females, cysts, and males, Mulvey and Stone (1976) proposed that *Globodera* be raised to generic rank and described a new genus *Punctodera* from the specific epithet of the type species name (Latin *punctum* = dot, and Greek (δερος) *deros* = skin), for species having extensive anal fenestration. An identification key for *Punctodera* was given by Golden (1986).

Genus *Punctodera* Mulvey & Stone, 1976

DIAGNOSIS (after Siddiqi, 2000)

Mature female and cyst: Spherical, pear-shaped or ovoid, with short projecting neck and heavy subcrystalline layer. Cuticle reticulate, subcuticle with punctations. D-layer present. Terminal region not cone-shaped; cyst light to dark brown. Vulval slit extremely short (<5 μm), anus at a short distance from vulval fenestra. Circumfenestrate, fenestra surrounding vulva 16-40 μm (*ca* 20 μm in type species) in diam., anus offset toward ventral margin of fenestra, an anal fenestra of similar shape and size to vulval fenestra present. Underbridge and perineal papilla-like tubercles absent. Bullae present or absent. Egg retained in body, no egg sac. *Male*: Vermiform, less than 1.5 mm long. DGO 2.6-4.6 μm. Spicules 31-33 μm long, distally pointed. Tail less than 0.5 anal body diam. long, cloacal lips not forming a tube. *J2*: Body 0.35-0.49 mm, stylet 24-26 μm and tail conical, 63-78 μm long, hyaline region in type species 38-41 μm long. Lateral field with four incisures. Pharyngeal glands filling body diam. Phasmid openings punctiform, without a lens-like structure. Parasites of monocotyledonous plants.

TYPE SPECIES

1. *Punctodera punctata* (Thorne, 1928) Mulvey & Stone, 1976
   = *Heterodera punctata* Thorne, 1928
   = *Heterodera* (*Globodera*) *punctata* (Thorne, 1928) Skarbilovich, 1959

OTHER SPECIES

2. *P. chalcoensis* Stone, Sosa Moss & Mulvey, 1976
3. *P. matadorensis* Mulvey & Stone, 1976
4. *P. stonei* Brzeski, 1998

Key to species of the genus *Punctodera*

1 – Cyst pear- or oval-shaped, without bullae, parasitising wheat and grasses . . . . *P. punctata*
– Cyst spherical- or subspherical-shaped, with bullae, parasitising maize or grasses 2

2 – Bullae small and scattered, parasitising maize in Mexico . . . . *P. chalcoensis*
– Bullae massive, between vulva and anal fenestra, parasitising grasses in Canada and Europe 3

3 – Stylet knobs strongly concave anteriorly, hyaline region < 42 μm in J2 . . . . *P. matadorensis*
– Stylet knobs flattish to slightly concave anteriorly, hyaline region $\geqslant 42$ μm in J2 . . . . *P. stonei*

Punctodera punctata differs from the closely related *P. matadorensis* and *P. stonei* by cyst shape, absence of bullae and some morphometric characters of the J2, as well as by the ability to infect roots of wheat plants (Table 24). However, some observations revealed that these features might be the subject of intraspecific variation for *P. punctata.* It is likely that the present descriptions of this species have been made from samples containing more than one species. Wouts and Baldwin (1998) believed that five different species can be recognised within the genus *Punctodera*: *P. punctata* from wheat, with short juveniles,

Table 24. *Morphological and morphometric characters useful for identification of* Punctodera *species (means are given in μm).*

Species	Cyst				J2			
	Vulval fenestral diam.	Anal fenestral diam.	Vulva to anus distance	Bullae	Body length	Stylet	Tail	Hyaline region
P. punctata	22-29	18-33	27-39	Absent	383-620	24-28	62-88	38-59
P. chalcoensis	19-30	21-22	42	Present	522-533	24-25	63-66	38-39
P. matadorensis	20	20	?	Present	526-540	24-27	62-66	37-41
P. stonei	27	23	59	Present	520	25	78	50

P. matadorensis and *P. stonei* from native grasses and with distinct bullae, *P. chalcoensis* from maize, and an undescribed species from grasses. More detailed molecular and morphological studies are required to estimate the reliability of these characters for diagnostics and to further test the validity of *Punctodera* species.

SPECIES DESCRIPTIONS

1. *Punctodera punctata* (Thorne, 1928) Mulvey & Stone, 1976
(Figs 17D; 61, 62)

Common name: grass cyst nematode.

This species was briefly described by Thorne (1928) based on type specimens from heavily infested wheat roots from a field in the Humboldt area, Saskatchewan, Canada. Several subsequent attempts to collect topotype specimens failed (Baker, 1957; Thorne & Malek, 1968; Mulvey & Stone, 1976). Subsequently, *P. punctata* was also reported as a common species infecting grasses from Europe (Franklin, 1951), USA (Chitwood, 1949b; Spears, 1956; Horne & Thames, 1966; Brzeski *et al.*, 1971) and Canada (Baker, 1957; Mulvey & Stone, 1976). However, all attempts to infect wheat or other cereals by these nematodes did not give any positive results. Redescriptions and measurements of *P. punctata* were based on several original slides and provided by Horne (1965) and Mulvey and Golden (1983). Morphological descriptions and measurements of populations from different countries and identified as *P. punctata* were provided by Franklin (1951), Horne (1965), Mulvey (1972), Solovjeva and Vasiljeva (1973), Mulvey and Stone (1976), Webley and Lewis (1977), Wouts and Weischer (1977), Chizhov and

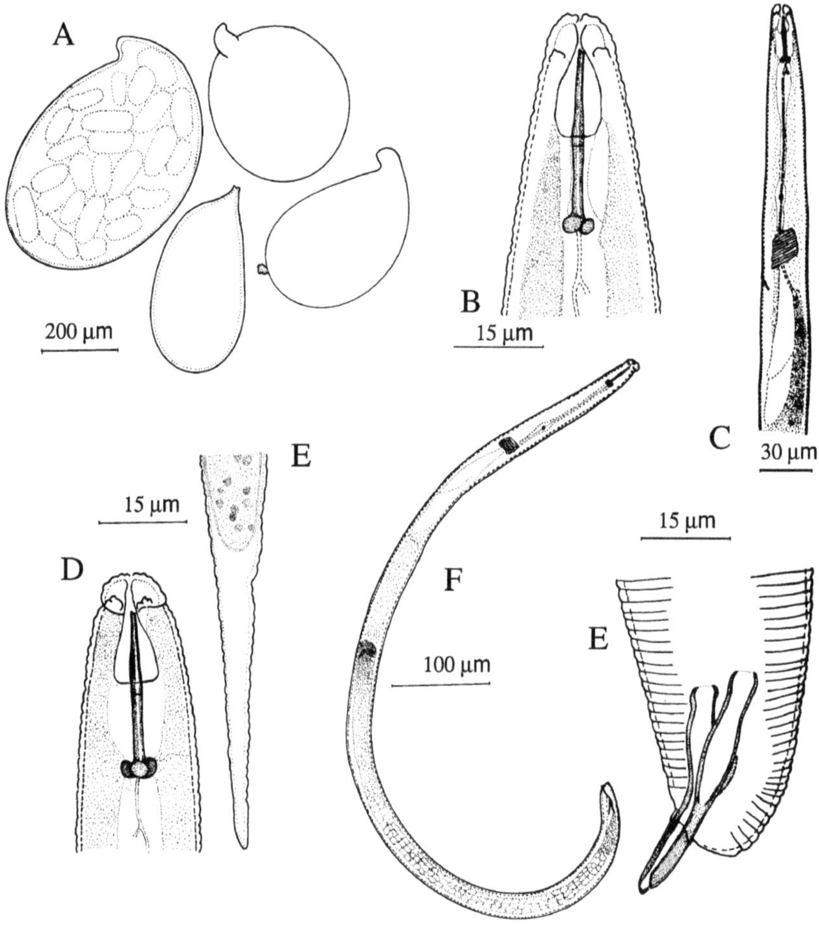

Fig. 61. Punctodera punctata. *A: Cysts; B: Anterior region of male; C: Pharyngeal region of male; D: Anterior region of J2; E: Tail of J2; F: Male. After Webley and Lewis (1977).*

Idekh (1980), Riggs (1982) and Wouts *et al.* (1986). Several authors (Oostenbrink, 1960; Mulvey, 1972; Solovjeva & Vasiljeva, 1973; Wouts *et al.*, 1986) remarked that *P. punctata* might represent a complex of several closely related species.

MEASUREMENTS

After Thorne (1928).

- *Paratype female* = 520 μm.

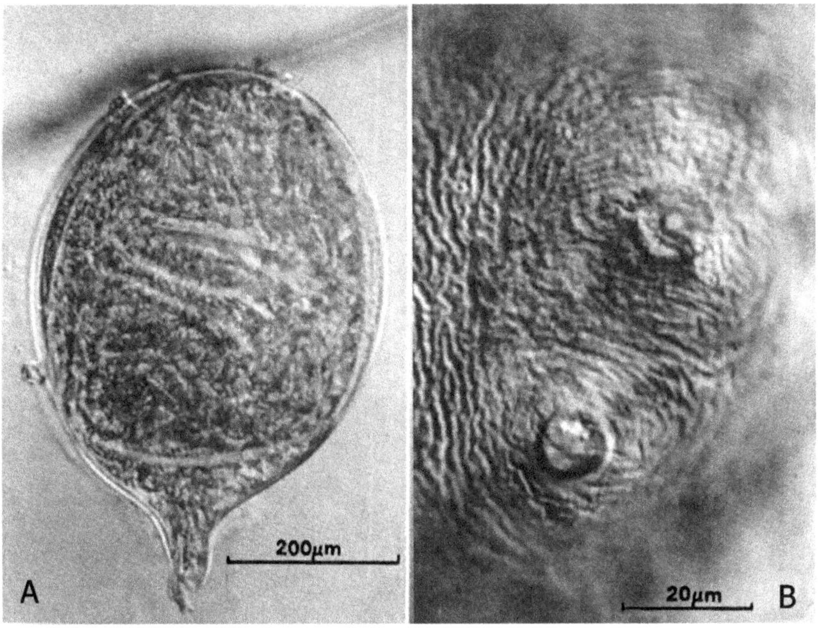

Fig. 62. Punctodera punctata. *A: Newly formed cyst; B: Terminal area of cyst showing vulval and anal areas. Material from Thorne's type collection. After Mulvey and Stone (1976).*

– *Paratype J2*: L = 350-470 μm; pharynx = 17-22% of body length.

See Table 25 for measurements of J2.

After Horne (1965).

– *Female* (n = 20): L (excluding neck) = 370 (330-420) μm; W = 236 (170-320) μm; neck = 123 (110-150) μm; L/W ratio = 1.59 (1.22-2.05); stylet = 27 (20-25) μm.

– *Male* (n = 20): L = 1100 (970-1270) μm; W = 27 (25-32) μm; stylet = 30 (27-33) μm; DGO = 5.5 (5-6) μm; spicules = 33.6 (28-36) μm.

– *Egg* (n = 50): L = 118 (110-113) μm; W = 45.6 (40-50) μm; L/W ratio = 2.6 (2.0-3.0).

After Solovjeva and Vasiljeva (1973).

– *Female* (n = 150): L = 540 (219-901) μm; W = 282 (159-682) μm; L/W ratio = 1.41; vulval slit (n = 58) = 22 (16-38) μm; vulva

Table 25. *Morphometrics of J2 of* Punctodera punctata punctata *from different locales (all measurements in μm).*

Population / Characters	Paratype slides from Thorne's collection, Mulvey & Golden (1983)	*Poa annua*, Horne (1965)	*Triticum aestivum*, Horne (1965)	After Solovjeva & Vasiljeva (1973)	Webley & Lewis (1977)	Münster, BBA, Germany, Wouts & Weischer (1977)	Münster, Venne, Germany, Wouts & Weischer (1977)	Moscow region, Russia, Chizhov & Idekh (1980)	Davert, Germany, Wouts *et al.* (1986)
J2 (n)	–	128	–	56	–	20	10	10	30
L	464 (438-490)	620 (520-680)	542 (520-570)	383 (208-538)	540 (510-570)	592 ± 14.0	583 ± 7.0	490-590	598 (548-643)
W	-	22 (20-24)	20 (19-21)	18.4 (11-22)	22 (20-24)	20.6 ± 0.2	21.1 ± 0.3	-	21.2 (19-23)
Stylet	25 (24.2-25.8)	27 (24-32)	25 (23-27)	28.2 (26.6-31.1)	27 (24-32)	28.1 ± 0.1	28.4 ± 0.2	26-28	28.3 (27-30)
DGO	4.5	–	–	–	5.4 (4-7)	–	–	–	5.1 (3.5-6.5)
Anterior end to median bulb	–	–	–	–	–	81.5 ± 0.8	82.3 ± 0.9	–	–
Tail	70.1 (62.7-77.5)	80 (68-93)	78 (75-83)	–	79.8 (68-93)	87.6 ± 0.8	86.9 ± 2.0	–	82.5 (75-92)
Hyaline region	40 (38.5-41.4)	–	–	44.6 (39.9-50.5)	45 (51-60)	55 ± 0.6	59.1 ± 1.4	–	53 (37-64)

to anus distance (n = 58) = 39 (22-77) μm; anal fenestral diam. (n = 58) = 18 (11-33) μm.

– *Egg* (n = 150): L = 105.8 (88-132) μm; W = 49.6 (43-55) μm; L/W ratio = 2.14.

After Webley and Lewis (1977).

– *Female*: L = 370 (330-420) μm; W = 236 (170-320) μm; stylet = 23.2 (20-25) μm.

– *Male*: L = 1087.6 (970-1270) μm; W = 27 (25-32) μm; a = 40.3 (32-47); stylet = 30.4 (27-33) μm; spicules = 33.6 (28-36) μm; DGO = 5.5 (5-6) μm.

– *Egg*: L = 118 (100-123) μm; W = 54.6 (40-60) μm; L/W ratio = 2.6 (2-3).

After Chizhov and Idekh (1980).

– *Cyst*: L = 565 (392-823) μm; W = 387 (298-710) μm; L/W ratio = 1.52 (1.16-1.89); vulval fenestral diam. = 29 (23.1-33.6) μm; anal fenestral diam. = 33 (25.2-42.0) μm; vulva to anus distance = 27.0 (16.8-42.0) μm.

– *Male* (n = 10): L = 910-1050 μm; a = 29.7-40.0; b = 4.8-6.2; c = 162-226; stylet = 26-29 μm; spicules = 28-34 μm; gubernaculum = 8-10 μm.

After Mulvey and Golden (1983).

– *Paratype females* (specimens from Thorne's 1927 collection): L = 499 $\pm$ 64 μm; W = 345 $\pm$ 38 μm; fenestral length = 16.5 μm; fenestral width = 20.3 μm; anal length = 17.8 μm.

– *Paratype males* (specimens from Thorne's 1927 collection): L = 900 $\pm$ 99 μm; stylet = 23.7 $\pm$ 1.0 μm; spicules = 32.4 $\pm$ 0.7 μm.

DESCRIPTION

After Webley and Lewis (1977).

Female

Ovoid, pear- or flask-shaped without vulval cone, white. Vulval slit *ca* 4 μm long, bordered by thickened ridges, set in a subcircular translucent area of cuticle. Anal slit less than 4 μm long, positioned towards ventral side of a similar subcircular area. Newly formed cysts with conspicuous subcrystalline layer. Vulval and anal fenestrae appearing as two pores penetrating cyst surface near rounded posterior end. Fenestrae usually of about equal size, separated by a distance of 1-3 vulval fenestral widths. Vulval fenestra occasionally twice diam. of anal fenestra. Bullae absent.

Male

Body tapering markedly anterior to stylet base. Labial region offset with 5-7 annuli. Lateral field with four incisures. Tail bluntly rounded, slightly angular dorsally. Stylet strong with well developed basal knobs projecting anteriorly. Median pharyngeal bulb ellipsoid, curved ventrally with rounded blunt tips. Cloacal opening thickened, prominent. Spicules with smooth tip (Fig. 11B). Single, unornamented, gubernaculum.

J2

Vermiform with long, pointed tail. Labial region offset with our annuli and robust labial framework. Body annuli distinct, the first seven or eight sometimes thickened. Stylet robust with basal knobs projecting anteriorly. Lateral field with four incisures. Median pharyngeal bulb well developed, ellipsoid, not filling body diam. Hemizonid one annulus anterior to excretory pore, no phasmid openings observed although small refractive bodies often present in hyaline region. Conspicuous hyaline region at least twice as long as stylet, distal third of tail tapering, ending in a rounded point.

TYPE HOST AND LOCALITY

Wheat, *Triticum aestivum* L. (Poaceae, Poales), in the Humboldt area located about 161 km east by north-east of Saskatoon, Saskatchewan, Canada.

DISTRIBUTION

Europe: Austria, Belgium, Czech Republic, Denmark, Finland, France, Germany, Hungary, Ireland, The Netherlands, Norway, Poland, Russia, Slovenia, Sweden, Yugoslavia (former); *North America*: Canada (Saskatchewan, Labrador, Newfoundland, Sable Island, Manitoba, Ontario, Quebec, and British Columbia), USA (Michigan, South Dakota, Texas, New Jersey and Arkansas); *Asia*: China (Chitwood, 1949b; Spears, 1956; Baker, 1957; Horne & Thames, 1966; Mulvey, 1972; Mulvey & Stone, 1976; Radice *et al.*, 1983; Stelter, 1984).

Other hosts

Although this species was originally described from wheat roots, extensive studies of the host range indicate that many grasses are the preferred hosts of this nematode. *Poa annua* L. and *P. pratensis* were listed as the most common natural hosts for *P. punctata* by Baker (1957), Horne (1965) and Chizhov and Idekh (1980). After a study of a Texas population, Horne (1965) and Horne and Thames (1966) found annual bluegrass to be a good host and, although wheat, oat and barley were infected by nematodes, they were successively poorer hosts. In host range tests by Radice *et al.* (1985) with a population from New Jersey, *P. annua* was the most suitable host followed by *P. pratensis*, *Lolium perenne* and *Festuca rubra*. Seidel (1973), Stone *et al.* (1976), Chizhov and Idekh (1980), Riggs (1982) and Radice *et al.* (1985) also found that the tested populations did not infect wheat and other cereals.

Life cycle

Horne (1965), Chizhov and Idekh (1980) and Radice *et al.* (1985) found that only a single generation occurs each year. *Punctodera punctata* developed from J2 to immature white females in 21 days on roots of *P. pratensis*. After 40 days, eggs, J2 and males were detected (Radice *et al.*, 1985). In Texas, USA, mature females producing eggs were found 70 days after planting hosts in the field (Horne, 1965). In the Moscow region, Russia, the first mature females with eggs were found during the end of June and cysts appeared at the end of July (Chizhov & Idekh, 1980).

Pathogenicity

A distinct chlorosis and unhealthy appearance of heavily infected *Poa* species growing on lawns and in natural habitats were noticed by Horne (1965), Brzeski *et al.* (1971) and Chizhov and Idekh (1980). In experiments conducted by Radice *et al.* (1985), inoculation with *P. punctata* on the most susceptible host, *P. annua*, had no effect on plant top dry weight, root dry weight or total dry weight despite reproduction of the nematode. Infection by moderate numbers of *P. punctata* appeared to cause little damage to *P. annua*.

RELATIONSHIPS

Punctodera punctata differs from other *Punctodera* species by the pear-shaped cysts and the absence of bullae (Table 24).

BIOCHEMICAL AND MOLECULAR CHARACTERISATION

Rumpenhorst (1985) provided the IEF protein profile for several populations of *P. punctata*. Fleming *et al.* (2000) showed that PCR-ITS-RFLP can be used routinely to discriminate among cyst nematodes of the genera *Heterodera*, *Punctodera* and *Globodera*. Subbotin *et al.* (2001a) and Sabo *et al.* (2002) sequenced the ITS-rRNA gene and studied the relationships of *P. punctata* with other cyst nematodes.

2. *Punctodera chalcoensis* Stone, Sosa Moss & Mulvey, 1976
(Figs 12C; 63, 64)

Common name: Mexican corn cyst nematode.

The Mexican corn cyst nematode was first observed in the late 1950s in maize fields in Huamantla, Tlaxcala, Mexico, and was identified as *H. punctata* and a Mexican race of this species (Sosa Moss & Gonzales, 1973; Vazquez, 1976; Baldwin & Mundo-Ocampo, 1991). Further detailed morphological and morphometric observations revealed differences between these populations and *P. punctata* and *P. matadorensis* and led to the description of a new species, *P. chalcoensis*, by Stone *et al.* (1976).

MEASUREMENTS

After Stone *et al.* (1976).

– *Holotype cyst*: L (excluding neck) = 510 μm; W = 470 μm; neck = 88 μm; L/W ratio = 1.1.

– *Paratype cysts* (n = 20): L (excluding neck) = 441 $\pm$ 69 μm; W = 416 $\pm$ 61 μm; neck = 95 $\pm$ 26 μm; L/W ratio = 1.06 $\pm$ 0.1; vulva to anus plates (n = 25): vulval fenestral length = 18.1 $\pm$ 2.7 μm; vulval fenestral width = 19.8 $\pm$ 2.9 μm; diam. of clear zone surrounding vulval fenestra $\approx$ 30 μm; vulval slit = 4.2 $\pm$ 0.4 μm; anal fenestra length = 21.1 $\pm$ 3.4 μm; anal fenestral width = 22.4 $\pm$ 3.1 μm; length of anal slit = 2.8 $\pm$ 0.4 μm; distance between fenestrae = 42.3 $\pm$ 8.8 μm.

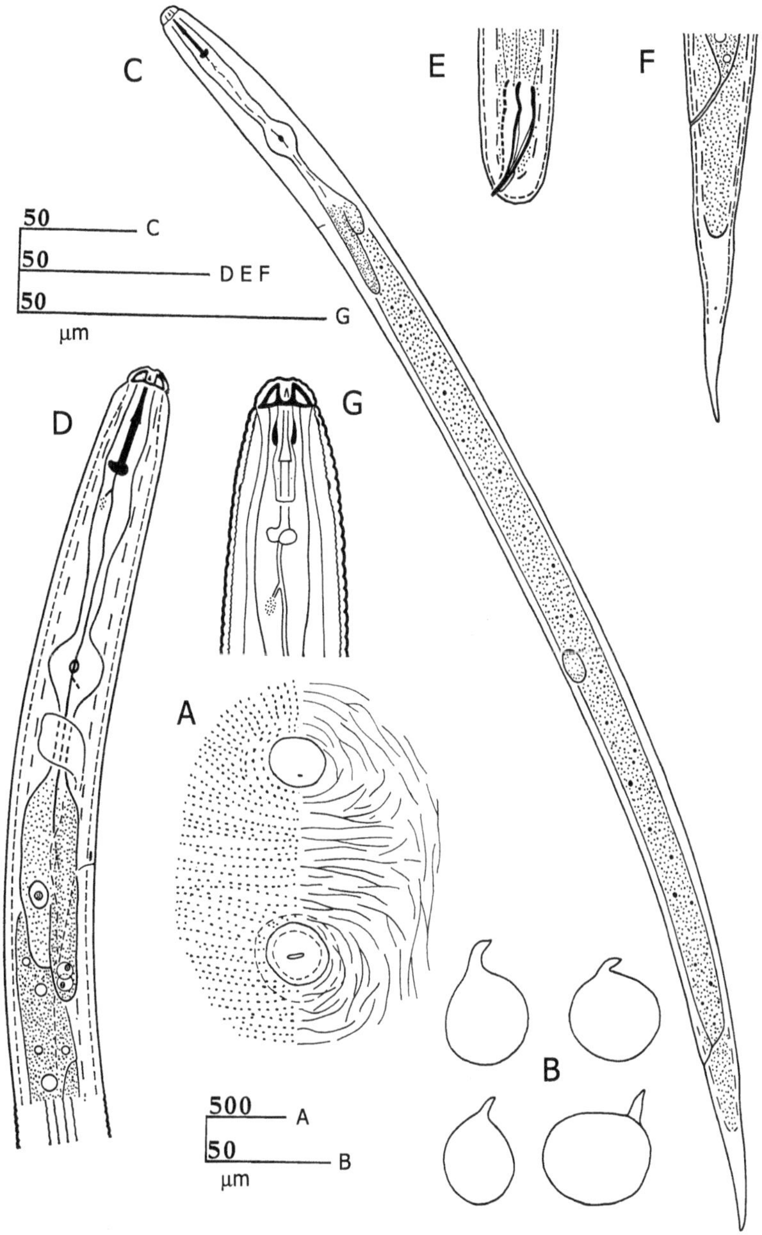

Fig. 63. Punctodera chalcoensis. *A: Perineum of female; B: Females; C: J2; D: Pharyngeal region of J2; E: Male tail; F: Tail of J2; G: Head of J2. After Stone* et al. *(1976).*

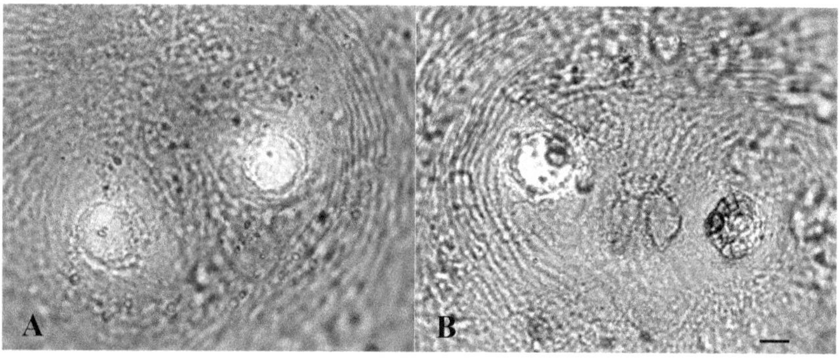

Fig. 64. Punctodera chalcoensis. *A, B: Vulval plates. (Original.) (Scale bar = 10 μm.)*

– *Paratype females* (n = 20): L (excluding neck) = 473 ± 105 μm; W = 429 ± 120 μm; neck = 142 ± 24 μm; L/W ratio = 1.14 ± 0.18; stylet = 25.8 ± 0.9 μm; DGO = 5.7 ± 0.9 μm; median bulb length = 25.7 ± 1.9 μm; median bulb width = 24.9 ± 2.4 μm; anterior end to median bulb valve = 77.4 ± 8.4 μm, excretory pore = 131.0 ± 0.6 μm; vulval fenestral length = 30.7 ± 7.0 μm; vulval fenestral width = 32.1 ± 7.0 μm; vulval slit = 4.0 ± 0.5 μm; anal fenestral length = 29.5 ± 5.7 μm; anal fenestral width = 1.1 ± 5.3 μm; anal slit = 3.0 ± 1.0 μm; distance between fenestrae = 67.7 ± 14.5 μm.

– *Paratype males* (n = 12): L = 985 ± 69 μm; W = 25.8 ± 1.7 μm; stylet = 26.8 ± 0.8 μm; DGO = 4.0 ± 0.6 μm; anterior end to: median bulb valve = 82.7 ± 3.7 μm, excretory pore = 131.3 ± 7.7 μm; tail = 2.8 ± 1.4 μm; spicules = 31.9 ± 1.8 μm; gubernaculum = 7.3 ± 0.8 μm; overall gonad length = 476 ± 107 μm.

– *Paratype J2* (unfixed in water, n = 20): L = 522 ± 26 μm; W = 21.6 ± 0.7 μm; stylet (S) = 24.4 ± 0.2 μm; DGO = 5.5 ± 0.8 μm; anterior end to: median bulb valve = 70.4 ± 6.3 μm, excretory pore = 110.1 ± 7.9 μm, pharyngeal glands = 176.0 ± 15.1 μm; tail = 66.2 ± 3.5 μm; hyaline region (H) = 38.6 ± 3.5 μm; H/S ratio = 1.6 ± 0.1.

– *Paratype J2* (fixed in 4% formaldehyde, n = 50): L = 533 ± 29 μm; W = 20.1 ± 0.6 μm; stylet = 24.7 ± 0.6 μm; DGO = 5.1 ± 0.7 μm; anterior end to: median bulb valve = 73.7 ± 2.7 μm, excretory pore = 107.7 ± 4.9 μm; tail = 63.2 ± 3.3 μm; hyaline region = 38.2 ± 3.1 μm; H/S ratio = 1.5 ± 0.1.

– *Paratype eggs* (n = 20): L = 114.4 ± 3.6 μm; W = 43.1 ± 1.7 μm; L/W ratio = 2.7 ± 0.1.

DESCRIPTION

Cyst

Spherical or subspherical with protruding neck, pale to dark brown, darkening with age. New cysts often retaining subcrystalline layer. Old cysts losing thin walls of vulval and anal fenestrae, younger cysts showing incomplete fenestration. Some specimens with small scattered bullae in perineal region or just anterior to vulval fenestra, but bullae may be lacking from many cysts. Cyst typically containing 200-400 eggs.

Female

Mature females white, spherical or subspherical with protruding neck. Labial region with one or two prominent annuli. Stylet slender, often curved, with rounded basal knobs. Excretory pore at base of neck. Cuticle greatly thickened except in head region. Vulval slit on slight circular prominence. Vulva a short transverse slit situated posteriorly, lying centrally in a circular zone lacking ridges and punctations and having cuticle reduced in thickness to form the vulval fenestra. Vulval fenestra appearing more transparent than surrounding body wall and surrounded by a narrow zone of cuticle with reduced thickness. Anus a transverse slit, shorter than vulval slit, lying 'dorsal' to vulva within a thin-walled cuticular zone, the anal fenestra. Anal fenestra of similar size and appearance to vulva fenestra but lacking surrounding clear zone. Anus located toward ventral side of fenestra. Clusters of elongate bodies associated with anal and vulval apertures. Thick white subcrystalline layer typically present.

Male

Body vermiform, heat-relaxed specimens with strong ventral curvature, tail frequently twisted through 180°. Labial region offset with 5-7 annuli. Tail bluntly rounded, length less than one-quarter of body diam. Cuticle with regular annulations, four lateral areolated incisures. Cephalids at level of second and eighth body annuli. Stylet well developed with flat to concave anteriorly basal knobs. Median pharyngeal bulb ellipsoidal, not filling body cavity. Hemizonid extending over two annuli, located one annulus anterior to excretory pore. Paired spicules flask-shaped proximally and tapering distally, curved ventrally with single distal point. Gubernaculum single, unornamented. Phasmids not observed.

J2

Vermiform body adopting a slight ventral curve on heat relaxation. Cuticle annulations distinct, lateral field with four intermittently areolated incisures reducing to three anteriorly and posteriorly (Fig. 9H). Labial region offset, apparently with 3-5, usually four annuli as seen by light microscopy. Labial disc elongated dorsoventrally by *ca* 1.5 times its diam., surrounded by distinct lateral lips bearing amphidial apertures, adjacent submedial lips fused with each other. Cephalids at level of second and seventh or eighth body annulus. Well developed stylet with massive basal knobs, knobs rounded posteriorly, flat to shallowly concave anteriorly. Pharyngeal gland lobes with ventral overlap, usually extending posteriorly *ca* one-third of body length but sometimes considerably shorter. Hemizonid 1-2 annuli anterior to excretory pore. Phasmid openings situated in posterior two-thirds of tail, small refractive bodies sometimes present in matrix of hyaline terminus.

TYPE HOST AND LOCALITY

Maize, *Zea mays* L. (Poaceae, Poales), near Chaco, Valley of Mexico, Mexico state, Mexico, at an altitude of 2300 m a.s.l.

OTHER HOSTS

Punctodera chalcoensis also parasitises teosinte, *Zea mexicana* (Schrad.) Kuntz, 1976. Under experimental conditions, several species of grasses and cereals were not infected by the nematode (Stone *et al.*, 1976).

DISTRIBUTION

North America: Mexico. Nematological surveys indicated a wide distribution of *P. chalcoensis* at elevations between 1500 and 2800 m a.s.l. throughout the maize-producing regions of Central Mexico from the Pacific Ocean to the Gulf of Mexico and in the following states: Mexico, Tlaxcala, Puebla, Veracruz, Michoacan and Jalisco (Baldwin & Mundo-Ocampo, 1991; Tovar Soto *et al.*, 2006). Climatic conditions at high altitude are temperate with marked seasonal rainfall. The nematode is most abundant in sandy soils (Stone *et al.*, 1976).

BIOLOGY

The life cycle spans the growing season of maize so that only one generation occurs per year (Baldwin & Mundo-Ocampo, 1991).

PATHOGENICITY

It is highly damaging to maize in Mexico, causing severe yellowing, stunting and even death of young seedlings (Baldwin & Mundo-Ocampo, 1991). Heavily attacked plants have stunted root systems with many short laterals, giving a bottle-brush effect, and the aerial parts of the plant appear unthrifty. Pathogenicity studies of *P. chalcoensis* on maize in the glasshouse showed an inverse correlation between root and top weight of mature plants and the number of nematode cysts in the inoculum (Hernandez Aragon, 1965).

RELATIONSHIPS

Punctodera chalcoensis most resembles *P. matadorensis* and *P. stonei*. It differs from *P. matadorensis* by the shape of the J2 stylet knobs, which are anteriorly flat to slightly concave in *P. chalcoensis vs* strongly concave. They also differ in the distribution of bullae in the cyst perineum, these being small scattered or absent in *P. chalcoensis vs* massive and always present in *P. matadorensis. Punctodera chalcoensis* differs from *P. stonei* by the shorter hyaline region of the J2 tail (38-39 *vs* 50 μm) (Table 24) and by its distinctive host range.

BIOCHEMICAL AND MOLECULAR CHARACTERISATION

Sabo *et al.* (2002) sequenced the ITS-rRNA gene and studied the relationships of *P. chalcoensis* with other cyst nematodes. The ITS-rRNA gene sequence distinguishes this species from *P. punctata*.

3. *Punctodera matadorensis* Mulvey & Stone, 1976
(Figs 65, 66)

In 1971, several cores of soil containing native grass roots from the Matador Ranch, Matador, Saskatchewan, Canada, were found to be infected with white females and cysts. These nematodes were tentatively identified as *Heterodera punctata* by Mulvey (1972, 1973). Further

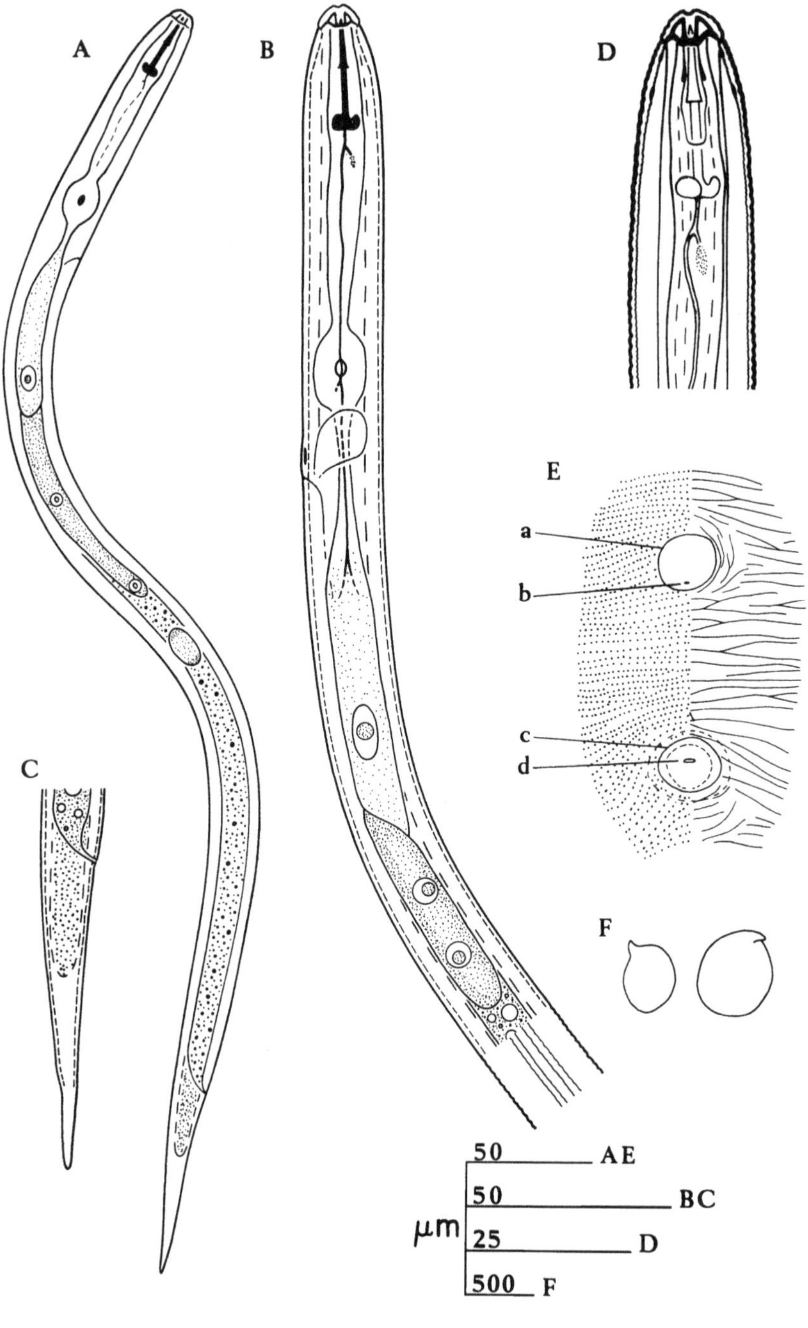
A
B
C
D
E
a
b
c
d
F
50 AE
50 BC
25 D
500 F
µm

detailed observations of the material led to the description of a new species, *P. matadorensis*, by Mulvey and Stone (1976).

MEASUREMENTS

After Mulvey and Stone (1976).

– *Holotype female*: L (excluding neck) = 440 μm; W = 400 μm; L/W ratio = 1.1; neck = 140 μm.

– *Paratype females* (n = 12): L(excluding neck)= 483 (430-560) μm; W = 403 (320-530) μm; L/W ratio = 1.21 (1.0-1.4); neck = 115-140 μm; stylet = 22 (21-23) μm; DGO = 5 μm.

– *Paratype J2* (n = 20) (fixed in 5% formalin and mounted in lactophenol): L = 526 (500-570) μm; W = 19-20 μm; stylet = 24.6 (24-25) μm; anterior end to pharyngeal gland base = 222 (200-250) μm; tail = 61.7 (58-65) μm; hyaline region = 39.1 (35-42) μm.

– *Paratype J2* (n = 20) (water mounts): L = 540 (510-580) μm; W = 21.2 (20-22) μm; stylet = 24 (23-25) μm; DGO = 5.5 (5-6) μm; stylet knob width = 5-6 μm; tail = 62 (60-65) μm; hyaline region = 36.5 (35-40) μm.

– *Paratype eggs* (n = 20): L = 105-128 μm; W = 43-50 μm.

After Wouts and Weischer (1977).

– *J2* (Saskatchewan, Canada) (n = 20): L = 535 ± 5.4 μm; stylet = 26.7 ± 0.13 μm; labial region height = 4.56 ± 0.07 μm; labial region diam. = 10.1 ± 0.09 μm; anterior end to median bulb = 76.4 ± 1.6 μm; tail = 65.8 ± 1.6 μm; hyaline region = 40.7 ± 0.98 μm.

DESCRIPTION

Female

Body basically ovoid, sometimes nearly spherical, white with heavy subcrystalline layer. Neck relatively long. Cephalids not observed. Stylet

Fig. 65. Punctodera matadorensis. *A-D: J2. A: Entire; B: Anterior showing pharyngeal glands, excretory pore and lateral field; C: Tail; D: Anterior showing robust stylet with large forward-directed basal knobs and dorsal pharyngeal gland duct opening; E: Terminal area of white mature female showing (a) anal fenestra, (b) anus, (c) vulval fenestra, (d) vulval slit, surface ridges, and subsurface punctuations; F: Cysts. After Mulvey and Stone (1976).*

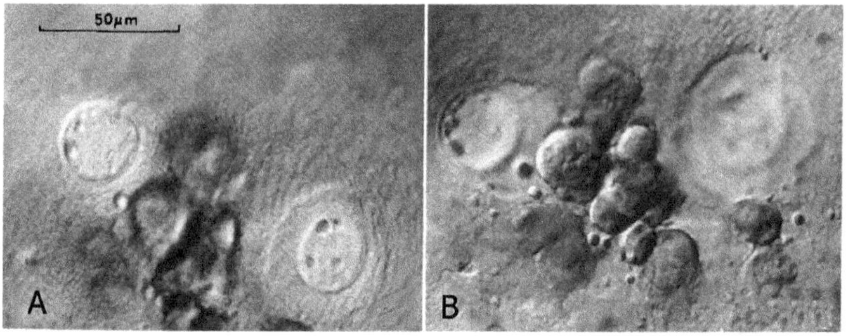

Fig. 66. Punctodera matadorensis. *A: Terminal area of mature cyst showing vulval and anal fenestrae and outline of underlying bullae; B: Terminal area of cyst showing distinct bullae. After Mulvey and Stone (1976).*

robust, dorsally curved, anterior surface of basal knobs sloping posteriorly. Median pharyngeal bulb large with well developed valve. Excretory pore at base of neck. Hemizonid and hemizonion not observed. Vulval slit *ca* 4 μm long, set in a circular, translucent area of cuticle – the vulval fenestra – which is 20 μm long and 20 μm wide. Anus a lateral slit, shorter than vulval slit, offset to ventral side of a circular translucent area of cuticle – the anal fenestra. Vulval and anal fenestra of about equal size. Fenestra in female not being broken through, obscured by covering of cuticle. Vagina distinct. Perineal tubercles and vulval bodies absent. Cuticle with zigzag pattern of ridges and distinct subcuticular punctations. Mature female containing as many as 200 eggs.

Cyst

Cyst basically ovoid or spherical with small protruding neck, light brown. All mature cysts with several heavy bullae located between vulval and anal fenestra. Both vulval and anal areas marked by circular fenestra. Vulval fenestra deteriorating rapidly, forming a round aperture with a distinct circumfenestral area which is absent for anal aperture.

Male

Not found.

J2

Labial region distinctly offset from rest of body, with four annuli, labial disc dorsoventrally elongate. Stylet massive with prominent, anteriorly directed knobs. Cephalids two and seven annuli posterior to head constriction. Hemizonid just anterior to level of excretory pore, itself located six or seven annuli posterior to median pharyngeal bulb. Lateral field with four areolated incisures, reducing to three anteriorly and posteriorly. Pharyngeal glands extending to nearly half body length. Tail tapering to rounded terminus, phasmid openings small but distinct, located in anterior half of tail.

TYPE HOST AND LOCALITY

Unknown natural grasses (Poaceae, Poales) from the Matador Ranch, Matador, Saskatchewan, Canada.

DISTRIBUTION

North America: Canada.

RELATIONSHIPS

Punctodera matadorensis differs from *P. punctata* in the shape of the cyst, which is ovoid to spherical for the former and pear-shaped for the latter, and also by the presence of massive bullae. It differs from *P. stonei* by J2 stylet knob shape (strongly concave anteriorly *vs* flattish to slightly concave anteriorly) and shorter hyaline region (35-42 *vs* 41-66 μm) (Table 24).

BIOCHEMICAL AND MOLECULAR CHARACTERISATION

No available data.

4. *Punctodera stonei* Brzeski, 1998

(Figs 67, 68)

Common name: Stone's cyst nematode.

This species was described from a grass lawn from Skirniewice, Poland.

MEASUREMENTS

After Brzeski (1998).

– *Paratype cysts* (n = 25): L = 0.59 ± 0.06 (0.47-0.74) mm; W = 0.44 ± 0.05 (0.37-0.55) mm; neck = 80-130 μm; L/W ratio =

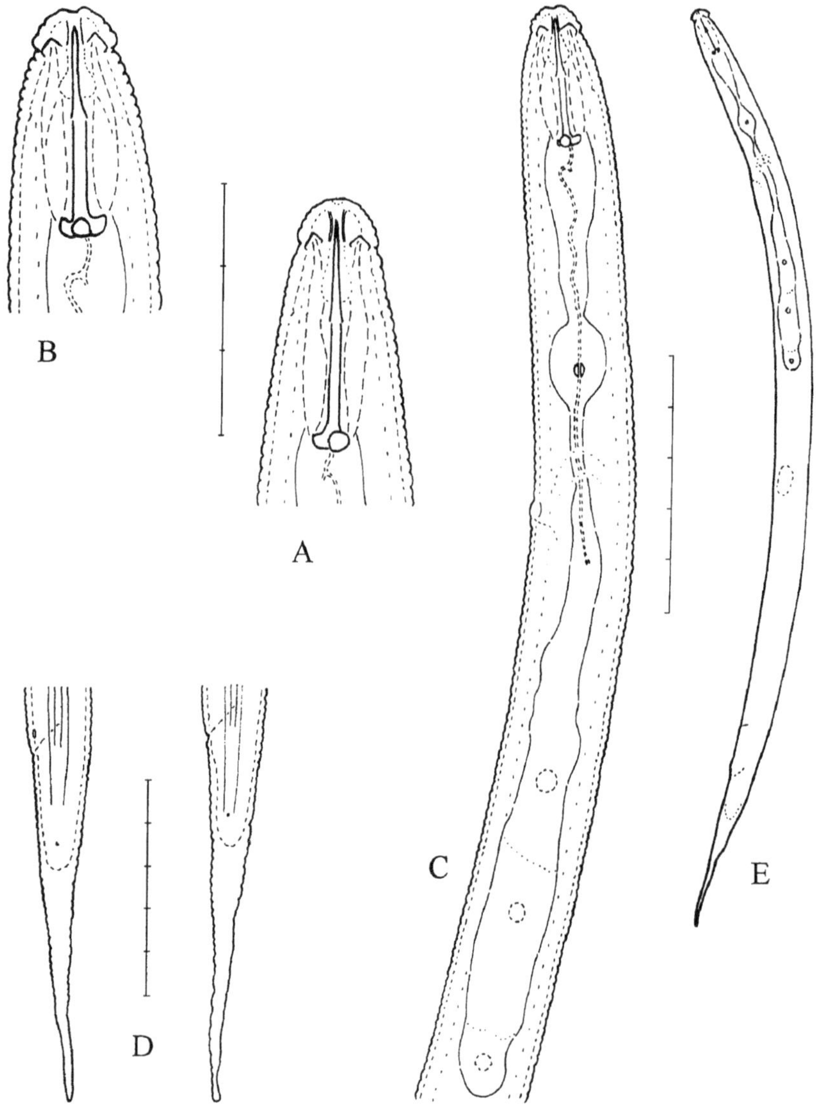

Fig. 67. Punctodera stonei. *A: Male anterior region; B: J2 anterior region; C: J2 pharyngeal region; D: J2 tail; E: J2 entire. After Brzeski (1998). (Smallest section of each scale bar = 10 μm.)*

1.32 ± 0.1 (1.01-1.50); vulval fenestral diam. = 26.6 ± 3.4 (22-36) μm; anal fenestral diam. = 23.0 ± 3.0 (18-32) μm; distance between fenestrae = 59 ± 26 (32-140) μm.

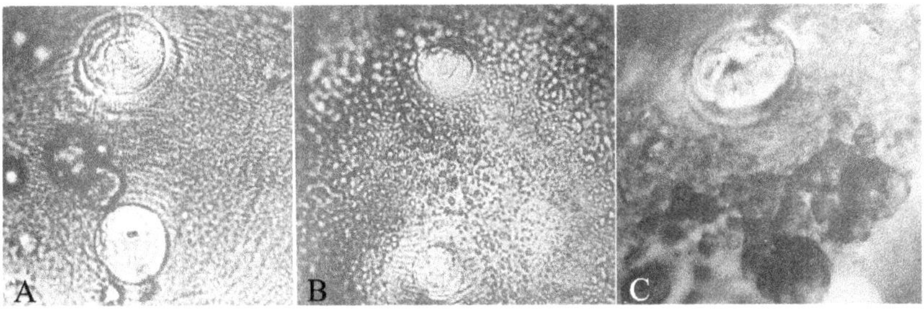

Fig. 68. Punctodera stonei. *A-C: Vulval plates. After Brzeski (1998).*

– *Paratype males* (n = 13): L = 0.88 ± 0.07 (0.84-0.96) mm; a = 31 ± 1.7 (28-33); b = 7.3 ± 0.5 (6.5-8.2); b′ = 4.8 ± 0.4 (3.9-5.3); stylet = 27.4 ± 0.8 (26.5-28) μm; labial region diam. = 9-11 μm; labial region height = 5 μm; DGO = 2.5-4 μm; anterior end to excretory pore = 123 (117-132) μm; posterior end of pharyngeal glands = 185 ± 9.1 (167-197) μm; pharynx = 120 ± 8.2 (106-135) μm; tail = 5-8 μm; spicules = 33 (29-36) μm.

– *J2* (summarised measurements of several populations): L = 520 (470-590) μm; a = 26 (23-29); b′ = 2.6 (2.1-3.4); c = 6.7 (5.9-7.9); c′ = 5.7 (4.5-6.7); stylet = 25.4 (24-26.5) μm; labial region diam. = 12 μm; labial region height = 5-6 μm; DGO = 3-6 μm; pharynx = 118 (95-132) μm; anterior end to: median bulb valve = 73 (68-82) μm, excretory pore = 99 (94-111) μm; posterior end of pharyngeal glands = 201 (148-257) μm; tail = 78 (60-95); hyaline region = 50 (41-66) μm.

– *Paratype eggs* (n = 70): L = 146 ± 7.0 (134-168) μm; W = 61 ± 5.0 μm; L/W ratio = 2.4 ± 0.2 (1.8-3.0).

Description

Cyst

Ovoid, posterior end rounded, light to medium brown in colour. Cuticle covered by transverse rows of punctations. Fenestral area with distinct punctations, sometimes cuticle near fenestra with ridges. Dark bullae present in large cysts, may be lighter in colour or rarely absent in small cysts. Bullae usually large, located just under surface between fenestra, sometimes smaller and scattered around both fenestra.

Male

Usually C-shaped upon relaxation. Cuticular annulation distinct, cuticle slightly swollen from head to posterior cephalids. Lateral field with four incisures, not areolated. Labial region offset by distinct groove, bearing 4-6 annuli plus labial disc. Stylet robust, cone comprising 47-50% of stylet, knobs mostly anteriorly flattened, sometimes slightly concave or sloping obliquely, 5-6 μm in diam.

J2

Usually very slightly arcuate upon relaxation. Cuticle annulation distinct, cuticle slightly swollen from head to posterior cephalids. Lateral field with four incisures, not areolated, inner lines extending to about anus level and thereafter only outer lines extending for a few annuli. Labial region offset by distinct groove, bearing three or four annuli. Basal head annulus always distinct, others not always well demarcated, sometimes annuli discontinuous or anastomosing. Stylet strong, knobs anchor-shaped 6-8 (7) μm diam. Phasmid openings small, 7-13 annuli posterior to anus. Tail tapering evenly in anterior part, becoming thin and almost cylindrical in posterior part. Terminal part often slightly clavate, tip rounded. Hyaline region forming 64 (56-82)% of tail.

Type host and locality

Unidentified lawn grasses (Poaceae, Poales) on north-eastern side of the Research Institute of Vegetable Crops, Skierniewice, Poland.

Distribution

Europe. This species was found in several provinces (Katowice, Bialystok) of Poland (Brzeski, 1998) and Slovakia (Sturhan & Lisková, 2004). Brzeski (1998) considered that populations from Toppheide, Germany, Rothamsted, Harpenden, UK (Wouts *et al.*, 1986), and a population from Spain (Castillo *et al.*, 1989) belonged to this species. Cysts identified as *P. punctata* with bullae-like structure from Estonia and Latvia might belong to *P. stonei* (Krall, 1977).

Relationships

Punctodera stonei most resembles *P. matadorensis*, differing from this species by the J2 stylet knobs which are flattish to slightly concave anteriorly *vs* strongly concave anteriorly and by the hyaline region length of 41-66 *vs* 35-42 μm (Table 24).

Biochemical and molecular characterisation

Holterman *et al.* (unpubl.) provided sequence of the 18S-rRNA gene for *P. stonei*.

Nematology Monographs & Perspectives, 2010, Vol. 8A, 203-259

Chapter 8

Systematics of the genus *Cactodera*

History

Mulvey (1972) separated *Heterodera* species into five groups based on cyst characteristics. In group 2, characterised by cysts with a posterior protuberance and circumfenestrate vulva region, he included the species *Heterodera betulae* Hirschmann & Riggs, 1969, *H. cacti* Filipjev & Schuurmans Stekhoven, 1941, *H. estonica* Kirjanova & Krall, 1963 and *H. weissi* Steiner, 1949. Krall and Krall (1978) erected the genus *Cactodera* from the specific epithet of the type species name (Greek (κάκτος) *kaktos* = prickly plant and Greek (δερος) *deros* = skin) for these species and added *H. amaranthi* Stoyanov, 1972 and *H. aquatica* Kirjanova, 1971. Subsequently, fivc more *Cactodera* species were described and three previously known *Heterodera* species were transferred to this genus. The description of the genus *Cactodera* was emended by Mulvey and Golden (1983). Diagnoses were later presented by Siddiqi (1986, 2000), Luc *et al.* (1988), Baldwin and Mundo-Ocampo (1991) and Wouts and Baldwin (1998). After erection of the genus *Betulodera* for *H. betulae* and transferring *C. johanseni* (Sharma *et al.*, 2001) to *Heterodera*, Sturhan (2002) proposed a new diagnosis for *Cactodera*, which is given here. Identification keys for *Cactodera* are given by Golden and Raski (1977), Golden (1986), Graney and Bird (1990) and Cid del Prado and Miranda (2008).

Genus *Cactodera* Krall & Krall, 1978

DIAGNOSIS (after Sturhan, 2002)

Mature female and cyst lemon-shaped to spherical, with posterior protuberance. Vulva terminal, vulval slit < 30 μm, fenestra circumfenestrate. Anus without fenestration. Bullae and underbridge absent, vulval denticles usually present. Cuticle with D-layer. J2 have lateral field with four incisures, phasmid openings punctiform. Labial disc and six labial sectors present.

TYPE SPECIES

1. *Cactodera cacti* (Filipjev & Schuurmans Stekhoven, 1941) Krall & Krall, 1978
   = *Heterodera cacti* Filipjev & Schuurmans Stekhoven, 1941

OTHER SPECIES

2. *C. acnidae* (Schuster & Brezina, 1979) Wouts, 1985
   = *Heterodera acnidae* Schuster & Brezina, 1979
3. *C. amaranthi* (Stoyanov, 1972) Krall & Krall, 1978
   = *Heterodera amaranthi* Stoyanov, 1972
4. *C. eremica* Baldwin & Bell, 1985
5. *C. estonica* (Kirjanova & Krall, 1963) Krall & Krall, 1978
   = *Heterodera estonica* Kirjanova & Krall, 1963
6. *C. evansi* Cid del Prado & Rowe, 2000
7. *C. galinsogae* Tovar Soto, Cid del Prado, Nicol, Evans, Sandoval Islas & Martinez Garza, 2003
8. *C. milleri* Graney & Bird, 1990
9. *C. radicale* Chizhov, Udalova & Nasonova, 2008
10. *C. rosae* Cid del Prado & Miranda, 2008
11. *C. salina* Baldwin, Mundo-Ocampo & McClure, 1997
12. *C. thornei* (Golden & Raski, 1977) Mulvey & Golden, 1983
    = *Heterodera thornei* Golden & Raski, 1977
13. *C. weissi* (Steiner, 1949) Krall & Krall, 1978
    = *Heterodera weissi* Steiner, 1949

MORPHOLOGICAL IDENTIFICATION

Diagnostics of *Cactodera* species is based on several characters: shape of cyst, fenestral diam., presence of vulval denticles, structure of eggshell, body, tail and hyaline region length for J2 (Table 26).

Key to species of the genus *Cactodera*

1 – Cyst generally twice or more longer than wide, mean L/W ratio = 2.3 . . . . *C. estonica*
– Cyst usually less than twice as long than wide, mean L/W ratio = 1.1-1.8 2

Table 26. *Morphological and morphometric characters useful for identification of the* Cactodera *species (means are given in μm).*

Stage	Cyst			Egg	J2			
Species	Size	L/W ratio	Fenestral diam.	Shell surface	Body length	Stylet	Hyaline region of tail	Tail
C. acnidae	698 × 417	1.5	30-33	smooth	411	21	22	44
C. amaranthi	642 × 472	1.4	31	smooth	368-374	21	14-15	35
C. cacti	442-587 × 328-475	1.1-1.4	26-38	punctate	368-504	22-26	16-25	41-55
C. eremica	620 × 434	1.5	21	punctate	480	27	18	40
C. estonica	852-922 × 383-391	2.3	18-30	smooth	428-440	23	17-19	34-40
C. evansi	459 × 334	1.4	18-23	punctate	387	20	21	40
C. galinsogae	523 × 384	1.4	41	punctate	401	22	18	37
C. milleri	632 × 506	1.2-1.4	13-18	punctate	426	22	18	43
C. radicale	682 × 438	1.6	24	smooth	488	25	20	52
C. rosae	654 × 433	1.5	18	punctate	397	20	6	39
C. salina	603 × 375	1.6	24	smooth	458	24	20	35
C. thornei	548-656 × 432-448	1.3-1.5	29-34	punctate	457-554	25-27	25-27	52-56
C. weissi	524-598 × 350-394	1.7	34	smooth	407-489	20-21	20-24	46

2 – Eggshell punctate 3
 – Eggshell smooth 9

3 – Mean stylet length of J2 $\geqslant$ 26 μm 4
 – Mean stylet length of J2 < 26 μm 5

4 – Tail = 48-64 μm, J2 hyaline region = 23-28μm, fenestral diam. = 23-41 μm .... *C. thornei*
 – Tail = 37-48 μm, J2 hyaline region = 17-24 μm, fenestral diam. = 14-25 μm .... *C. eremica*

5 – Mean body length $\geqslant$ 411 μm, mean tail length $\geqslant$ 42 μm for J2 6
 – Mean body length < 411 μm, mean tail length < 42 μm for J2 7

6 – Fenestral diam. = 23-35 μm, vulval slit = 10-15 μm, stylet of J2 = 21-27 μm .... *C. cacti*
 – Fenestral diam. = 7-22 μm, vulval slit = 14.3-17.8 μm, stylet of J2 = 21-23 μm .... *C. milleri*

7 – Fenestral diam. < 25 μm 8
 – Fenestral diam. $\geqslant$ 25 μm .... *C. galinsogae*

8 – Hyaline region length for J2 = 4-8 μm .... *C. rosae*
 – Hyaline region length for J2 = 16-23 μm .... *C. evansi*

9 – Mean tail of J2 < 40 μm 10
 – Mean tail of J2 $\geqslant$ 40 μm 11

10 – Body length = 340-406 μm, stylet = 20.2-21.3 μm for J2, fenestral diam. = 25-38 μm .... *C. amaranthi*
 – Body length = 410-514 μm, stylet = 23.4-25 μm for J2, fenestral diam. = 20.4-27.6 μm .... *C. salina*

11 – J2 stylet knobs anterior surface concave, DGO = 4.5-5.6 μm .. 12
 – J2 stylet knobs anterior surface convex, DGO = 2.5-3.0 μm .... *C. acnidae*

12 – Vulval denticles present, tail of J2 = 43-50 μm .... *C. weissi*
 – Vulval denticles absent, tail of J2 = 46-60 μm .... *C. radicale*

MOLECULAR IDENTIFICATION

PCR-ITS-RFLP (Table 27) and sequences of ITS-rRNA gene are used for molecular diagnostics of *Cactodera* species.

Table 27. *Approximate sizes (in bp)* of restriction fragments generated by eight diagnostic enzymes after digestion of PCR products of the ITS-rRNA regions amplified by TW81 and AB28 primers for cyst nematodes of the genus* Cactodera.

Species	Unrestricted	*Alu*I	*Ava*I	*Bsh*1236I	*Bsu*RI	*Cfo*I	*Mva*I	*Pst*I	*Rsa*I	References
Cactodera cacti	**981**	**552, 239, 190**	**706, 275**	**644, 337**	**458, 214, 151, 72, 40, 24, 22**	**343, 301, 185, 145, 7**	**331, 249, 198, 128, 75**	**643, 338**	***924*, 545, 379, 36, 21**	Tanha Maafi *et al.* (2003)
Cactodera rosae	977	893, 84, 4	977	875, 102	547, 197, 158, 75	344, 196, 178, 102, 81, 76	462, 270, 245	642, 335	586, 368, 14, 9	This study
Cactodera estonica	976	839, 81, 52, 4	976	976	527, 277, 172	348, 327, 301	457, 273, 246	638, 338	485, 265, 112, 54, 51, 9	This study

*Fragments obtained in results of virtual sequence digestion using Webcutter2; **Bold** font – verified by PCR-RFLP study; Normal font – without RFLP verification; *Italics* font – additional fragments, sometimes poorly visualised on a gel.

SPECIES DESCRIPTIONS

1. *Cactodera cacti* (Filipjev & Schuurmans Stekhoven, 1941) Krall & Krall, 1978
(Figs 69, 70)

Common name: *cactus* cyst nematode.

A cyst nematode infecting cacti, *Discocactus akkermannii* (Lindl) Barthlott and *Cereus speciosus* K. Schumacher, both of which were expressing declining symptoms, was first recorded and described by Adam (1932) from Maartensdijk, near Utrecht, The Netherlands. Goffart (1936) also provided several measurements of different stages of this nematode. However, Adam (1932) and Goffart (1936) considered this nematode as *H. schachtii*. Filipjev and Schuurmans Stekhoven (1941), on the basis of the description by Adam (1932), named this species *Heterodera cacti*. In 1978, Krall and Krall (1978) erected the genus

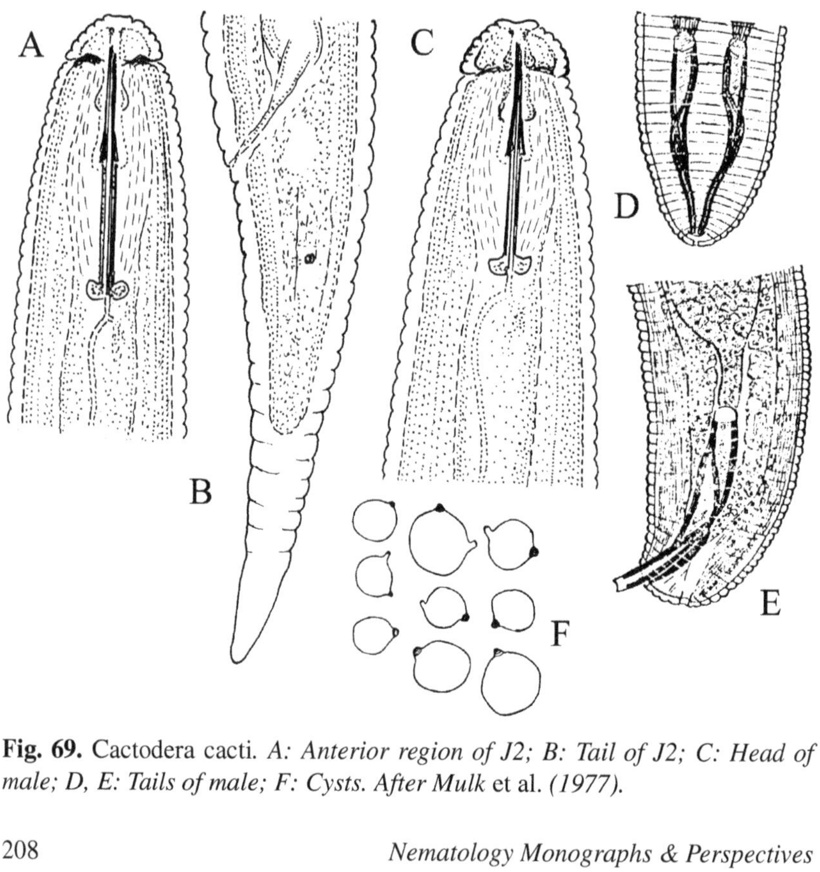

Fig. 69. Cactodera cacti. *A: Anterior region of J2; B: Tail of J2; C: Head of male; D, E: Tails of male; F: Cysts. After Mulk* et al. *(1977).*

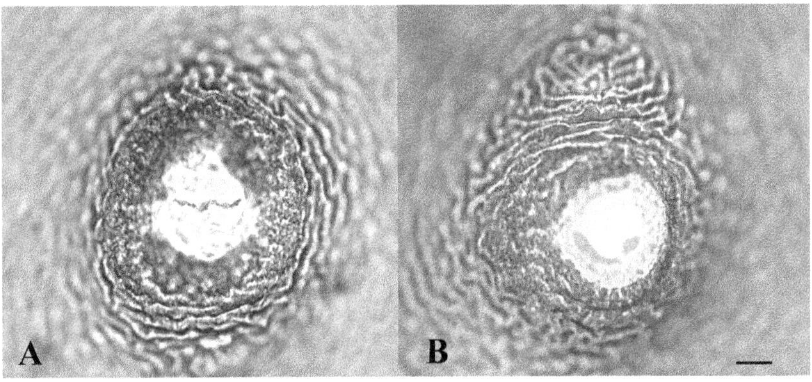

Fig. 70. Cactodera cacti. *A, B: Vulval cones. (Courtesy of V.N. Chizhov, Russia.) (Scale bar = 10 μm.)*

Cactodera to accommodate *C. cacti* and other species. The cactus cyst nematode is distributed worldwide, mainly on plants of the family Cactaceae grown in glasshouses as ornamentals. Several authors provide descriptions and morphometrics of different stages of this species (Kirjanova & Terenteva, 1961; Ambrogioni, 1969; Narbaev, 1973; Golden & Raski, 1977; Mulk, 1977; Mulvey & Golden, 1983; Othman *et al.*, 1988; Graney & Bird, 1990; Cordero *et al.*, 1991; Pan *et al.*, 1997; Tanha Maafi *et al.*, 2004).

MEASUREMENTS

– *Cyst*: see Table 28.

After Graney and Bird (1990).

– *Female* (n = 30): L (excl. neck) = 550 ± 59 (401-652) μm; W = 450 ± 62 (323-574) μm; L/W ratio = 1.2 ± 0.1 (1.0-1.4); stylet = 33.2 ± 1.2 (30.2-35.6) μm; DGO = 6.9 ± 1.2 (5.4-9.8) μm; vulval slit = 12.6 ± 1.4 (9.5-15.1) μm; vulva to anus distance = 40.6 ± 4.2 (34.5-50.0) μm; anterior end to excretory pore 124.9 ± 21.2 (90.2-159.0) μm.

– *Egg* (n = 30): L = 107.2 ± 3.2 (101.1-113.1) μm; W = 49.1 ± 2.4 (44.0-54.7) μm; L/W ratio = 2.2 ± 0.1 (1.9-2.4).

Table 28. *Morphometrics of cysts of* Cactodera cacti *from different locales (all measurements in µm).*

Population Characters	The Netherlands, *Phyllocactus* (Adam, 1932)	Italy, Ambrogioni (1969)	Yugoslavia (former), Golden & Raski (1977)	USA, Michigan, Graney & Bird (1990)	Russia, Kazachenko (1993)	Iran, Mahalat, Tanha Maafi *et al.* (2004)
n	50	25	15	30	10	21
Length	497 (328-617)	536 (384-780)	587 (498-662)	558 ± 66 (389-658)	442 (369-516)	490 ± 62 (380-600)
Width	447 (309-598)	378 (240-588)	475 (387-550)	470 ± 72 (323-598)	328 (270-405)	382 ± 59 (280-490)
Length/width	1.1	1.4 (1.1-2.0)	1.3 (1.1-1.4)	1.2 ± 0.1 (1.0-1.4)	1.4 (1.2-1.8)	–
Fenestral diam.	–	26-28	38 (29-48)	26.2 ± 3.6 (15.5-29.8)	18.2-28.6	28 ± 3.3 (23-35)
Vulva slit-anus distance	–	–	–	–	–	41 ± 6.4 (35-55)

After Adam (1932).

– *Male* (n = 6): L = 1010 (972-1048) μm; a = 34.8 (33.5-37.1); b = 9.7 (9.3-9.8); c = 200; stylet = 28 (26-29) μm; spicules = 33.5 (32.3-36) μm; gubernaculum = 11.4 (10.2-12) μm.

After Mulk (1977).

– *Male* (n = 5): L = 910-1113 μm; a = 33-42; b = 5.0-5.7; b′ = 6.7-8.3; c = 113.7-146.4; stylet = 22-28 μm; spicules = 30-37 μm; gubernaculum = 13-15.

– *J2*: see Table 29.

DESCRIPTION

Female

Body lemon-shaped to almost spherical, pearly white, yellow, or golden, maturing to light brown. Neck offset, vulval cone prominent. Subcrystalline layer may be present (Fig. 6F). Gelatinous egg sac not observed. Labial disc elevated. Stylet and knobs well developed. Excretory pore located near median bulb.

Cyst

Cyst usually lemon-shaped, but may be rounded with protruding neck and vulva, light or medium brown, sometimes reddish brown. Vulval denticles generally present, visible beneath fenestral surface, sometimes inconspicuous, occurring individually at random or sometimes in small clusters. Cone tops abullate, circumfenestrate. Cuticular surface pattern at mid-body comprising straight to wavy lines extending around cyst at right angles to cyst axis, often interrupted by short oblique or vertical lines. Outer cyst surface often granular, with heavy, random punctations.

Male

Labial region nearly continuous with body, usually with six annuli. Dorsal stylet knob anterior surface concave. Hemizonid 1-3 annuli anterior to excretory pore. Spicules with bifid tips. Testis single, sperm present.

J2

Body vermiform, tapering anteriorly and posteriorly. Labial region slightly offset, bearing five annuli, with slightly dorsoventrally elongated

Table 29. *Morphometrics of J2 of* Cactodera cacti *from different locales (all measurements in μm).*

Population Characters	The Netherlands, *Phyllocactus*, Adam (1932)	Italy, Golden & Raski (1977)	Mexico, Golden & Raski (1977)	Georgia, USA, Golden & Raski (1977)	Wageningen, The Netherlands, Wouts & Weischer (1977)	Michigan, USA, Graney & Bird (1990)	Russia Kazachenko (1993)	Mahalat, Iran, Tanha Maafi *et al.* (2004)
n	–	50	25	23	20	30	25	10
L	486 (428-584)	409 ± 18 (364-453)	418 ± 38 (376-451)	432 ± 17 (412-451)	487 ± 4.1	504 (456-540)	368 (344-382)	448 ± 38.6 (402-513)
a	22 (19.2-29.4)	18 ± 1.5 (15-20)	19 ± 2.2 (17-21)	19.4 ± 1.0 (18-21)	–	–	23.4 (19.6-27.5)	25.4 ± 1.7 (22.3-27.4)
b	6 (5.6-6.8)	2.6 ± 0.3 (2.1-3.1)	2.7	2.8	–	–	2.6 (2.4-2.9)	–
c	9.4 (8.4-10.5)	8.9 ± 0.6 (7.8-10.1)	9.9 ± 0.4 (9.5-10.7)	9.9 ± 0.4 (9.5-10.7)	–	–	9 (8.5-10.1)	10.3 ± 0.8 (9.1-11.2)
Stylet	25.2 (22-28.2)	23.5 ± 0.6 (22.4-24.6)	23.4 ± 0.4 (22.4-24.0)	22.9 ± 0.5 (22.4-24.0)	26.1 ± 0.2	25.1 (24.4-26.3)	21.9 (21-22.8)	24.3 ± 1.0 (23-26)
Labial region height	–	–	3.5 ± 0.5 (2.8-4.5)	2.6 ± 0.4 (2.2-3.4)	4.6 ± 0.1	4.6 (3.9-4.9)	3.4 (2.4-3.5)	4
Labial region diam.	–	–	7.0 ± 0.5 (6.2-8.4)	7.2 ± 0.5 (6.2-7.8)	9.8 ± 0.1 (10.0-11.0)	10.5	8 (7.8-10.4)	8.5 ± 0.5 (8-9)

Table 29. *(Continued).*

Population Characters	The Netherlands, *Phyllocactus*, Adam (1932)	Italy, Golden & Raski (1977)	Mexico, Golden & Raski (1977)	Georgia, USA, Golden & Raski (1977)	Wageningen, The Netherlands, Wouts & Weischer (1977)	Michigan, USA, Graney & Bird (1990)	Russia Kazachenko (1993)	Mahalat, Iran, Tanha Maafi *et al.* (2004)
DGO	–	3.8 ± 0.3 (3.4-4.6)	5.0 ± 0.5 (4.5-6.1)	3.8 ± 0.3 (3.4-4.5)	–	3.8 (2.9-4.4)	2.1	4.5 ± 0.5 (4-5)
Anterior end to median bulb valve (MB)	–	–	–	–	79.8 ± 0.8	77.6 (73.6-82.0)	61 (58.5-65)	70 ± 4.6 (64-76)
Anterior end to excretory pore	–	–	–	–	–	112.9 (103.9-121.5)	85.6 (80.6-89.7)	101 ± 10 (91-117)
Body diam. at mid-body	–	–	–	–	19.7 ± 0.2	–	–	17.6 ± 1.4 (16-20)
Body diam. at anus level	–	–	–	–	–	–	10.9 (9.1-11.7)	12 ± 0.4 (12-13)
Tail	-	46 ± 3.6 (36-52)	43 ± 3.6 (34-48)	44 ± 2.2 (39-48)	53.0 ± 0.9	55.0 (49.3-59.5)	40.7 (37.7-45.5)	44 ± 3.8 (39-48)
Hyaline region	–	15.7 ± 1.8 (12-19)	19 ± 2.2 (14-21)	17.2 ± 1.2 (14.5-19.6)	24.8 ± 0.5	17.6 (13.7-20.5)	19.1 (15.6-20.8)	16 ± 2.5 (12-19)
L/MB	–	–	–	–	–	–	–	6.4 ± 0.5 (5.7-7.1)

labial disc, two small lateral and four large submedial lips. Frequent fusion between adjacent submedial lips, lateral lips and oral disc, and between lateral lips and submedial lips. Lateral field with four incisures, outer field sometimes areolated. Dorsal stylet knob anterior surface sloping posteriorly, may be slightly concave. Cephalids not observed. Hemizonid just anterior or superimposed at level of excretory pore. Transition to hyaline region demarcated with outline usually V-shaped and U-shaped, rarely sloping ventrally. Tail tapering, with hyaline region often shorter than stylet.

Egg

Eggshells heavily punctuate (Fig. 6C) as seen under optical microscope with oil immersion.

TYPE HOST AND LOCALITY

Collected by Adam (1932) from cactus plants, *Discocactus akkermannii* (Lindl) Barthlott (Cactaceae, Caryophyllales) and *Cereus speciosus* K. Schumacher, in a glasshouse in Maartensdiijk near Utrecht, The Netherlands.

DISTRIBUTION

The dispersal of *C. cacti* from native regions in the Americas is beyond doubt associated with the international trade of infested ornamental cactus plants around the world. *Europe*: Austria, Belgium, Czech Republic, Denmark, Estonia, France, Germany, Hungary, Italy, Lithuania, Malta, The Netherlands, Poland, Portugal, Romania, Russia, Slovenia, Spain, Sweden, Switzerland, UK, Ukraine, Yugoslavia (former); *Africa*: Algeria; *Asia*: Armenia, China, Georgia, India, Iran, Israel, Japan, Kazakhstan, South Korea, Turkmenia, Uzbekistan, Vietnam; *North America:* Canada, USA (Michigan, California, Maryland, Arkansas, Georgia, Florida, Colorado, Minnesota, New York), Mexico; *South America and Caribbean*: Argentina, Brazil, Cuba, Colombia; *Oceania*: Australia, New Zealand (Golden & Raski, 1977; Mulk, 1977; Mulvey & Golden, 1983; Stelter, 1984; Sosa-Moss, 1987; Baldwin & Mundo-Ocampo, 1991; Esser, 1992).

OTHER HOSTS

The cactus cyst nematode has been associated with or infecting plants belonging to three families: Cactaceae (Caryophyllales): *Cereus, Cleistocactus, Coryphantha, Discocactus, Echinocactus, Echinopsis, Echinocereus, Epiphyllum, Gymnocalycium, Hatiora, Heliocereus, Hylocereus, Leuchtenbergia, Mammillaria, Melocactus, Notocactus, Nopalea, Notocactus, Opuntia, Oreocereus, Rebutia, Rhipsalis, Schlumbergera, Selenicereus, Thelocactus*; Umbelliferae (order Apiales): *Apium* and Euphorbiaceae (order Malpighiales): *Euphorbia* (Kirjanova & Krall, 1971; Mulk, 1977; Esser, 1992). Reports of *C. cacti* infection on other plant families: Crassulaceae (Saxifragales), Moraceae (Rosales), Apocynaceae (Gentianales), Asclepiadaceae (Gentianales), Asteraceae (Asterales) and Oleaceae (Lamiales) (Shmalko, 1959; Scognamiglio *et al.*, 1985; Paneque & Sampedro, 1986; Narbaev & Mirsalimowa, 1989); and from members of the monocotyledonous genera *Agave, Aloe* and *Haworthia* (Asparagales) (Scognamiglio *et al.*, 1985; Paneque & Sampedro, 1986) require confirmation.

RELATIONSHIPS

Cactodera cacti resembles *C. weissi, C. acnidae, C. milleri* and *C. galinsogae.* It differs from *C. weissi* and *C. acnidae* in having eggshells heavily punctate *vs* shells without visible markings, J2 with longer stylet (22-26 *vs* 20, 21 μm, respectively). It differs from J2 *C. galinsogae* by a longer tail (41-55 *vs* 37 μm) and from *C. milleri* by cysts with larger fenestral diam. (26-38 *vs* 13-18 μm) (Table 26).

PATHOGENICITY

Infected plants may exhibit various symptoms included branched roots and increased numbers of rootlets. Plants become reddish brown to yellow in colour, wilted, stunted, with reduced flower production and shortening of the flowering period. With high infection the host may die (Mulk, 1977; Esser, 1992).

BIOCHEMICAL AND MOLECULAR CHARACTERISATION

Tanha Maafi *et al.* (2003) sequenced the ITS-rRNA gene and studied relationships of *C. cacti* with other cyst nematodes. The ITS-rRNA gene sequence enables this species to be distinguished from other *Cactodera*. PCR-ITS-RFLP profile for this species is given in Table 27.

2. *Cactodera acnidae* (Schuster & Brezina, 1979) Wouts, 1985
(Fig. 71)

Common name: water hemp cyst nematode.

A cyst nematode was found in soybean fields in eastern Nebraska associated with roots of a weed identified as water hemp, *Acnida altissima.* Morphological differences with similar species suggested a new species, which was described by Schuster and Brezina (1979).

MEASUREMENTS

After Schuster and Brezina (1979).

– *Holotype female*: L (including neck) = 745 μm; W = 370 μm; L/W ratio = 2.0; L (excl. neck) = 672 μm; W = 370 μm; L/W ratio = 1.8.

– *Paratype females* (n = 18): L (excluding neck) = 647 ± 46 (560-722) μm; W = 413±47 (358-498) μm; L/W ratio = 1.6±0.2 (1.3-1.8); L (including neck) = 744 ± 74 (633-874) μm; L/W ratio = 1.8 ± 0.2 (1.4-2.6); neck: 108 ± 34 (78-196) μm; stylet = 23.3 (19.5-25.0) μm; vulval slit length = 17 (12-20) μm.

– *Paratype cysts* (n = 35): L (including neck) = 698 ± 81.8 (504-857) μm; W = 417±55 (319-493) μm; L/W ratio = 1.8±0.3 (1.2-2.4); L (excluding neck) = 635 ± 75.4 (543-756) μm; L/W ratio = 1.5 ± 0.2 (1.1-2.1); neck = 69 ± 28 (56-140) μm. circumfenestra *ca* 33 μm long by 30 μm wide; vulval slit = 17 (12-20) μm.

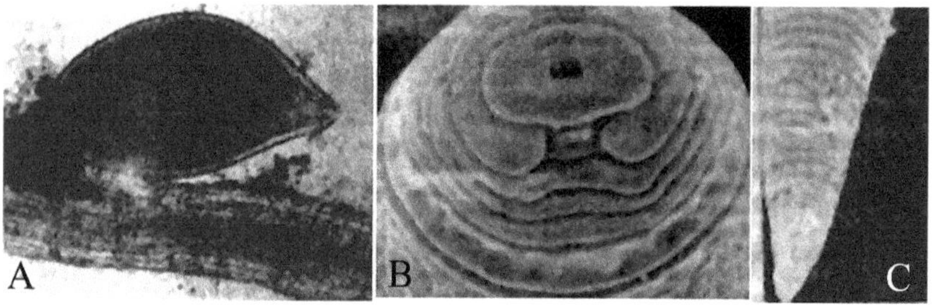

Fig. 71. Cactodera acnidae. *A: White female on root; B: Lip region of J2; C: Tail of J2. After Schuster and Brezina (1979).*

– *Allotype male*: L = 1120 μm; W = 28 μm; a = 40; stylet = 22.2 μm; spicules = 25.4 μm; gubernaculum = 10 μm.

– *Paratype males* (n = 13): L = 1127 (1120-1148) μm; a = 39-45; b = 13.3-13.7; c = 87-228; stylet = 21.7 (19.5-23.4) μm; spicules = 25.7 (24.7-27.8) μm; gubernaculum = 9-11 μm.

– *Paratype J2* (n = 100): L = 411 (361-448) μm; a = 20.3-29.9; b = 4.2-6.9; c = 8.9-9.9; stylet = 21.2 (18.9-24.7) μm; labial region diam. = 10.7 (8.3-12.2) μm; labial region height = 3.3 (2.8-5.6) μm; DGO = 2.5-3.0 μm; hyaline region = 21.6 (17.5-26) μm; tail = 44 (43-48) μm.

– *Paratype eggs* (n = 41): L = 99 (88-112) μm; W = 48 (34-52) μm; L/W ratio = 2.0.

DESCRIPTION

Female

Body lemon-shaped, pearly white, with protruding neck and vulva. Cuticular pattern at mid-body, consisting of wavy, straight, to zigzag lines. Granular appearance in dorsal and ventral margins of vulva (under LM). Anus opening in small depressed area, located *ca* 5% of body length from vulva. Subcrystalline layer present. Egg sac not observed.

Cyst

Lemon-shaped, with protruding neck and vulva, circumfenestrate, light to dark brown. Cuticular pattern at mid-body with straight to wavy lines extending transversely around cysts and often broken by oblique or vertical lines. Subcrystalline layer present mainly on young cysts. Anus opening in a small depressed area among cuticular ridges. Bullae absent. Punctations are often present on terminal area of cyst. Vulval cone end with surface ridge patterns characterised as two zones in anterior and posterior halves of vulval area, one resembling a broad band of cuticle forming smooth lips and other looking like cheeks with punctations. Beyond this area some parallel ridges around cone.

Male

Body filiform, cylindrical, tapering at both ends. Cuticle distinctly annulated, annulation *ca* 1.1 μm apart at mid-body. Labial region offset from body-contour, bearing six annuli on labial region. Tail short, obtuse with a slight protuberance, bluntly round dorsally. Spicules, curved, well developed.

J2

Body filiform, elongate tapering anteriorly and posteriorly. Labial framework heavily sclerotised, somewhat offset; labial region slightly offset with four annuli, labial disc oval surrounded by six labial sectors including four submedial and two small lateral sections. Amphidial apertures conspicuous. Stylet well developed with semi-spherical knobs, convex anteriorly and posteriorly. Lateral field with four lines extending posteriorly to *ca* eighth annulus from tail tip. Tail conoid, with thin rounded terminus. Phasmids, not reported.

Egg

Cylindrical to kidney-shaped. Eggshell hyaline, without visible surface markings or punctations.

TYPE HOST AND LOCALITY

Water hemp, *Acnida altissima* (Riddell) Moq. ex Standl (Amaranthaceae, Caryopyllales), Platte Valley, in eastern Nebraska, USA.

OTHER HOSTS

Host range studies demonstrated that edible pigweed variety (*Amaranthus* sp.) and rough pigweed (*A. retroflexus* L.) were susceptible and that the nematode reproduced in their roots. Erect knotweed (*Polygonum erectum* L.) roots also showed J2 and immature females 4 weeks after transplanting seedlings into infected soil. Maize (*Zea mays* L.), soybean (*Glycine max* (L.) Merr.), sugar beet (*Beta vulgaris*), tobacco (*Nicotiana tabacum* L.), knotweed (*Polygonum aviculare* L.), lambsquarters (*Chenopodium album* L.), smartweed (*Polygonum persicaria* L.), horsetail (*Equisetum arvense* L.), sandburr (*Centhrus pauciflorus* Benth.) and white clover (*Trifolium repens* L.) were not hosts (Schuster & Brezina, 1979).

DISTRIBUTION

North America: USA (Nebraska).

BIOLOGY

Two or more generations occur in the field during the summer in Nebraska, USA.

RELATIONSHIPS

Cactodera acnidae is most similar to *C. weissi*, *C. amaranthi* and *C. salina*. The mean J2 length is about the same for *C. acnidae* and *C. weissi* (411 and 407 μm) but it is longer than for *C. amaranthi* (411 *vs* 368-374 μm) and smaller than that for *C. salina* (411 *vs* 458 μm) (Table 26). The shape of tail in *C. acnidae* is conoid with a rounded terminus while that of *C. weissi* tapers to a fine rounded terminus with annulations continuing to the end of the tail. Tail of *C. acnidae* J2 (43-48 μm) is longer than that of *C. amaranthi* (32-40 μm). In J2 of *C. acnidae* the stylet knob anterior surface is convex and DGO = 2.5-3.0 μm, whereas in J2 of *C. weissi* the anterior surface is concave and DGO = 4.5-5.6 μm.

BIOCHEMICAL AND MOLECULAR CHARACTERISATION

No available data.

3. *Cactodera amaranthi* (Stoyanov, 1972) Krall & Krall, 1978
(Figs 72, 73)

Common name: amaranth cyst nematode.

In 1957, G.J. Rau found a new cyst nematode associated with spiny amaranth, *Amaranthus spinosus* L., in the Zellwood area, near Sanford, Florida, USA, but did not describe it. This nematode was subsequently described by Stoyanov (1972) from roots of *A. viridis* L. collected at La Habana, Cuba. Golden and Raski (1977) provided additional description and measurements of the amaranth cyst nematode from Florida and Cuba.

MEASUREMENTS

After Golden and Raski (1977).

– *Female* (n = 2): L (including neck) = 563 ± 98 (404-688) μm; W = 366 ± 87 (181-464) μm; L/W ratio = 1.6 ± 0.3 (1.1-2.2); stylet = 25.3 ± 1.5 (24.9-26.6) μm; DGO = 4.8 ± 0.5 (4.3-5.2) μm; vulval slit = 15 ± 1.2 (14-16) μm.

– *Cyst* (n = 25): L (including neck) = 642 ± 62 (525-774) μm; W = 472 ± 54 (370-550) μm; L/W ratio = 1.4 ± 0.12 (1.1-1.7); diam.

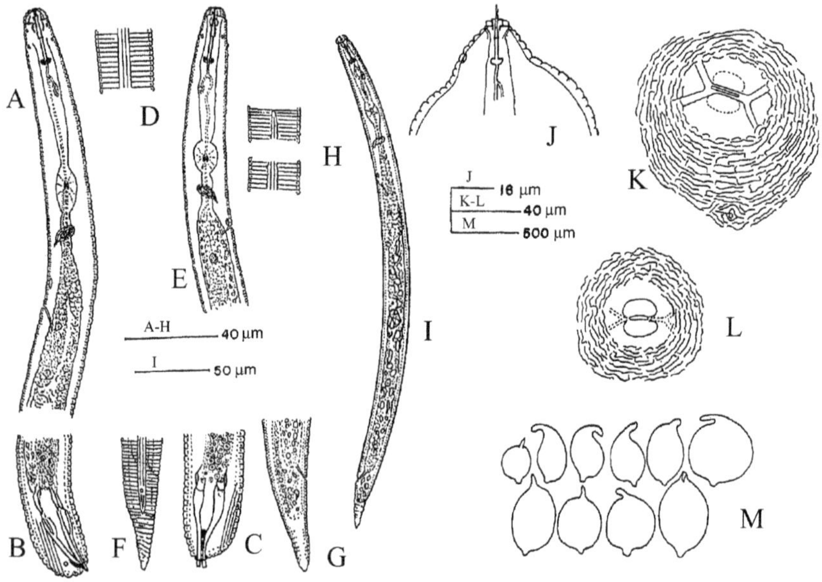

Fig. 72. Cactodera amaranthi. *A-D: Male. A: Anterior region; B, C: Tail; D: Lateral field; E-I: J2. E: Anterior region; F, G: Tail; H: Lateral field; I: Entire body; J: Anterior end of female; K, L: Vulval plates; M: Cysts. After Stoyanov (1972).*

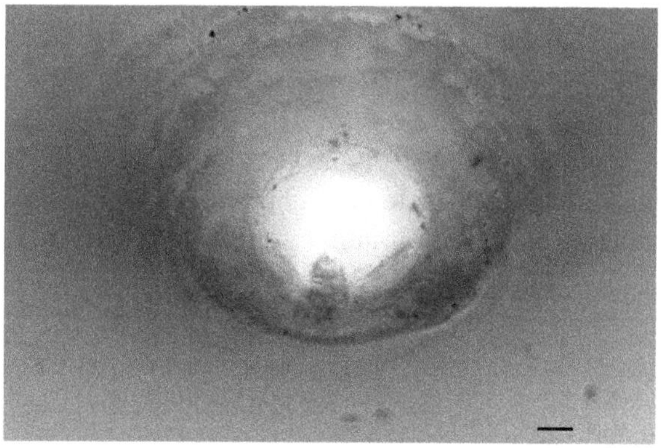

Fig. 73. Cactodera amaranthi. *Vulval cones. (Original.) (Scale bar = 10 μm.)*

of fenestra (on longest axis of fully fenestrated cysts) = 31 ± 3.6 (25-38) μm.

– *Male* (n = 25): L = 974 ± 85 (792-1195) μm; a = 38 ± 5 (31-47); b = 6.2 ± 0.1 (5.4-6.8); c = 558 ± 274 (302-1244); stylet = 25.6 ± 0.5

(24.7-26.4) μm; DGO = 3.7 ± 0.6 (2.6-4.3) μm; spicules = 30.5 ± 1.4 (27-33) μm; gubernaculum = 10.0 ± 1.4 (8-12) μm.

– *J2* (Florida; n = 25): L = 374 ± 15 (340-406) μm; a = 18 ± 1 (16-20); b = 2.8±0.2 (2.5-3.0); c = 10.6±0.6 (9.7-11.8); stylet = 20.7±0.4 (20.2-21.3) μm; labial region diam. = 8.7 ± 0.4 (8.5-9.4) μm; labial region height = 3.6 ± 0.4 (3.4-4.3) μm; DGO = 3.9-5.1 μm (4.5 ± 0.4); tail = 35 ± 2.2 (32-40) μm; hyaline region = 15 ± 1.9 (11-18) μm; anterior end to median bulb valve = 58 ± 3 (52-64) μm.

– *Paratype J2* (Cuba, n = 17, cracked from eggs of two cysts): L = 368 ± 18 (343-402) μm; a = 18 ± 2.4 (15-23); b = 2.7 ± 0.1 (2.5-2.9); c = 10.5±1.0 (9.1-12.4); stylet = 20.7±0.4 (20.4-21.3) μm; labial region diam. = 8.5 μm; labial region height = 3.4 μm; DGO = 4.6±0.4 (4.0-5.2) μm; tail = 35 ± 2.5 (31-40) μm; hyaline region = 14 ± 1.0 (12-16) μm; anterior end to median bulb valve = 56 ± 1.9 (54-60) μm.

– *Egg* (n = 25): L = 96±5 (88-108) μm; W = 48±2.7 (43-52) μm; L/W ratio = 2.0 ± 0.2 (1.8-2.3).

Description

Female

Lemon-shaped with protruding neck and vulvar cone. Neck relatively long, amphidial aperture not observed. Stylet slender, knobs 2-3 μm wide, anterior surface concave. Pharyngeal bulb large with well developed valve. Vulval cone circumfenestrate. Anus small, without a fenestra, located 10-13 μm dorsal to vulval fenestra margin. Mature females containing more than 100 eggs in various stages of development.

Cyst

Lemon-shaped, elongate-ovoid with long protruding neck and vulval cone. Length (including neck): 525-774 μm; width: 370-550 μm; L/W ratio: 1.1-1.7. Circumfenestrate. No bullae reported. Cyst varying from light to dark brown, occasionally very dark (almost black). Cuticular ornamentation comprising straight to wavy lines surrounding mid-body, the primarily transverse pattern may be broken by short oblique or vertical lines. Vulval denticles present (*ca* 20 μm beneath upper surface of fenestra), either individually, in small groups or in a partial or complete ring, denticles *ca* 5-10 μm long and 3-5 μm wide at their base. Anus distinct, without surrounding pattern, located *ca* 10% of body length from posterior end.

J2

Lateral field with four incisures. Labial disc present but not conspicuous; generally six sectors present. Pharyngeal glands filling body cavity. Tail conical, hyaline region comprising *ca* half of tail. Phasmid opening pore-like, without a lens-like structure.

Egg

Eggshell often somewhat yellowish, without visible markings as seen with both LM and SEM. J2 often folded four times within egg.

TYPE HOST AND LOCALITY

Slender amaranth *Amaranthus viridis* L. (Amaranthaceae, Caryopyllales), at Terrenos del Instituto de Investigaciones Tropicales, Santiago de las Vegas, La Habana, Havana, Cuba.

OTHER HOSTS

Amaranthus spinosus L., and *A. retroflexus* L. (Golden & Raski, 1977). Stoyanov (1973) found that, of 100 plant species from 36 families, the only hosts for *C. amaranthi* were *Amaranthus* spp. and *Acnida cuspidata*, all members of the family Amaranthaceae. Reports that this nematode infects spinach and some Chenopodiaceae in Mexico (Tovar Soto *et al.*, 2006) and potato in Argentina (Chaves, 1987) require confirmation.

DISTRIBUTION

North America and Caribbean: Cuba, Mexico, USA (Florida, Colorado, Illinois, Kansas, Missouri, Tennessee) (Golden & Raski, 1977). A cyst nematode resembling *C. amaranthi* and heavily parasitising *Amaranthus putulus* Bertoloni was reported in a citrus orchard near Maputo, Mozambique (Luc, 1986a).

RELATIONSHIPS

Cactodera amaranthi resembles *C. weissi* and *C. acnidae*. It differs from these *Cactodera* species by having the J2 with a shorter body (368-374 *vs* 407, 411 μm, respectively) and shorter hyaline region (14-15 *vs* 20, 22 μm, respectively) (Table 26). The cysts of *C. amaranthi* are generally darker in colour than those of *C. weissi*, the former being rather dark brown with some appearing almost black, although these colour changes alone are not reliable for diagnostics.

BIOCHEMICAL AND MOLECULAR CHARACTERISATION

No available data.

4. *Cactodera eremica* Baldwin & Bell, 1985
(Figs 74, 75)

This cyst nematode was found and described by Baldwin and Bell (1985) from roots of *Atriplex confertifolia* (Tort. and Frem.) Wats. growing in a desert habitat in Utah.

MEASUREMENTS

After Baldwin and Bell (1985).

– *Holotype female*: L (including neck) = 380 μm; W = 250 μm; neck = 49.2 μm; stylet = 23.0 μm; DGO = 3.3 μm; anterior end to excretory pore = 106.5 μm; vulva to anus distance = 49.6 μm.

– *Paratype females* (n = 20): L (including neck) = 546.3 ± 36.3 (373-676) μm; W = 352 ± 36.9 (233-495) μm; neck = 101.61 ± 7.28 (70.0-132.8) μm; stylet = 25.53 ± 0.82 (23.7-27.4) μm; DGO = 3.9 ± 0.4 (3.0-4.8) μm; vulva to anus distance (lateral view) = 51.2 ± 3 (43.2-63.0) μm; vulva length = 15.8 ± 1.3 (14.0-19.0) μm.

– *Paratype cysts* (n = 10): L (including neck) = 620.0 ± 70.1 (530-810) μm; W = 434.0 ± 78.7 (290-590) μm; L/W ratio = 1.5 ± 0.2 (1.2-1.9); circumfenestral diam. = 20.6 ± 3.2 (14.4-25.2) μm.

– *Paratype male* (n = 15): L = 1170 ± 73 (950-1320) μm; W = 26.6 ± 1.6 (22.0-29.0) μm; a = 43.4 ± 1.3 (39.3-48.9); b = 7.6 ± 0.5 (6.4-9.8); b′ = 6.7 ± 0.6 (5.3-10.0); T = 63.8 ± 2.0 (57.4-68.8); stylet = 28.5 ± 0.8 (25.5-30.5) μm; DGO = 3.4 ± 0.4 (3.0-4.5) μm; pharynx = 183.3 ± 7.9 (165.0-205.0) μm; anterior end to excretory pore = 142.6 ± 6.8 (125.0-155.0) μm; testis = 745.5 ± 37.8 (605.7-757.1) μm; spicules = 38.2 ± 1.5 (33.0-42.0) μm; gubernaculum = 10.6 ± 0.7 (9.5-13.5) μm.

– *Paratype J2* (n = 20): L = 480 ± 0.02 (440-510) μm; W = 20.9±1.1 (18.5-22.5) μm; a = 23.1±0.9 (20.9-23.8); b = 2.7±0.2 (2.5-3.1); b′ = 3.2 ± 0.3 (2.6-3.7); c = 11.9 ± 0.8 (10.1-12.5); c′ = 3.2 ± 0.2

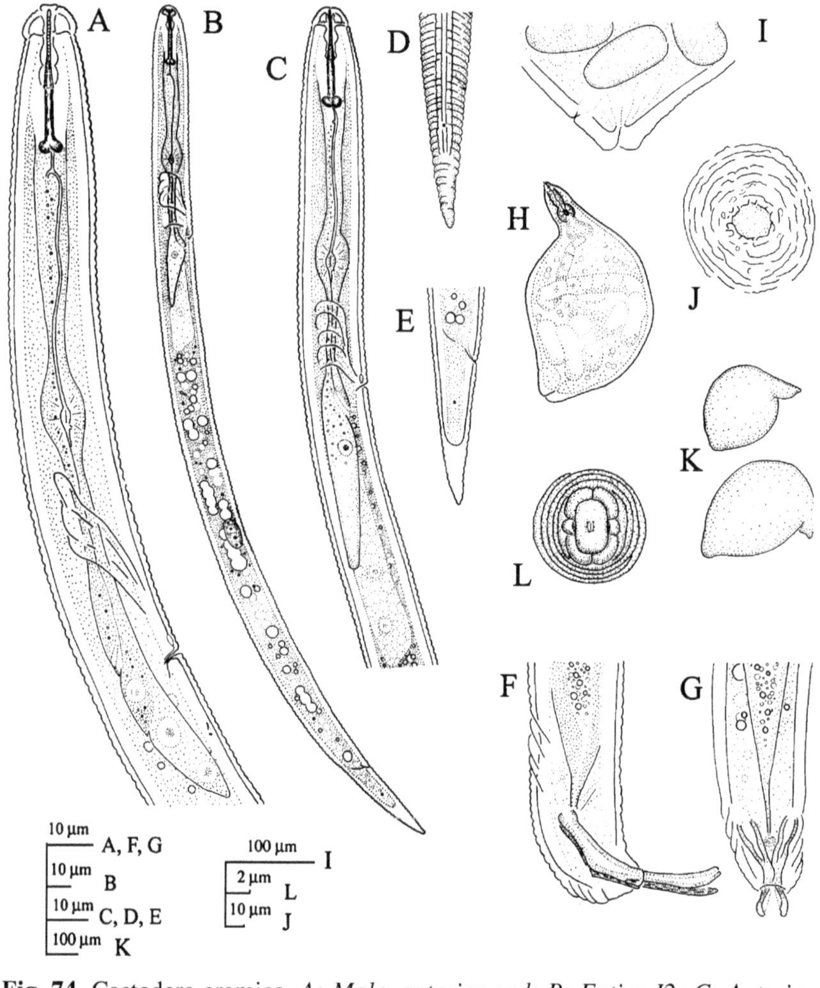

Fig. 74. Cactodera eremica. *A: Male, anterior end; B: Entire J2; C: Anterior region of J2; D, E: Tails of J2; F, G: Tails of males; H: Entire female; I: Vulval cone, lateral view; J: End view of cone; K: Cysts; L: Face view of J2 as observed with SEM. After Baldwin and Bell (1985).*

(2.8-3.6); stylet = 26.5 ± 0.7 (25.0-28.0) μm; DGO = 4.4 ± 0.7 (3.5-6.0) μm; pharynx = 175.8 ± 10.6 (155.0-205.0) μm; anterior end to: excretory pore = 103.8 ± 5.6 (95.0-115.0) μm, genital primordium = 295.6 ± 16.1 (260.0-320.0) μm; tail = 40.4 ± 2.9 (36.5-47.5) μm; hyaline region = 18.1 ± 1.8 (17.0-23.5) μm.

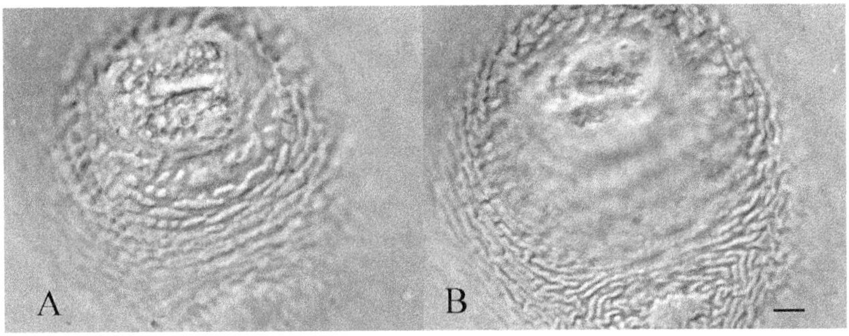

Fig. 75. Cactodera eremica. *Vulval cones. (Original.) (Scale bar = 10 μm.)*

– *Paratype eggs* (n = 15): L = 104.5 ± 1.5 (99-115) μm; W = 50.3 ± 0.5 (48-54) μm; L/W ratio = 2.1 ± 0.6 (1.9-2.4).

DESCRIPTION

Female

Body shape varying in large specimens from lemon to nearly spherical, smaller individuals asymmetric with a strong dorsal curvature. Young females pearly white with small terminal vulval cone. An amorphous cement-like material surrounding head up to neck region. Cuticle with subsurface punctations and wavy transverse striations, head region relatively smooth without striations. Labial disc elevated with four submedial lobes. Anterior most labial annulus continuous with six sectors fused. Stylet, slightly curved dorsally. Two ovaries. Anus small but conspicuous. Surface of perineal region wrinkled with coarse discontinuous striae. Vulval slit narrow.

Cyst

Lemon-shaped, typically brown with protruding neck and vulval cone; surface pattern of transverse striae becoming broken and somewhat wavy posteriorly. Bullae absent. Denticles not observed, anus inconspicuous.

Male

Body cylindrical, tapering anteriorly, posterior half of heat-killed specimens twisted *ca* 90°. Labial region slightly offset, consisting of round labial disc, surrounded by squarish first labial annulus and 7-9 discontinuous head annuli. First labial annulus only rarely separated into full complement of four large submedial lips (sectors) and a pair of small lateral lips. In most males, first annulus continuous (not separated into lips), in other individuals, various intermediate degrees of fusion and separation occur. Amphidial aperture a very small lateral slit at periphery of labial disc. Lateral field with four incisures, centre ridge not areolated, outer ridges often areolated anteriorly and posteriorly. Cephalids not observed. Subventral stylet knobs sloping posteriorly, dorsal knob large, flattened anteriorly. Testis highly variable in length, filled with large, irregularly-shaped, sperm. Tail very short or absent. Spicules bifid (Fig. 11A). Short spicule sheath distinctive by absence of annulation. Phasmid openings not observed.

J2

Body cylindrical and tapering, especially posteriorly. Labial region nearly continuous with body contour, having about five annuli. *En face* pattern consistent among individuals, in form of dorsoventrally elongate labial disc surrounded by six lips of first labial annulus. Lateral lips greatly reduced. Four lateral lines, outer two ridges partially areolated, particularly anteriorly and posteriorly. Hemizonid about one annulus anterior to excretory pore. Stylet knobs robust, rounded. Pharyngeal gland lobe distinct dorsally, subventral region elongate but inconspicuous; pharyngo-intestinal junction ventral, typically located just posterior to level of excretory pore. Phasmid opening pore-like.

Egg

Surface with very fine punctations visible with LM; punctations resolved as minute tubercles (*ca* 0.2 × 0.5 μm) with SEM.

TYPE HOST AND LOCALITY

Roots and surrounding soil of shadscale, *Atriplex confertifolia* (Tort. and Frem.) Wats. (Amaranthaceae, Caryopyllales), near Fairfield, Cedar Valley, Utah, USA.

DISTRIBUTION

North America: USA (Utah).

RELATIONSHIPS

Cactodera eremica most nearly resembles *C. thornei*. J2 of *C. eremica* have a shorter tail (40 *vs* 56 μm) and shorter hyaline region (18 *vs*

25 μm) (Table 26). The egg surface in both species is covered with minute tubercles.

BIOCHEMICAL AND MOLECULAR CHARACTERISATION

No available data.

5. *Cactodera estonica* (Kirjanova & Krall, 1963) Krall & Krall, 1978

(Fig. 76)

Common name: Estonian cyst nematode.

This species was originally described from a few cysts collected from soil around the roots of several different plants in Estonia (Kirjanova & Krall, 1963).

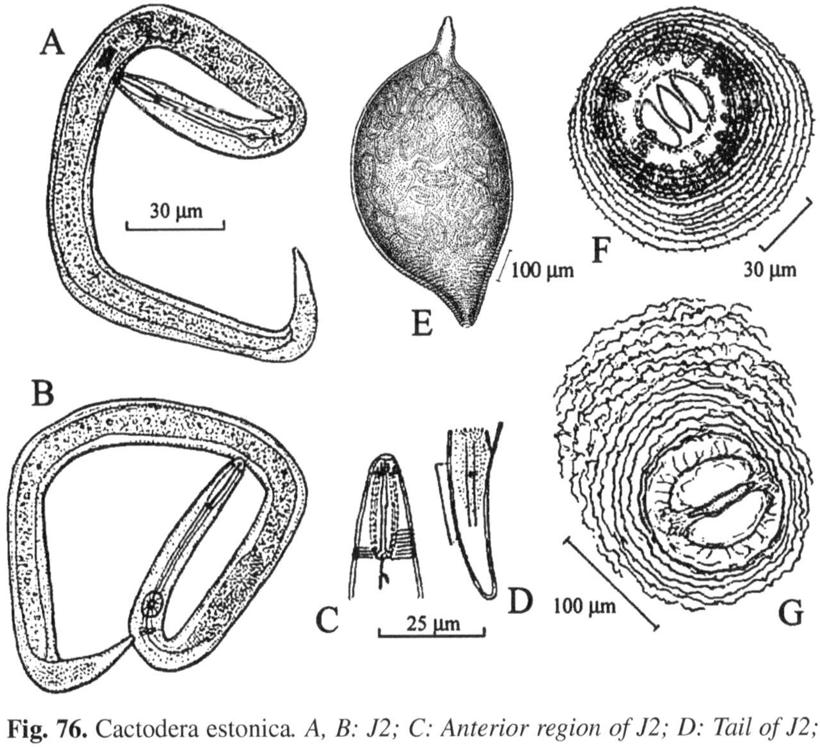

Fig. 76. Cactodera estonica. *A, B: J2; C: Anterior region of J2; D: Tail of J2; E: Cyst; F, G: Vulval plate. After Kirjanova and Krall (1963, 1971).*

MEASUREMENTS

After Kirjanova and Krall (1963).

– *Paratype females* (n = 12): L = 922 (536-1178) μm; W = 391 (207-535) μm; L/W ratio = 2.3; stylet = 21 μm.

– *Paratype J2*: L = 428 (392-456) μm; stylet = 22.8 (21-25) μm; tail = 33.5 (31.5-36) μm; hyaline region = 19.2 (18-21) μm.

– *Paratype eggs*: L = 124.9 (108.5-133) μm; W = 48.2 (42-52) μm; L/W ratio = 2.58.

After Golden and Raski (1977).

– *Cyst* (Poland; n = 8): L (including neck) = 852 ± 130 (686-1014) μm; W = 383 ± 57 (312-468) μm; L/W ratio = 2.3 (2.0-2.4); vulva to anus distance = 71-107 μm.

– *Paratype J2* (n = 8): L = 440 ± 12 (426-465) μm; a = 22.8 ± 1.0 (22-25); b = 2.8 ± 0.1 (2.7-2.9); c = 11.2 ± 0.7 (10.5-11.9); stylet = 22.8±0.4 (22.1-23.6) μm; labial region diam. = 8.3±0.4 (7.7-8.5) μm; labial region height = 3.7 ± 0.5 (3.4-4.3) μm; DGO = 3.6 ± 0.4 (3.4-4.3) μm; tail = 40 ± 2.8 (36-44) μm; hyaline region = 17 ± 2.5 (14-21) μm; anterior end to median bulb valve = 67 ± 3.4 (62-72) μm.

After Sturhan (2010).

– *Paratype J2*: labial region diam. = 9.0-9.5 μm, labial region height = 4.0-4.8 μm; stylet = 22.7 (22.2-23.6) μm; tail = 38.5 and 40 μm; hyaline region = 19 (17-21.5); c′ = 2.6 and 3.0.

DESCRIPTION

Female

Mostly lemon-shaped, some highly elongated and asymmetrically spindle-shaped. Cuticle 5-7 μm thick. Cuticular superficial ornamentation with fine zigzag lines. Size of vulval cone 52-63 μm. Anus located in centre of a distinct area. Subcrystalline layer not reported.

Cyst

Lemon-shaped, some elongated, with protruding neck and vulva, circumfenestrate, light to dark brown. Cyst wall pattern consists of basically straight to wavy transverse lines encircling cyst and commonly broken by short oblique or vertical lines. Cuticle thickness *ca* 6-10 μm.

Fenestral diam. *ca* 20-24 μm. Vulval denticles usually present, located in fenestra, *ca* 6-12 μm in length and 3-6 μm wide at base. Denticles may occur singly at random, in small clusters, or as a partial or complete ring. Anus distinct.

Male

Not found.

J2

Overall morphology similar to other species of *Cactodera*. Tail, including hyaline region, rather short and blunt. Original authors reported three or four head annuli but Sturhan reported six (rarely five) annuli in paratypes. Lateral field composed of four incisures. Stone (1986) and Sturhan (2002) pointed out that the originally reported number of five incisures in lateral fields was incorrect. Tail bluntly rounded.

Egg

Surface without visible markings as seen under oil immersion.

TYPE HOST AND LOCALITY

Pigweed, *Polygonum aviculare* L. (Polygonacea, Caryopyllales), Village of Makas, Tartu district, Estonia.

DISTRIBUTION

Europe: Estonia, Poland, Sweden, The Netherlands, Ukraine, Yugoslavia (former); *Asia*: China, Turkey; *North America*: USA (Kirjanova & Krall, 1963; Mulvey, 1972; Stoyanov, 1977; Norgren & Goldon, 1986; Sturhan, 2006; Peng *et al.*, unpubl.).

RELATIONSHIPS

Cactodera estonica can be distinguished from other species of the genus by the shape of the cysts (L/W ratio averaging 2.3 *vs* a maximum value < 1.8 for the other species) (Table 26). Mulvey (1972) suggested that this species is close to *C. weissi*, which is known to parasitise a polygonaceous host, although in North America. However, these species differ in cyst shape. Molecular analysis rejects the hypothesis of a close relationship with *C. weissi*.

BIOCHEMICAL AND MOLECULAR CHARACTERISATION

The ITS rRNA gene of *C. estonica* was sequenced and analysed by Subbotin *et al.* (2001a). PCR-ITS-RFLP profile is given in Table 27.

6. *Cactodera evansi* Cid del Prado & Rowe, 2000
(Fig. 77)

Common name: Evans' cyst nematode.

In 1977, cysts, males and J2 belonging to an unknown *Cactodera* species were recovered from soil and roots of carnation plants, *Dianthus caryophyllus* L., in Mexico. Further examination of specimens led to the authors describing it as a new species.

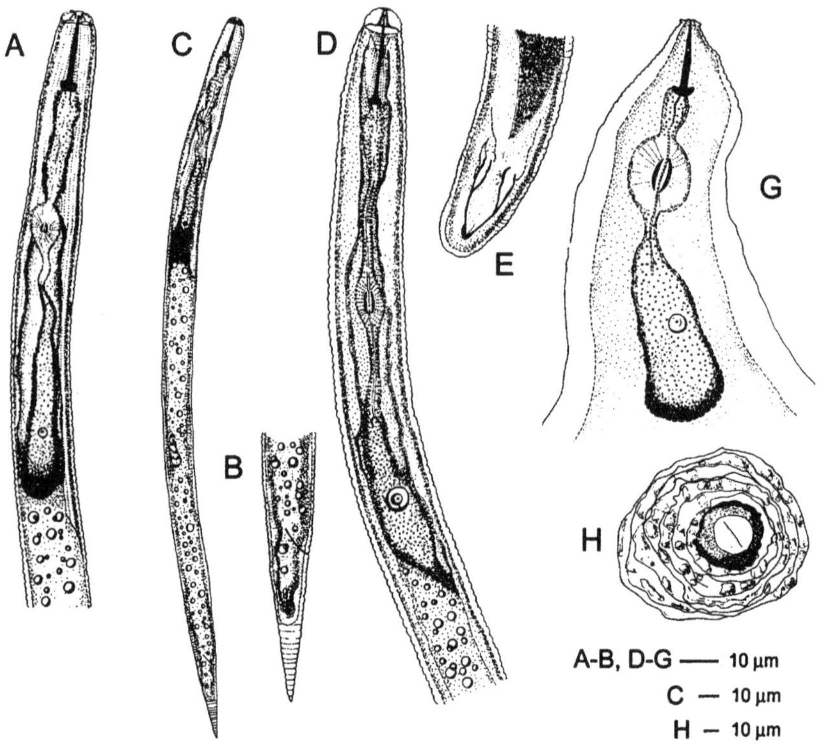

Fig. 77. Cactodera evansi. *A: Anterior region of J2; B: Tail of J2; C: Entire body of J2; D: Anterior end of male; E: Tail of male; H: Vulval cone; G: Anterior end of female. After Cid del Prado and Rowe (2000).*

MEASUREMENTS

After Cid del Prado and Rowe (2000).

– *Holotype female*: L (including neck) = 390 μm; W = 200 μm; L/W ratio = 1.9; vulva to anus distance = 36 μm.

– *Paratype females* (n = 15): L (including neck) = 433 ± 3.4 (356-560) μm; W = 239 ± 3.9 (148-395) μm; L/W ratio = 1.8 ± 4.1 (0.6-3.8); neck = 83 ± 1.3 (52-128) μm; stylet = 20-25 μm; vulva to anus distance = 43 ± 8.1 (32-80) μm; fenestral diam. = 18-23 μm; vulval slit = 17-19 μm.

– *Paratype cysts* (n = 31): L (including neck) = 459 ± 10.7 (416-528) μm; W = 334±9.3 (284-384) μm; L/W ratio = 1.4±0.4 (1.2-1.7); neck = 98.3±1.8 (92-116) μm; fenestral diam. = 18.0-22.8 μm; vulval slit = 17.1-19.0 μm; vulva to anus distance = 43 ± 8.1 (32-80) μm.

– *Allotype male*: L = 1010 μm; W = 26 μm; a = 38.8; b = 8.7; b′ = 6.3; c = 280; stylet = 26 μm; pharynx = 160 μm; anterior end to excretory pore = 142 μm; spicules = 30 μm.

– *Paratype males* (n = 11): L = 979 ± 6.5 (780-1090) μm; W = 26 ± 1.6 (24-28) μm; a = 36.7 (26.4-46.1); b = 9.2 ± 1.0 (8.3-11.3); b′ = 6.6 ± 5.1 (5.8-8.0); c = 266 ± 3.2 (116-340); stylet = 24.9 (20.8-26.4) μm; DGO = 3.4 ± 1.0 (2.8-4.0) μm; labial region diam. = 10.6±9.7 (8.6-12.4) μm; pharynx = 159±1.1 (152-180) μm; spicules = 24.7 ± 2.8 (20.0-30.0) μm; tail = 3.8 (2.8-4.4) μm.

– *J2* (n = 8): L = 387 ± 1.3 (358-420) μm; a = 22.1 ± 6.9 (21.1-23.3); b = 4.9 ± 0.2 (4.8-5.2); b′ = 2.6 ± 0.6 (2.2-2.9); c = 9.6 (8.9-11.5); stylet = 20 ± 1.1 (20-24) μm; DGO = 3.0 (2.8-4.0); tail = 40±2.8 (34-44) μm; hyaline region = 21.1±2.0 (16.4-23.2) μm.

– *Paratype embryonated eggs* (n = 14): L = 89.0 ± 2.4 (82-94) μm; W = 35 ± 9.2 (32-38) μm; L/W ratio = 2.5 ± 1.3 (2.1-2.8).

DESCRIPTION

Female and cyst

Body spherical to lemon-shaped, some oval, with short neck. Vulval cone small but distinct. Distinct cuticular pattern comprised of irregular annulations. Excretory pore at level of posterior end of pharynx gland lobe. Labial disc with first annulus wider than others. Subcrystalline layer absent. Well developed knobs projecting anteriorly. In lateral view, vagina short and uterus oval to round.

Male

Vermiform body, slightly curved ventrally. Lateral field with four lines, outer lateral lines partially areolated. Spicules slightly curved with bifid tips. Gubernaculum and phasmids not observed. Cloacal *tubus* conspicuous.

J2

Body filiform, tapering anteriorly and posteriorly. Labial region with four annuli, offset from body. Labial framework heavily sclerotised. Cuticular annulation of body distinct, annuli measuring *ca* 1.4 μm at mid-body, subcuticular annuli less distinct and twice as numerous. Lateral field without areolation except anteriorly, composed of three incisures and averages 3.5 μm wide at mid-body. Stylet strong with prominent knobs that are concave on anterior surface. Hemizonid prominent, located anterior and adjacent to excretory pore, hemizonion usually *ca* 8-10 annuli posterior to latter. Tail tapering to a fine, rounded terminus, annulation continuing to terminus. Phasmids small, indistinct; located in centre of lateral field about midway on tail.

Egg

External egg surface with distinct punctations visible with both LM and SEM as micro-protuberances.

TYPE HOST AND LOCALITY

Carnation, *Dianthus caryophyllus* L. (Caryophyllaceae, Caryopyllales), at Predio las Parvas, Villa Guerrero, Mexico State, Mexico.

DISTRIBUTION

North America: Mexico.

RELATIONSHIPS

Cactodera evansi is morphologically similar to *C. galinsogae* and *C. rosae*. It differs from *C. galinsogae* by smaller cysts (459 × 334 *vs* 523 × 384 μm) and a smaller fenestral diam. (18-23 *vs* 41 μm), and from *C. rosae* by a longer hyaline region of the J2 tail (21 *vs* 6 μm) (Table 26).

BIOCHEMICAL AND MOLECULAR CHARACTERISATION

No available data.

7. *Cactodera galinsogae* Tovar Soto, Cid del Prado, Nicol, Evans, Sandoval Islas & Martinez Garza, 2003

(Figs 78, 79)

This cyst nematode was found in 1999-2000 when taking soil samples from fields that had grown barley, wheat, oats, and maize in the high valleys of the states of Hidalgo and Tlaxcala, Mexico, and in the Central Valley of Mexico (Tovar Soto *et al.*, 2003, 2008).

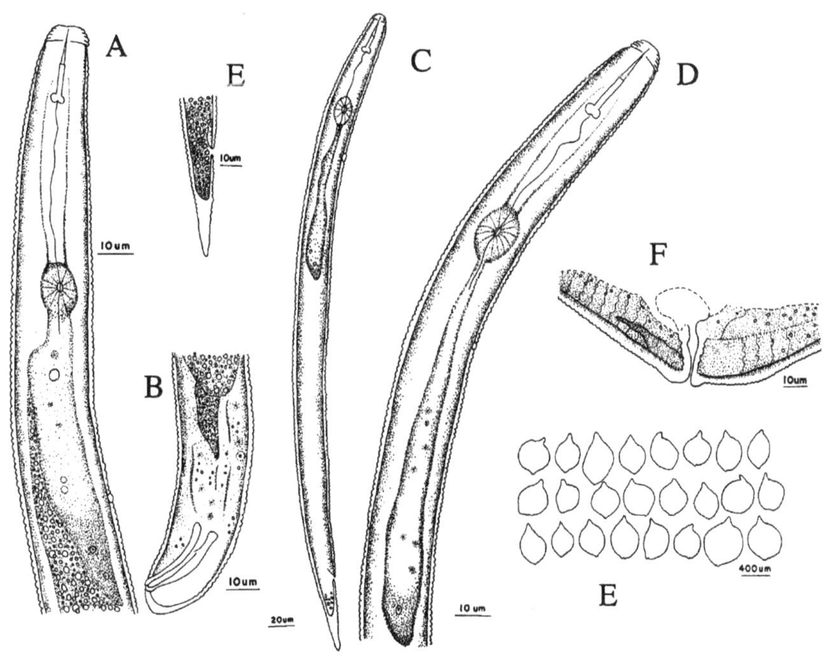

Fig. 78. Cactodera galinsogae. *A: Anterior end of male; B: Tail of male; C: Entire J2; D: Anterior region of J2; E: Cysts; F: Vulval cone (lateral view). After Tovar Soto* et al. *(2003).*

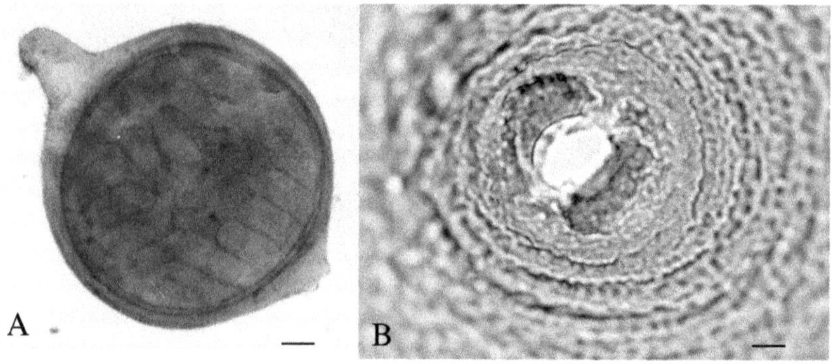

Fig. 79. Cactodera galinsogae. *A: Cyst; B: Vulval cone. (Courtesy I. Cid del Prado Vera, unpubl.) (Scale bars: A = 50 μm; B = 10 μm.)*

MEASUREMENTS

After Tovar Soto *et al.* (2003).

– *Holotype female*: L (including neck) = 480 μm; W = 300 μm; stylet = 20 μm; L/W ratio = 1.6.

– *Paratype females* (n = 49): L (excluding neck) = 483 (373-709) μm; W = 323 (197-471) μm; L/W ratio = 1.5 (1.2-2.1); stylet = 19.5 (16.2-24.3) μm; DGO = 5.7 (4.4-9.3) μm; vulval slit = 29.6 (20.1-40.3) μm.

– *Paratype cysts* (n = 34): L (excluding neck) = 523 (453-675) μm; W = 384 (284-508) μm; L/W ratio = 1.4 (1.1-1.7); fenestral diam. = 41.2 (32.8-56.4) μm.

– *Paratype males* (n = 19): L = 830 (680-960) μm; stylet = 21.6 (20.0-24.0) μm; spicules = 9.5 (10.0-32.0) μm.

– *J2* (n = 31): L = 401 (358-443) μm; a = 23.1 (16.7-26.2); b′ = 2.7 (2.4-3.4); c = 11.1 (9.2-16.0); stylet = 21.7 (18.9-31.3) μm; labial region height = 4.1 (3.1-5.6) μm; hyaline region = 17.9 (10.0-24.0) μm; tail = 36.5 (26.0-45.0) μm.

– *Paratype eggs* (n = 50): L = 108.9 (99.2-121.6) μm; W = 40.3 (36.8-45.7) μm; L/W ratio = 2.7 (2.3-3.1).

DESCRIPTION

Female

Body spherical to subspherical with protruding neck and small vulval cone. Colour changing from pearly white-creamy to light brown, as females mature. Cuticular surface pattern in mid-body region with transverse branching striae forming an interlaced pattern. Strong stylet with well developed rounded knobs. Vulva located on a small terminal projection or vulval cone. Egg retained inside female. Egg sac, when present, without eggs.

Cyst

Body spherical to subspherical with protruding straight neck and small vulvar cone. Light to dark brown, circumfenestra present, bullae and vulval denticles absent.

Male

Male smaller than other *Cactodera* species; body cylindrical, C-shaped when heat killed. Curved and twisted at posterior end. Lateral field with four lines, outer lines partially areolated. Labial framework, heavily sclerotised. Labial region morphology typical of *Cactodera* with four or five annuli. Stylet robust with rounded and well developed knobs. Phasmid openings not resolved. Spicules slightly curved and bifid, gubernaculum not observed.

J2

Body cylindrical, slightly ventrally curved when heat-killed. Labial framework strongly sclerotised. Labial region with three or four indistinct annuli. Robust stylet with well developed, slightly rounded knobs. Length of hyaline region less than half total tail.

Egg

Surface of eggs with a punctate pattern.

TYPE HOST AND LOCALITY

Roots and associated soil of *Galinsoga parviflora* Cav. (Asteraceae, Asterales), the high valleys, 'La Raya', municipality of Singuilucan, State of Hidalgo, Mexico.

OTHER HOSTS

Cactodera galinsogae is also a parasite of the dicotyledons *Bidens odorata* Cav., *B. ballsii* Sherff. (Asteraceae) as well as barley (*Hordeum vulgare* L.) and wild oats (*Avena fatua* L.) (Poaceae) (Tovar Soto *et al.*,

2008). *Cactodera galinsogae* reproduced better on *G. parviflora* and *B. odorata* than on *H. vulgare* and *A. fatua*. Other weeds (*B. serrulata* Desf. and *B. ballsii* Sherff.) that grew in the associated barley fields are also hosts of this species (Tovar Soto *et al.*, 2003).

DISTRIBUTION

North America: Mexico.

RELATIONSHIPS

Cactodera galinsogae differs from *C. evansi* in several respects: cysts of *C. evansi* are smaller, light to dark brown, with conspicuous vulval cone and cuticular pattern with irregular annulations; fenestral diam. is 18.0-22.8 μm and vulval slit is 17.1-19.0 μm. In *C. galinsogae*, cysts have a straight neck; with the vulval cone smaller than in the other two species, the cuticular pattern at mid-body is comprised of 'branching' transverse striae that form an interlaced pattern; vulval denticles absent; fenestral diam. is 32.8-56.4 μm and vulval slit is 20-40 μm (Table 26).

BIOCHEMICAL AND MOLECULAR CHARACTERISATION

Sequence of the ITS-rRNA gene was provided by Subbotin *et al.* (unpubl.).

8. *Cactodera milleri* Graney & Bird, 1990
(Figs 80, 81)

Common name: Miller's cyst nematode.

In the autumn of 1982, a nematode population with lemon-shaped cysts belonging to the genus *Cactodera* was isolated from soil samples collected from a former asparagus field in Mattawan, Michigan. The host was not identified at that time as the soil samples were taken when the field had just been cultivated and planted to rye. Detailed morphological examination of the cysts, eggs, and J2 and testing of host range revealed significant differences of this nematode relative to known species. It was considered to be a new species and was described as *C. milleri* by Graney and Bird (1990).

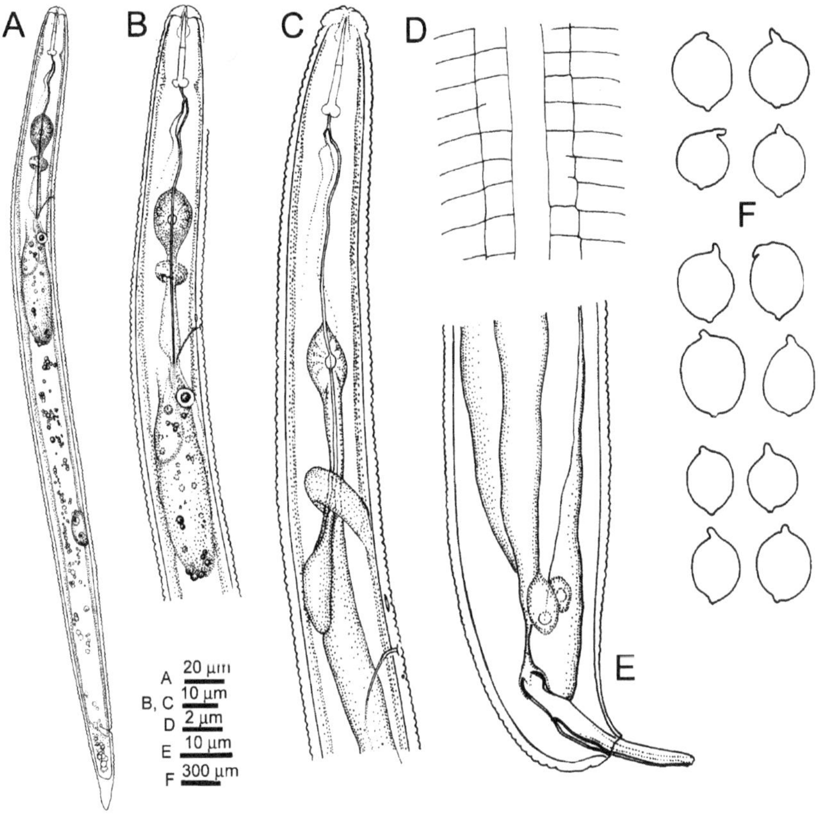

Fig. 80. Cactodera milleri. *A: Entire body of J2; B: Anterior region of J2; C: Anterior end of male; D: Lateral field of male; E: Tail of male; F: Cysts. After Graney and Bird (1990).*

MEASUREMENTS

After Graney and Bird (1990).

– *Holotype female*: L (including neck) = 658 μm; L (excl. neck) = 598 μm; W = 466 μm; stylet = 23.9 μm; DGO = 5.4 μm; anterior end to excretory pore = 138.0 μm; vulval slit = 16.1 μm.

– *Paratype females* (n = 50): L (including neck) = 718 ± 88.2 (538-885) μm; L (excl. neck) = 625 ± 84.5 (436-789) μm; W = 491 ± 83.7 (335-640) μm; L (including neck)/W = 1.5 ± 0.2 (1.2-1.9) μm; L (excluding neck)/W = 1.3±0.1 (1.1-1.5) μm; stylet = 24.1±0.6 (22.9-25.6) μm; DGO = 5.7 ± 1.2 (4.4-9.3) μm; anterior end to excretory

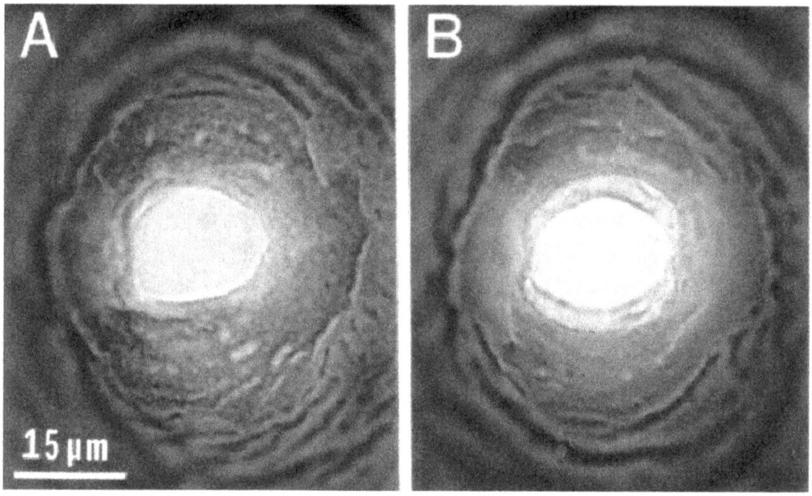

Fig. 81. Cactodera milleri. *A, B: Vulval cone. After Graney and Bird (1990).*

pore = 124.8 ± 16.8 (100.0-180.5) μm; vulval slit = 16.2 ± 0.8 (14.3-17.8) μm; vulva to anus distance = 47.4 ± 6.2 (36.6-62.9) μm.

– *Paratype cysts* (n = 50): L (including neck) = 700 ± 59 (550-849) μm; L (excl. neck) = 632 ± 55 (514-730) μm; W = 506 ± 48 (419-598) μm; L (including neck)/W = 1.4 ± 0.1 (1.2-1.6) μm; L (excl. neck)/W = 1.2 ± 0.1 (1.1-1.4); fenestral diam. (dorsal-ventral) = 13.3 ± 3.2 (7.1-21.4) μm; fenestral diam. (lateral) = 18.7 ± 2.2 (14.3-22.0) μm.

– *Allotype male* (measurements in water): L = 1368 μm; W = 31.2 μm; a = 43.8; b = 8.3; b′ = 7.1; c = 235.9; stylet = 25.3 μm; labial region height = 5.8 μm; labial region diam. = 11.7 μm; DGO = 4.4 μm; pharynx = 164.7 μm; tail = 5.8; spicules = 35.1 μm; gubernaculum = 11.7 μm; (measurements in glycerin): L = 1202 μm; W = 24.9 μm; a = 48.3; b = 10.0; b′ = 8.2; c = 222.6; stylet = 24.4 μm; labial region height = 5.1 μm; labial region diam. = 10.2 μm; DGO = 2.4 μm; pharynx = 120.0 μm; tail = 5.4 μm; spicules = 34.6 μm; gubernaculum = 11.7 μm.

– *Paratype males* (n = 30): L = 1316 ± 71.4 (1190-1499) μm; W = 31.1 ± 1.7 (27.1-33.6) μm; a = 42.5 ± 2.7 (37.8-48.7); b = 8.3 ± 0.4 (8.0-9.1); b′ = 7.1 ± 0.3 (6.9-7.7); c = 219.9 ± 39.6 (156.1-291.1); stylet = 25.0 ± 0.6 (23.4-26.1) μm; labial region height = 5.9±0.3 (5.1-6.4) μm; labial region diam. = 12.0±0.4 (11.5-13.1) μm;

DGO = 3.6 ± 0.9 (2.0-5.1) μm; tail = 6.2 ± 1.0 (4.6-8.0) μm; spicules = 36.2 ± 1.0 (34.6-37.6) μm; gubernaculum = 12.7 ± 0.7 (11.7-13.7) μm; pharynx = 163.3 ± 10.7 (144.7-178.2) μm.

– *Paratype J2* (n = 30): L = 426 ± 24.4 (370-479) μm; W = 22.4 ± 1.5 (19.5-25.4) μm; a = 19.0 ± 0.8 (17.0-21.2); b = 2.9 ± 0.1 (2.6-3.1); b′ = 2.5 ± 0.1 (2.3-2.6); c = 10.0 ± 0.4 (9.0-10.8); c′ = 3.2 ± 0.2 (2.8-3.6); stylet = 21.8 ± 0.5 (21.0-22.9) μm; labial region height = 4.0 ± 0.3 (3.6-4.4) μm; labial region diam. = 9.2 ± 0.2 (8.8-9.7) μm; DGO = 4.2 ± 0.4 (3.2-5.1) μm; anterior end to: median bulb valve = 66.4 ± 1.9 (62.9-70.1) μm, excretory pore = 100.1 ± 5.3 (89.7-110.7) μm; pharynx = 146.8 ± 6.8 (137.6-166.6) μm; tail = 42.7 ± 2.6 (36.6-48.8) μm; hyaline region = 18.2 ± 1.7 (14.6-21.2) μm; anal body diam. = 13.3 ± 0.6 (12.2-15.4) μm.

– *Paratype eggs* (n = 50): L = 102.1 ± 4.8 (93.4-113.7) μm; W = 43.6 ± 2.7 (38.1-51.2) μm; L/W ratio = 2.4 ± 0.2 (1.9-2.8).

Description

Female

Body shape variable, lemon-shaped to nearly spherical. Neck usually distinctly offset. Some specimens with prominent vulval cone reduced on larger spherical females. Newly formed females pearly white, maturing to yellow or golden, becoming light brown. Subcrystalline layer and gelatinous egg sac not observed. Labial disc slightly elevated. Stylet and knobs well developed. Excretory pore posterior to median bulb.

Cyst

Cyst commonly lemon-shaped, some rounded. Colour variable, dark brown to almost black. Cone top abullate, circumfenestrate. Some fenestra surrounded by thin, semitransparent, cuticle-like material. Vulval denticles often present. Anus usually distinct. Cuticular pattern at mid-body consisting of straight to wavy transverse striae.

Male

Body vermiform, tapering at extremities, more so anteriorly. Lateral field with four incisures, outer two bands areolated. Labial region slightly offset, six or seven annuli. Stylet and knobs distinct. Stylet shaft curved slightly dorsal. Dorsal stylet knob sloping posteriorly. Amphidial aperture not observed. Hemizonid distinct, located five annuli anterior to excretory pore. Spicules curved, tips slightly notched (bifid). Tail short, phasmid openings not observed.

J2

Body vermiform, tapering at extremities, greater posteriorly. Lateral field with four lines, outer two bands areolated. Labial region offset, with five or six annuli. *En face* pattern with slightly elongated labial disc, four submedial and two lateral lips. Lateral lips greatly reduced. Stylet and knobs well developed. Anterior surface of dorsal stylet knob varying from slightly to deeply concave, only a few without concavity. Amphidial aperture not observed. Hemizonid *ca* one body annulus anterior to excretory pore.

Egg

Exterior shell with small punctations forming a distinct pattern, visible with light microscope and SEM.

TYPE HOST AND LOCALITY

Common lambsquarter, *Chenopodium album* L. (Amaranthaceae, Caryophyllales), in a former vineyard of the Francis Ryan farm adjacent to Country Road 652 and the Francis Ryan residence in Section 13 of Antwerp Township in Van Buren County near Mattawan, Michigan, USA.

OTHER HOSTS

Results of the host range study indicated a narrow host preference with reproduction on two species of *Chenopodium*, *viz*., *C. amaranticolor* Cofte. Reyn. and *C. quinoa* Willd. (Graney & Bird, 1990).

DISTRIBUTION

North America: USA (Michigan, Wisconsin) (Graney & Bird, 1990; Schroeder *et al.*, 2008).

RELATIONSHIPS

Cactodera milleri differs distinctly from *C. cacti*, the most closely related species, in the following morphometric characters: female stylet

(22.9-25.6 *vs* 29.3-35.6 μm) and vulval slit length (14.3-17.8 *vs* 9.5-15.1 μm); male body length (1190-1499 *vs* 869-1166 μm), stylet length (23.4-26.1 *vs* 26.1-29 μm) and a smaller fenestral cyst diam. (13-18 *vs* 26-38 μm) (Table 26).

MOLECULAR CHARACTERISATION

The ITS-rRNA gene of *C. milleri* was sequenced and analysed by Ferris *et al.* (1999a).

9. *Cactodera radicale* Chizhov, Udalova & Nasonova, 2008
(Fig. 82)

During a nematological survey along bank of the Usta River, Novgorod region, Russia, an unknown cyst nematode was found and described as *C. radicale* (Chizhov *et al.*, 2008).

MEASUREMENTS

After Chizhov *et al.* (2008).

– *Holotype female*: L = 622 μm; W = 380 μm; L/W ratio = 1.6; fenestral diam. = 23 μm; vulval slit = 18 μm; vulva to anus distance = 28 μm.

– *Paratype cysts* (n = 18): L = 682 ± 27.6 (553-986) μm; W = 438 ± 23.8 (220-626) μm; L/W ratio = 1.6 ± 0.07 (1.3-2.6); fenestral diam. = 24 ± 0.7 (17-28) μm; vulval slit = 18 ± 2.0 (14-22) μm; vulva to anus distance = 30 ± 2.7 (20-58) μm.

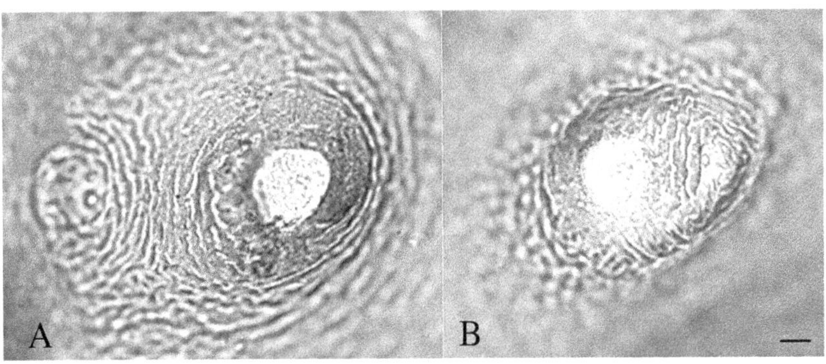

Fig. 82. Cactodera radicale. *A, B: Vulval plates. After Chizhov* et al. *(2008). (Scale bar = 10 μm.)*

– *Paratype J2* (n = 15): L = 488 ± 3.9 (467-520) μm; W = 25 ± 0.4 (23-27) μm; a = 19 ± 0.2 (17.9-21.3); b = 3.7 ± 0.04 (3.3-3.9); b′ = 3.0 ± 0.05 (2.7-3.4); c = 9.3 ± 0.2 (8.5-10.3); c′ = 3.6 ± 0.1 (2.9-4.1); stylet = 25 ± 0.4 (20-27) μm; labial region diam. = 10-11 μm; labial region height = 4-5 μm; DGO = 4.5 (3.6-5.8) μm; anterior end to: median bulb valve = 75 ± 1.2 (67-84) μm, nerve ring = 94 ± 1.2 (88-97) μm, excretory pore = 107 ± 1.2 (100-117) μm; pharyngo-intestine valve = 131 ± 1.5 (125-144) μm; pharyngeal gland base = 165 ± 2.8 (150-187) μm; tail = 52 ± 1.0 (46-60) μm; anal body diam. = 15 ± 0.3 (13-17) μm; hyaline region = 20 ± 1.2 (15-28) μm; ratio of hyaline region to stylet = 0.8 ± 0.1 (0.6-1.3) μm.

– *Paratype eggs*: L = 117 (100-129) μm; W = 48 (44-55) μm.

DESCRIPTION

Cyst

Lemon-shaped, yellowish with small vulval cone. In some cysts vulval cone absent. Anus situated at base of vulval cone. Vulval denticles absent.

Male

Not found.

J2

Body cylindrical, elongate, tapering to posterior end. Labial region with four annuli. Length of hyaline region shorter than stylet. Stylet knobs rounded to slightly projecting anteriorly. Tail terminus blunt.

Egg

Surface smooth.

TYPE HOST AND LOCALITY

Host unknown, Zavod village, Krasnobakovskii district, Novgorod region, Russia. The following plant species (with percentages representing their relative density) were found in the sampling areas: *Elytrigia repens* (71.6%), *Achillea millefolium* (7.4%), *Hypericum maculatum* (3.9%), *Leucanthemum vulgare* (3.5%), *Polygonum aviculare* (2.6%), *Festuca pratensis* (2.0%), *F. rubra* (1.5%), *Prunella vulgaris* (1.8%), *Agrostis tenuis* (1.8%), *Rhinanthus angustifolius* (1.7%), *Stellaria media* (0.7%),

Vicia cracca (0.5%), *Trifolium repens* (0.4%), *T. pratense* (0.3%) and *Carex* sp. (0.3%).

DISTRIBUTION

Europe: Russia.

RELATIONSHIPS

Cactodera radicale is distinguished from all other species of *Cactodera* by a combination of morphological and morphometric characters. It differs from *C. estonica* by lower L/W ratio for cyst (1.6 *vs* 2.3) and absence *vs* presence of vulval denticles; from *C. thornei* by smaller fenestral diam. (24 *vs* 34 μm) and shorter J2 body (488 *vs* 554 μm); from *C. weissi* by absence *vs* presence of vulval denticles; J2 with longer stylet (20-27 *vs* 19.6-21.8 μm) and longer tail (46-60 *vs* 43-50 μm) (Table 26).

BIOCHEMICAL AND MOLECULAR CHARACTERISATION

No available data.

10. *Cactodera rosae* Cid del Prado & Miranda, 2008

(Figs 83, 84)

In 2001-2002, during a nematological survey in the high valleys, Hidalgo State, Mexico, lemon-shaped cysts were found on barley roots and in the surrounding soil and were later described as a new species (Cid del Prado & Miranda, 2008).

MEASUREMENTS

After Cid del Prado and Miranda (2008).

– *Paratype cysts* (n = 40): L (including neck) = 654 ± 26.6 (460-840) μm; W = 433±22 (280-560) μm; neck = 78.2±4.8 (48-120) μm; L/W ratio = 1.5 ± 0.1 (1.2-2.1); fenestral diam. = 17.6 ± 1.0 (10-21) μm; distance from anus to fenestra edge = 27.5 ± 1.0 (25-60) μm.

– *Paratype J2* (n = 30): L = 397 ± 10.7 (348-472) μm; a = 21.3±1.1 (15-26.5); c = 10.5±0.7 (5.8-14.5); c′ = 3.8±0.5 (2.5-6.8); stylet = 20±1.0 (16-26) μm; anterior end to excretory pore = 87±12.0

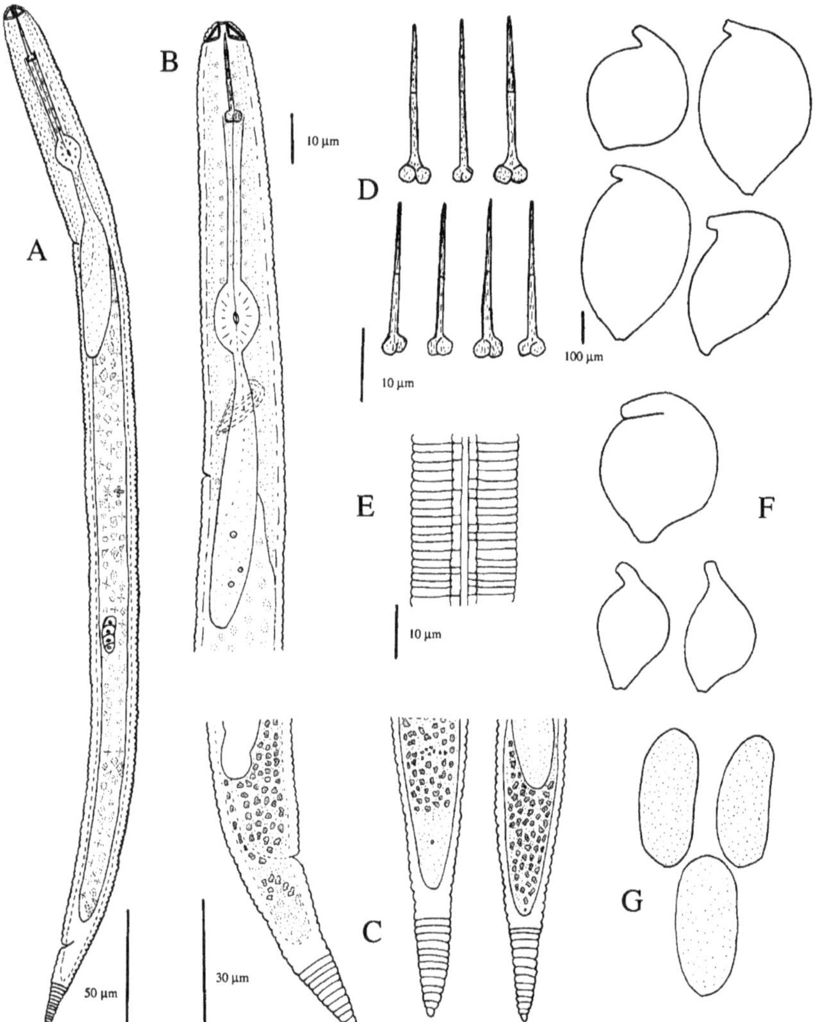

Fig. 83. Cactodera rosae. *A: J2 entire body; B: Anterior region of J2; C: J2 tail; D: Stylets of J2; E: Lateral field; F: Cysts; G: Eggs. After Cid del Prado and Miranda (2008).*

(76-92) μm; hyaline region = 6.3 ± 0.6 (4.0-8.0) μm; tail = 39.2 ± 3.2 (31-68) μm.

– *Paratype eggs* (n = 55): L = 109±3.0 (92-156) μm; W = 48±1.7 (37-62) μm; L/W ratio = 2.3 ± 0.1 (1.7-2.7).

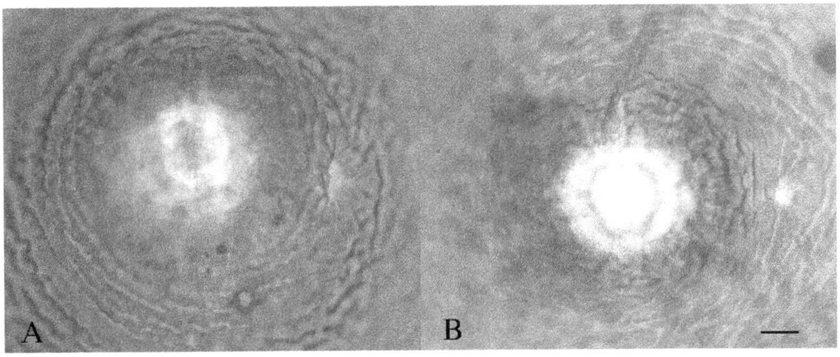

Fig. 84. Cactodera rosae. *A, B: Vulval plates. (Courtesy of I. Cid del Prado, Mexico.) (Scale bar = 10 μm.)*

DESCRIPTION

Cyst

Mostly large and lemon-shaped. Colour varying from dark brown to black with some translucent brown. Neck curved and orientated posteriorly, vulval cone very conspicuous with rounded end. Cuticular pattern varying with age of cyst. Vulval cone with parallel vertical lines. Denticles present around fenestra in some cysts. Anus conspicuous, located in a clear circular or oval depression.

Female

Not found.

Male

Not found.

J2

Body slightly curved, with anterior end rounded and posterior end with an acute terminus. Labial region with four or five annuli, separated from body by a constriction. Labial disc elongated rectangular shape, with six pseudolips, ventral and dorsal lips rounded, lateral lips reduced. Stylet knobs nearly round but anteriorly concave. Lateral field with four incisures and incomplete areolation. Tail terminus conical.

Egg

Oval with rounded ends. Eggshell ornamented with fine punctations over entire surface.

TYPE HOST AND LOCALITY

Collected from barley roots, *Hordeum vulgare* L. (Poaceae, Poales), in San Juan Ixtilmaco, Municipio de Apan, Hidalgo State, Mexico.

DISTRIBUTION

North America: Mexico.

RELATIONSHIPS

Cactodera rosae is similar to *C. evansi*. It differs from this species by a shorter hyaline region (4-8 *vs* 16-23 μm) in the J2 (Table 26).

BIOCHEMICAL AND MOLECULAR CHARACTERISATION

The ITS rRNA gene of *C. rosae* was sequenced and analysed by Subbotin *et al.* (unpubl.) and the PCR-ITS-RFLP profile is provided in Table 27.

11. *Cactodera salina* Baldwin, Mundo-Ocampo & McClure, 1997 (Figs 85, 86)

This species was found in Sonora, Mexico, from in the rhizosphere of *Salicornia bigelovii* Torr. and was described by Baldwin *et al.* (1997).

MEASUREMENTS

After Baldwin *et al.* (1997).

– *Holotype female*: L (including neck) = 446 μm; W = 182 μm; L/W ratio = 2.5; neck = 124 μm; stylet = 24.4 μm; DGO = 2.0 μm; anterior end to excretory pore = 114 μm; vulva to anus distance (lateral view) = 33.2 μm; vulva length = 12.4 μm; cuticle thickness = 6.1 μm.

– *Paratype females* (n = 6): L (including neck) = 447 ± 28.3 (336-515) μm; W = 194.7 ± 19.7 (149-272) μm; L/W ratio = 2.4 ± 0.2 (1.9-3.4); neck = 128.7 ± 12.1 (95-178) μm; stylet = 23.8 ± 0.2 (23.5-24.4) μm; DGO = 1.6 ± 0.2 (1.0-2.0) μm; anterior end to excretory pore = 92.6±3.2 (84-101) μm; vulva to anus distance (lateral view) = 36.9 ± 2.6 (27.5-40.8) μm; vulva length = 11.2 ± 0.5 (10-12.7) μm; cuticle thickness = 8.2 ± 0.5 (6.6-9.2) μm.

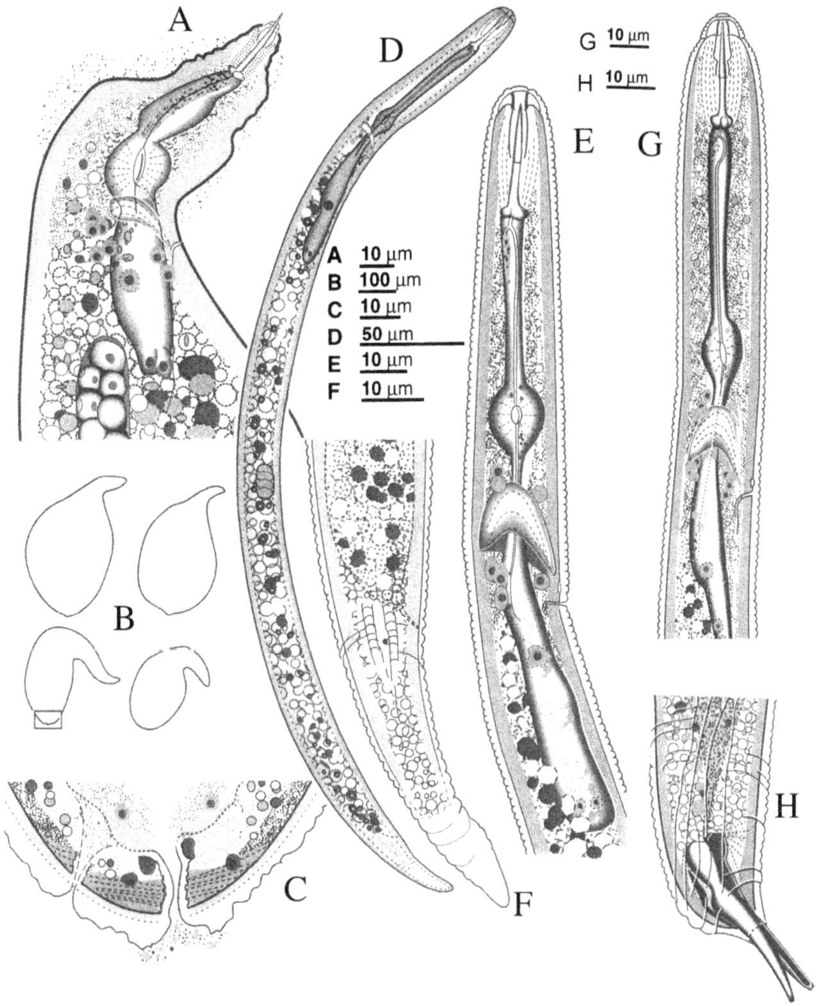

Fig. 85. Cactodera salina. *A: Anterior end of female; B: Outlines of females; C: Terminal protuberance with anus and vulva; D: Entire J2; E: Anterior end of J2; F: Tail of J2; G: Anterior end of male; H: Tail of male. After Baldwin* et al. *(1997).*

– *Paratype cysts* (n = 20): L (including neck) = 603 ± 20 (415-742) μm; W = 375 ± 14 (193-475) μm; L/W ratio = 1.6 ± 0.04 (1.4-2.2) μm; cuticle thickness = 6.6 ± 0.3 (4.9-9.0) μm; fenestral diam. = 23.8 ± 0.4 (20.4-27.6) μm.

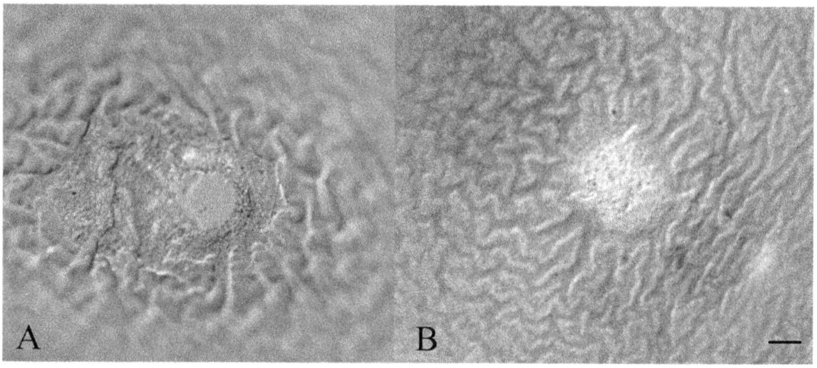

Fig. 86. Cactodera salina. *A, B: Vulval plates. (Original.) (Scale bar = 10 μm.)*

– *Paratype male* (n = 15): L = 1001 ± 21 (906-1118) μm; W = 28.7 ± 0.4 (27-31) μm; a = 34.9 ± 0.7 (31.7-37.8); b = 8.0 ± 0.9 (4.7-11.2); b′ = 6.3 ± 0.3 (5.5-7.1); stylet = 25.8 ± 0.4 (24-28) μm; DGO = 2.48 ± 0.2 (2.0-3.1) μm; pharynx = 161.2 ± 7.1 (142-179) μm; anterior end to excretory pore = 117.8 ± 2.7 (108-133) μm; testis = 489 ± 29.8 (415-565) μm; spicules = 40.8 μm; gubernaculum = 13.3 μm.

– *Paratype J2* (n = 22): L = 458 ± 6.8 (410-514) μm; W = 21.9 ± 0.2 (20.4-24.4) μm; a = 20.9 (18.3-23.3); b′ = 3.2 ± 0.03 (2.9-3.5); c = 11.7 ± 0.3 (9.9-16.0); stylet = 24.3 ± 0.1 (23.4-25.0) μm; DGO = 3.2 (2.5-4.0) μm; pharynx = 146 (130-163) μm; anterior end to excretory pore = 97.8 ± 1.3 (84-107) μm; tail = 34.8 ± 1.1 (31-48) μm; hyaline region = 19.6 ± 1.1 (10.2-30.6) μm; phasmid posterior to anus = 34.8 ± 1.1 (25.5-40.8) μm.

– *Paratype eggs* (n = 10): L = 110 ± 1.0 (107-119) μm; W = 49.5 ± 0.5 (45.4-51.4) μm.

Description

Female

Body asymmetrical, dorsally curved with imperceptible cone in younger individuals, oval or nearly spherical with a minute posterior protuberance in large specimens. Neck offset, usually reflexed ventrally. Young females pearly white, becoming increasingly opaque with age and after fixation. Cuticle with pattern of deep zigzag folds, anterior to excretory pore a gradual transition to wavy striations and finally a relatively smooth head region. Anterior to excretory pore a dense material covering head and neck region. Excretory pore slightly posterior to level of median bulb, pharyngeal gland region elongate. Stylet slightly curved dorsally Ovaries paired. Subcrystalline layer not observed.

Cyst

Oval to nearly spherical, often with minute posterior cone, creamy yellow to light brown. Dense material typically persistent, covering neck region. Cyst surface with pronounced zigzag pattern. Small cone region circumfenestrate, bullae and denticles not observed. Anus visible as minute pore within a smooth region; smooth region surrounding circumfenestra.

Male

Body vermiform with little anterior or posterior tapering, posterior end of heat-killed specimens with a nearly 90° twist. Labial region slightly offset with *ca* 4-7 discontinuous annulations. Labial disc elevated in SEM, labial sectors and labial disc together forming a square, labial sectors highly irregular in pattern among individuals but generally lateral lips fused with adjacent submedial sectors, lateral sectors much smaller than submedial sectors. Lateral field with four incisures, outer ridges often areolated anteriorly, pattern of lines distorted posteriorly. Hemizonid *ca* one annulus anterior to excretory pore, cephalids not observed. Stylet knobs robust, slightly flattened anteriorly. Spicules distinctly bifid; inconspicuous, elongate gubernaculum present. Phasmids not observed.

J2

Body cylindrical, tapering posteriorly. Labial region not offset, with *ca* five annuli. Labial region as viewed in SEM consisting of oval labial disc surrounded by a full complement of six separate sectors, size of lateral sectors greatly reduced relative to others. Four lateral lines with outer two ridges partially areolated at anterior and posterior ends. Hemizonid *ca* 1.5 annuli long, positioned one annulus anterior to excretory pore. Stylet knobs robust, sloping slightly posteriorly. Phasmids rarely visible with LM, with SEM often seen as a minute pore within an inner lateral line just posterior to anus level.

Egg

Surface lacking punctations or tubercles.

TYPE HOST AND LOCALITY

Roots and surrounding sand and silt of *Salicornia bigelovii* Torr. (Amaranthaceae, Caryophyllales) in experimental plots adjacent to the Centro de Estudios de Desiertos y Oceanos, Puerto Peñasco, Sonora, Mexico, adjacent to the northern tip of the Sea of Cortez.

DISTRIBUTION

North America: Mexico.

RELATIONSHIPS

Cactodera amaranthi, *C. weissi*, and *C. acnidae* share with *C. salina* the character of a smooth egg surface without tubercles, but *C. amaranthi* and *C. weissi* have a shorter J2 stylet (21 μm and 20 μm, respectively *vs* 24 μm) and a more posterior DGO (3.9-5.1 μm and 4.5-5.6 μm, respectively, *vs* 2.5-4.0 μm) compared to *C. salina. Cactodera acnidae* cysts tend to be longer (698 μm) than those of *C. salina* (603 μm) (Table 26) and the stylet knobs are convex *vs* posteriorly sloping in *C. salina. Cactodera acnidae* also has a much more pronounced posterior cone than *C. salina.* Mature cysts of *C. salina* are light coloured as are the cysts of *C. weissi*, in contrast to the dark cysts of *C. amaranthi* and *C. acnidae.*

12. *Cactodera thornei* (Golden & Raski, 1977) Mulvey & Golden, 1983
(Figs 87, 88)

Common name: Thorne's cyst nematode.

The specimens of this species were first noted in 1942 by Thorne (1943) from walnut orchards near San Jose and Santa Clara, California, USA. Further sampling of this field indicated that this cyst species was feeding on the roots of Miner's lettuce, *Montia perfoliata* (Donn ex. Willd.) Howell. In 1975, cysts collected from a cherry orchard at Linden, California by A. Weine (CDFA), and were described by Golden and Raski (1977) as a new species which was named in honour of the late eminent nematologist, Professor Thorne. Mulvey and Golden (1983) reported this nematode from several other locations in California. Peng

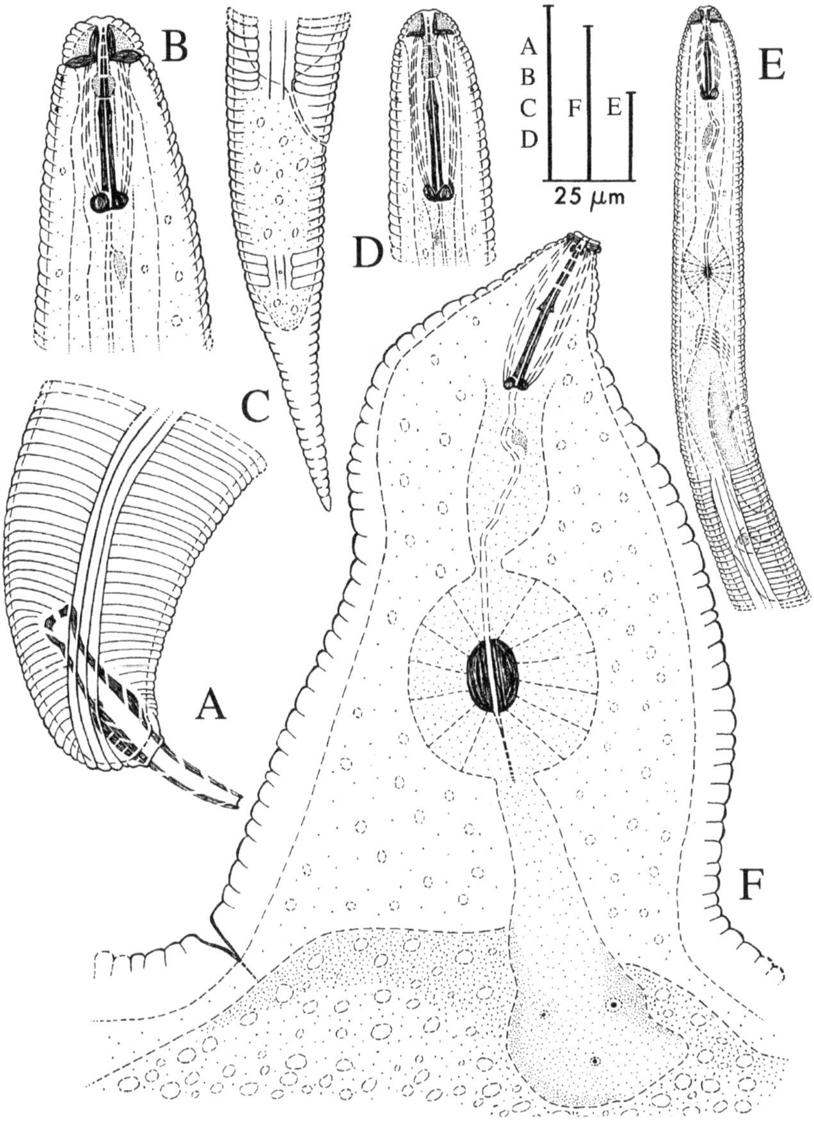

Fig. 87. Cactodera thornei. *A: Tail of male; B: Head of male; C: Tail of J2; D: Anterior region of J2; E: Pharyngeal region of J2; F: Anterior region of female. After Golden and Raski (1977).*

and Vovlas (1994) reported finding this species in cereals and grasses in China and provided a morphological description and morphometrics for this population.

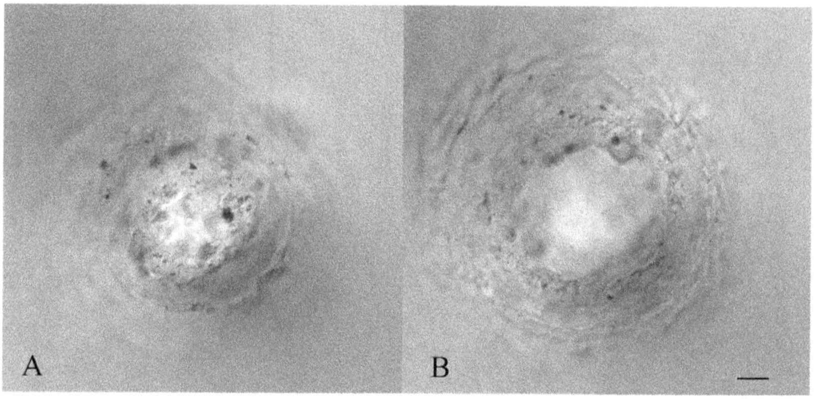

Fig. 88. Cactodera thornei. *A: Vulval plate at vulva level; B: Vulval plate at low level. (Original.) (Scale bar = 10 μm.)*

MEASUREMENTS

After Golden and Raski (1977).

– *Holotype female*: L (including neck) = 635 μm; width 330 μm; L/W ratio = 1.9; stylet = 29.0 μm; DGO = 5.8 μm.

– *Paratype females* (n = 25): L (including neck) = 561 ± 67 (456-704) μm; W = 326 ± 74.2 (194-459) μm; L/W ratio = 1.8 ± 0.3 (1.3-2.2); stylet = 29.4 ± 0.8 (28.4-30.7) μm; DGO = 5.6 ± 0.3 (5.2-6.0) μm; vulval slit = 17 (16-19) μm.

– *Paratype cysts* (n = 40): L (including neck) = 656 ± 83 (485-806) μm; W = 448 ± 81 (286-581) μm; L/W ratio = 1.5 ± 0.18 (1.2-1.9); fenestral diam. = 34 (31-36) μm.

– *Allotype male*: L = 1436 μm; a = 36; b = 6; c = 552; stylet = 29.8 μm; DGO = 6.4 μm; tail = 2.6 μm; spicules = 42.5 μm; gubernaculum = 12.6 μm.

– *Paratype males* (n = 20): L = 1503 ± 125 (1315-1708) μm; W = 33 ± 2.1 (29-36) μm; a = 37 ± 4.2 (30-45); b = 8.8 ± 0.7 (7.3-9.8); c = 321 ± 97 (199-505); stylet = 29.4 ± 1.4 (27.8-31.0) μm; DGO = 6.8 ± 1.1 (5.5-8.0) μm; lateral field width = 5.7 (5.6-6.2) μm; anterior end to median bulb valve = 51 ± 3.4 (45-60) μm; spicules = 43 ± 2 (39-46) μm; gubernaculum = 13 ± 2.2 (10-15) μm; tail = 5.0 ± 1.7 (2-8) μm.

– *Paratype J2* (n = 30): L = 554 ± 38 (446-620) μm; a = 24 ± 2.0 (20-28); b = 3.9 ± 0.7 (2.4-4.7); c = 9.8 ± 0.6 (8.8-11.2); stylet = 27 ± 0.9 (25-28) μm; DGO = 6 ± 0.6 (5-7) μm; anterior end to median bulb valve = 38 ± 1.6 (35-42) μm; hyaline region = 25 ± 1.5 (23-28) μm; tail = 56 ± 4.9 (49-64) μm.

– *Paratype eggs* (n = 20): L = 129 ± 2.6 (123-134) μm; W = 53 ± 2.1 (50-57) μm; L/W ratio = 2.4 ± 0.1 (2.3-2.6).

After Peng and Vovlas (1994).

– *Cyst* (n = 20): L = 548±78.8 (364-633) μm; W = 432.9±83.4 (287.9-568.0) μm; neck = 63.2 ± 19.3 (22.7-100) μm; L/W ratio = 1.28±0.17 (0.97-1.59); fenestral diam. (dorsal-ventral length) = 31.6± 5 (25.4-41.3) μm; fenestral diam. (lateral) width = 28.5 ± 5 (22.2-41.3) μm; anus-fenestral edge distance = 28.2 ± 12.7 (13-66.7) μm.

– *J2* (n = 20): L = 457 ± 20.0 (424-496) μm; W = 19.0 ± 0.5 (18.3-19.6) μm; a = 23.9 ± 1.1 (22.3-25.8); b = 3.2 ± 0.4 (2.5-3.7); c = 8.9 ± 0.6 (8.3-10.1); stylet = 25.1 ± 0.74 (23.5-26.1) μm; DGO = 3.3 ± 0.42 (2.6-3.9) μm; hyaline region = 26.8 ± 2.7 (22.8-29.4) μm; tail = 51.5 ± 2.2 (47.7-55.6) μm.

The Chinese population differs from the North American populations by smaller cyst size, shorter stylet and shorter DGO lengths for J2.

Description

Female

Body basically lemon-shaped with protruding neck and vulva; white to very light brown. Cuticle 12-15 μm thick with surface heavily punctate to granular, generally obscuring cuticular pattern. Labial region slightly offset, with two annuli, second annulus being larger. Stylet strong, curved slightly dorsal, with well developed stylet knobs. Anus small but distinct, located *ca* 10% of body length from vulva.

Cyst

Cyst basically lemon-shaped, with protruding neck and vulva, light brown to medium brown, abullate, circumfenestrate. Excretory pore near base of neck. Vulval denticles located *ca* 20 μm beneath upper surface of fenestra and occurring singly, in small clusters, or as a partial ring. Vulval denticles measuring *ca* 5-10 μm long × 3-6 μm wide. External cyst wall pattern at mid-body consisting of basically straight to wavy transverse lines and frequently broken by short oblique or vertical lines. Punctations often present on cysts, randomly arranged, sometimes heavy. Anus distinct, without surrounding pattern, located *ca* 10% of cyst length from posterior end.

Male

Body cylindrical, elongate, tapering gradually at each end. Labial region offset, bearing five or six annuli. Lateral field consisting of four incisures forming three bands of equal width on most of body, not areolated except in extreme anterior portion. Labial framework heavily sclerotised, stylet strong with rather rounded, prominent knobs. Hemizonid small but distinct, usually located 5-6 annuli anterior to excretory pore, hemizonion smaller, *ca* ten annuli posterior to excretory pore. Spicules arcuate, single notch at tips. Tail extremely short, somewhat variable in shape. Small phasmids near tail terminus.

J2

Body cylindrical, elongate, tapering especially to posterior end. Labial region bearing five indistinct annuli, offset from body. Cuticular annulation of body distinct, annuli measuring *ca* 1.4 μm at mid-body; subcuticular annuli less distinct and twice as numerous. Lateral field without areolation except anteriorly; composed of four incisures, averaging 3.5 μm wide at mid-body. Labial framework heavily sclerotised. Stylet strong, with prominent knobs, knobs concave on anterior surface. Hemizonid prominent, located anterior and adjacent to excretory pore, hemizonion usually *ca* 8-10 annuli posterior to latter. Tail tapering to a fine, rounded terminus, annulation continuing to terminus, sometimes giving a bead-like effect. Phasmid openings small, indistinct, located in centre of lateral field about midway on tail.

Egg

Eggshell as seen under optical microscope (oil immersion) covered with small but discrete punctations; under SEM, these resemble microprotuberances.

TYPE HOST AND LOCALITY

Roots of Miner's lettuce, *Montia perloliata* (Donn ex. Willd.) Howell (Portulaceae, Caryophyllales), growing in a Persian walnut orchard near San Jose, California, USA.

DISTRIBUTION

North America: USA (California); *Asia*: China (Yiaoshuitan town, Xining city, Qinhai province) (Golden & Raski, 1977; Peng & Vovlas, 1994).

RELATIONSHIPS

Cactodera thornei is readily distinguished from all other species of *Cactodera* by the following J2 characters: longer stylet (24-28 μm), longer body (424-620 μm) and longer tail (48-64 μm). However, the lower and upper values of the ranges may overlap with those of other species, making identification difficult when only a few specimens are available (Table 26). The stylet of male and female *C. thornei* is also far longer than those of other species in this group.

BIOCHEMICAL AND MOLECULAR CHARACTERISATION

No available data.

13. *Cactodera weissi* (Steiner, 1949) Krall & Krall, 1978
(Figs 12E; 89, 90)

Common name: knotweed cyst nematode.

This species was briefly described by Steiner (1949) from specimens parasitising *Polygonum pensylvanicum* L. at Beltsville, MD, USA. Later, Tarjan and Sasser (1953) and Golden and Raski (1977) provided additional information on all stages of this species and designated lectotypes.

MEASUREMENTS

After Golden and Raski (1977).

– *Hololectotype female*: L = 605 μm; W = 405 μm; L/W ratio = 1.5; stylet = 22.6 μm; DGO = 5.2 μm.

– *Female* (n = 20): L (including neck) = 503 ± 76 (408-688) μm; W = 278 ± 65 (201-408) μm; L/W ratio = 1.4 ± 1.9 (1.4-2.3) μm; stylet = 22.8 ± 0.53 (21.8-24.3) μm; DGO = 5.2 ± 0.3 (4.5-5.6) μm; vulval slit = 13-20 μm.

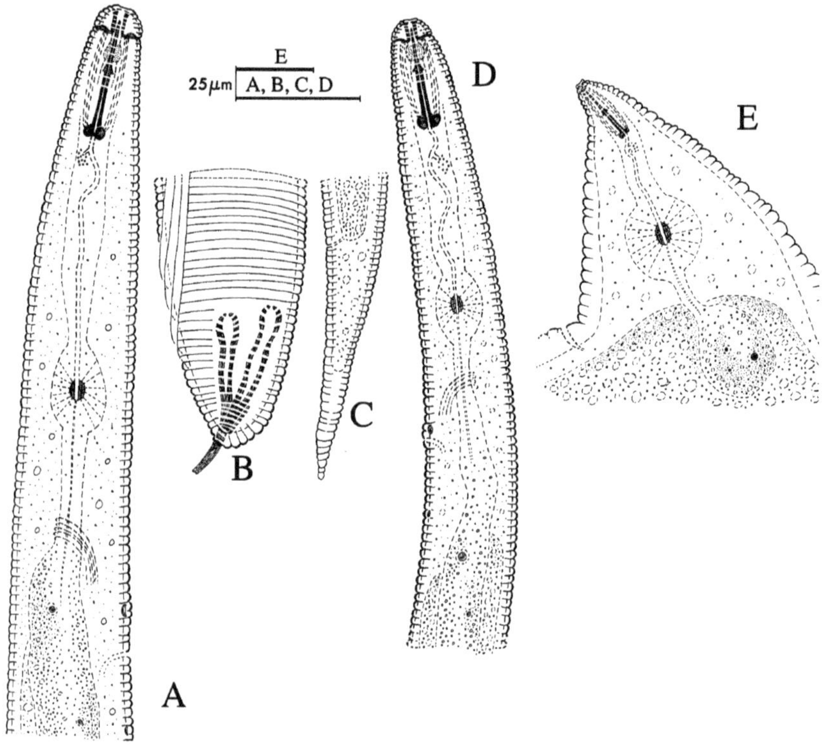

Fig. 89. Cactodera weissi. *A: Anterior end of male; B: Tail of male; C: Tail of J2; D: Pharyngeal region of J2; E: Anterior end of female. After Golden and Raski (1977).*

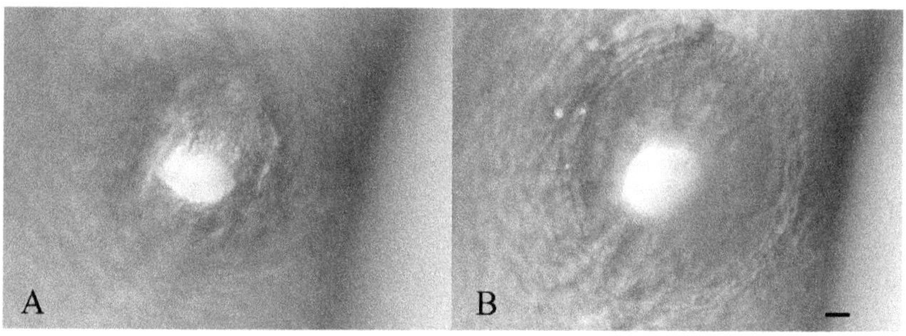

Fig. 90. Cactodera weissi. *A: Vulval plate at vulva level; B: Vulval plate at deeper level. (Original.) (Scale bar = 10 μm.)*

– *Cyst* (n = 20): L (including neck) = 598 ± 87 (385-750) μm; W = 350 ± 68 (230-460) μm; L/W ratio = 1.7 ± 0.3 (1.2-2.3); fenestral diam. (on longest axis) = 34 ± 2.8 (29-38) μm.

– *Allolectotype male*: L = 1040 μm; W = 31 μm; a = 32; b = 5.6; c = 928; stylet = 24.6 μm.

– *Male* (n = 20): L = 1034 ± 69 (943-1240) μm; W = 29-36 μm; a = 31 ± 2.1 (27-33); b = 6.0 ± 0.9 (5.1-6.7); c = 631 ± 112 (346-991); stylet = 24.2 ± 0.9 (22.5-25.2) μm; DGO = 4.8 ± 0.4 (4.5-5.6) μm; anterior end to median bulb = 45-60 μm; spicules = 32.4 ± 1.0 (30.8-33.6) μm; gubernaculum = 10 ± 0.5 (9.5-11) μm; tail = 1.6 ± 0.5 (1.1-2.8) μm.

– *J2* (n = 20): L = 407 ± 12 (372-420) μm; a = 20 ± 0.9 (18-22); b = 2.6 ± 0.1 (2.3-3.5); c = 8.9 ± 0.3 (8.4-9.5); stylet = 20.4 ± 0.5 (19.6-21.8) μm; anterior end to median bulb valve = 35-42 μm; DGO = 5.8 ± 0.3 (4.5-5.6) μm; tail = 45.9 ± 1.9 (43-50) μm; hyaline region = 20.2 ± 1.9 (17-24) μm.

– *Egg* (n = 20): L = 101 ± 2.9 (96-106) μm; W = 42 ± 0.8 (40-44) μm; L/W ratio = 2.4 ± 0.1 (2.3-2.6).

After Riggs *et al.* (1982).

From Illinois, USA, *P. pennsylvanicum.*

– *Cysts* (n = 25): L = 577 μm; W = 394 μm.

– *J2* (n = 4): L = 489 μm; W = 20.9 μm; stylet = 21.2 μm; DGO = 5.8 μm; hyaline region = 23.7 μm.

From Arkansas, USA, *P. pennsylvanicum.*

– *Cyst* (n = 25): L = 524 μm; W = 370 μm.

– *J2* (n = 4): L = 461 μm; W = 21.1 μm; stylet = 20.4 μm; DGO = 6.0 μm; hyaline region = 22.9 μm.

Description

Female

Body basically lemon-shaped, with protruding neck and vulva, pearly-white. Cuticle *ca* 12 μm thick; randomly arranged punctations on surface, often giving granular appearance, sometimes obscuring external cuticular pattern, which is essentially as for the cyst. Labial region with two annuli, second annulus larger, offset from body. Stylet rather delicate with small distinct knobs sloping posteriorly. Pharyngeal glands and excretory pore commonly located at base of neck. Anus small, inconspicuous, located *ca* 10% of body length from posterior end.

Cyst

Body lemon-shaped, with protruding neck and vulval cone, varying from light to dark brown, no bullae reported, circumfenestrate. Excretory pore situated near base of neck. Vulval denticles *ca* 16-20 μm beneath fenestral surface, measuring *ca* 6-11 μm in length and 3-5 μm in width at their base. Vulval denticles may occur singly at random, in small clusters, as a partial ring, or occasionally as a complete ring. External cyst wall pattern at mid-body consisting of basically transverse straight to wavy lines and commonly broken by short oblique or vertical lines. Rather fine punctations, arranged randomly, often present on cyst wall. Anus distinct though small, without a surrounding pattern; located *ca* 10% of cyst length from posterior end.

Male

Body cylindrical, elongate, tapering gradually at each end. Labial region offset, bearing five or six annuli. Cuticular annulation distinct, annuli *ca* 1.8 μm wide, subcuticle with 'annuli' about twice as numerous. Lateral field 5.6-6.2 μm wide, consisting of four incisures forming three bands of equal width on most of body, not areolated except in extreme anterior portion. Labial framework heavily sclerotised, stylet strong with rather rounded, prominent knobs. Centre of median bulb 45-60 μm from anterior end. Hemizonid small but distinct, usually located five or six annuli anterior to excretory pore, hemizonion smaller, *ca* ten annuli posterior to excretory pore. Testis single. Spicules arcuate, single notch at tips. Tail extremely short, somewhat variable in shape. Small phasmid openings near tail terminus.

J2

Body cylindrical, elongate, tapering especially to posterior end. Labial region bearing five indistinct annuli, offset from body. Cuticular annulation of body distinct, annuli measuring *ca* 1.4 μm at mid-body, subcuticular 'annuli' less distinct and twice as numerous. Lateral field without areolation except anteriorly, composed of three incisures, averaging 3.5 μm wide at mid-body. Labial framework heavily sclerotised. Stylet strong with prominent knobs anteriorly concave. Hemizonid prominent, located anterior and adjacent to excretory pore, hemizonion usually *ca* 8-10 annuli posterior to latter. Tail tapering to a fine, rounded terminus, with annulation continuing to terminus, sometimes giving a bead-like effect. Phasmid openings small, indistinct; in centre of lateral field about midway on tail.

Egg

Eggshells hyaline, without visible markings as seen with both optical microscope and SEM.

TYPE HOST AND LOCALITY

Roots of knotweed, *Polygonum pensylvanicum* L. (Polygonaceae, Caryophyllales), Beltsville, Maryland, USA.

OTHER HOSTS

Cactodera weissi is known to infect various species of *Polygonum*, including *P. punctatum* Ell. Buckwheat, *Fagopyrum esculentum* Moench, *Rumex coglomeratus* Murr. and *R. abtusifolius* L. were host plants for a population of this species from Nansemond County, Virginia (Golden & Raski, 1977; Riggs, 1982).

DISTRIBUTION

North America: Canada (British Columbia, Ontario and Quebec); USA (Arkansas, Florida, Illinois, Iowa, Maryland, Michigan, Missouri, Nebraska, New Jersey, New York, North Carolina, Pennsylvania, Virginia, West Virginia and Wisconsin) (Golden & Raski, 1977; Mulvey & Golden, 1983).

RELATIONSHIPS

Cactodera weissi is most closely related to *C. amaranthi* and *C. acnidae* but differs from the first species in longer J2 tail (43-50 *vs* 32-40 μm) and hyaline region (17-24 *vs* 11-18 μm). J2 of *C. weissi* have the anterior surface of the stylet knobs concave *vs* convex in *C. acnidae* and DGO = 4.5-6.0 *vs* 2.5-3.0 μm (Table 26).

BIOCHEMICAL AND MOLECULAR CHARACTERISATION

Sabo *et al.* (2002) sequenced the ITS-rRNA gene and studied relationships of *C. weissi* with other cyst nematodes.

Nematology Monographs & Perspectives, 2010, Vol. 8A, 261-265

Chapter 9

Systematics of the genus *Dolichodera*

History

This unusual cyst nematode was found in a sample from the Ottawa River near Deschenes, Quebec, Canada. Characters of the new taxon were unlike those of any described genera of cyst nematodes and were regarded as sufficiently distinct to justify proposing a new genus, *Dolichodera* (from the Greek *dolichos* = long, and *deros* = skin), by Mulvey and Ebsary (1980).

Genus *Dolichodera* Mulvey & Ebsary, 1980

DIAGNOSIS (after Siddiqi, 2000)

Female body elongate-oval, without terminal protuberance, white, swollen part 400-500 μm long, 140-270 μm wide, 2-2.8 times as long as wide, neck moderately long. Cuticle not annulated but with fine irregular striae. Vulval area terminal or just subterminal, circumfenestrate, fenestra *ca* 20 μm in diam., bullae present, perineal tubercles absent. Anus pore-like, lacking a fenestra, located 10-13 μm dorsal to vulval fenestral margin. *Cyst* light brown, elongate oval. Several large bullae present. Perineal tubercles absent. Vulva circumfenestrate, underbridge absent. Anus lacking fenestra. *Male* not found. *J2* with long tail (95-120 μm). Lateral field with three incisures, inner one faint. Labial region hemispherical, offset, with two annuli. Tail tip narrowly rounded. Phasmid openings lacking a lens-like ampulla, located about one anal body diam. posterior to anus.

Note

Siddiqi (2000) suggested that *Dolichodera* might be a junior synonym of *Punctodera* and that a primary difference between the two was the absence of anal fenestration in the former.

TYPE AND ONLY SPECIES

Dolichodera fluvialis Mulvey & Ebsary, 1980

Note

Wouts (1985) believed that *Thecavermiculatus andinus* Golden, France, Jatala & Astogaza, 1983 did not fit the description of the genus *Thecavermiculatus* but was closer to the genus *Dolichodera* and proposed to transfer this species. This species has a spherical body, the vulva and anus are *ca* 60 μm apart, the vulval slit is less than 10 μm long, there is the beginning of a circumfenestrate vulva and it has a rather wide pharyngeal gland which is not restricted to the ventral side of the body of the infective J2. However, as Wouts (1985) also noted, since no cyst is formed and the female cuticle is partially annulated, it does not seem completely identical to *Dolichodera*, suggesting it appears more primitive and may even represent an independent genus. Siddiqi (2000) considers this nematode as *Atalodera andina* (Golden, France, Jatala & Astogaza, 1983) de Souza & Huang, 1994.

SPECIES DESCRIPTION

***Dolichodera fluvialis* Mulvey & Ebsary, 1980**
(Figs 91, 92)

MEASUREMENTS

After Mulvey and Ebsary (1980).

– *Holotype female*: L (excluding neck) = 550 μm; W = 270 μm; L/W ratio = 2.04.

– *Paratype female* (n = 1): L (excluding neck) = 400 μm; W = 140 μm; L/W ratio = 2.8; neck = 120 μm; stylet = 15 μm; anterior end to median bulb valve = 60 μm; fenestral diam. = 20 μm.

– *Paratype cysts* (n = 12): L (excluding neck) = 575 (450-700) μm; W = 235 (140-300) μm; L/W ratio = 2.6 (1.8-4.0).

– *Paratype J2* (n = 5): L = 600 (570-610) μm; a = 29.3 (27-31); b = 2.2; c = 5.6 (5.1-6.3); stylet = 23 (22-24) μm; stylet knob width = 2-3 μm; DGO = 7-8 μm; tail = 103 (95-120) μm; anterior end to median bulb valve = 80-90 μm.

– *Paratype eggs* (n = 15): L = 76 (68-82) μm; W = 32 (29-40) μm.

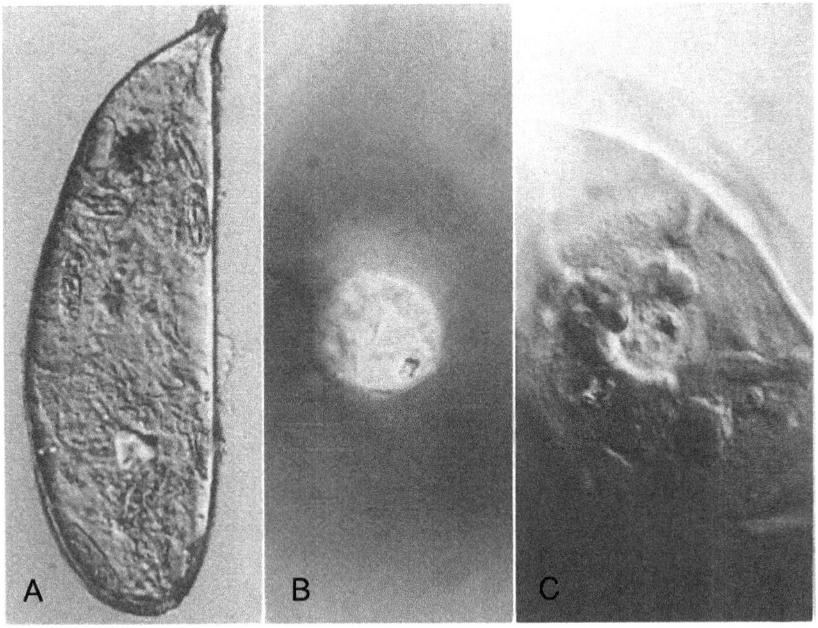

Fig. 91. Dolichodera fluvialis. *Cyst. A: Whole cyst with slightly subterminal vulva fenestra; B: Fenestral area which has deteriorated to form a round aperture; C: Bullae in fenestral area. After Mulvey and Ebsary (1980).*

DESCRIPTION

Female

Body elongate ovoid, white, cuticle 3-4 μm thick, non-annulated but with fine irregular striae. Subscrystalline layer absent. Neck relatively long, cephalids not observed. Stylet slender, knobs 2-3 μm wide, anterior surface concave. Median pharyngeal bulb large with well developed valve. Vulval area circumfenestrate. Anus small, without fenestra, located 10-13 μm dorsal to vulval fenestra margin. Mature female containing more than 100 eggs in various stages of development.

Cyst

Cyst elongate-ovoid with long protruding neck, light brown. Cuticle very thin (3-4 μm). Several large refractive bullae located slightly internal to fenestral area. Vulval area deteriorated to form terminal or subterminal aperture, round. Most cysts containing many empty eggshells indicating that J2 had hatched and emerged from cysts.

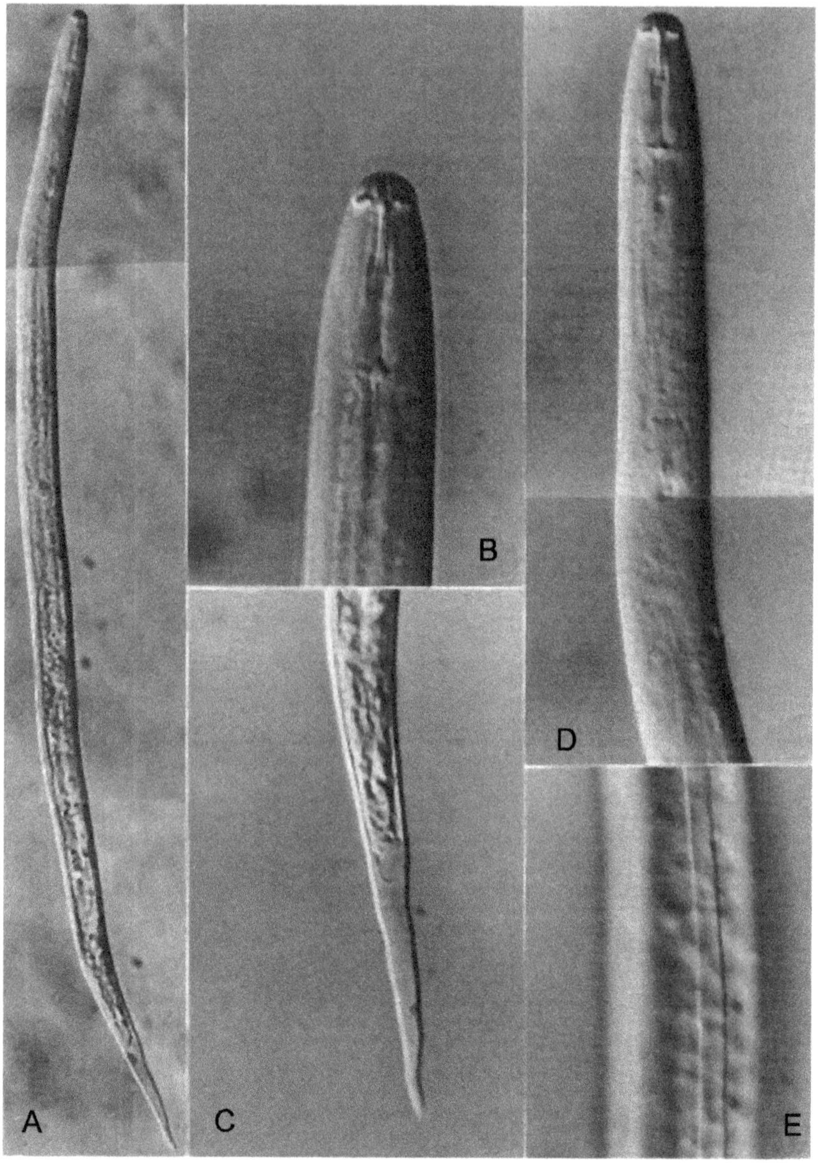

Fig. 92. Dolichodera fluvialis. *J2. A: Posteriorly located genital primordium; B: Head end showing sclerotised labial skeleton and stylet; C: Tail; D: Pharyngeal region; E: Lateral field with three incisures. After Mulvey and Ebsary (1980).*

Male

Unknown.

J2

Cuticle moderately annulated, lateral field with three incisures, median incisure faint. Labial region with two annuli, offset from body. Stylet slender, anterior surfaces of knobs concave. Labial framework heavily sclerotised. Cephalids not observed. Median bulb valve prominent Excretory pore 30-35 μm posterior to level of median bulb valve, hemizonid immediately anterior to pore. Tail long, terminus narrowly rounded. Phasmid openings small, lacking lens-like ampulla, located one anal body diam. posterior to anus.

TYPE HOST AND LOCALITY

The host plant remained unknown (see Mulvey & Ebsary, 1980) until 1981 when root and soil samples revealed that *Spartina pectinata* Link (Poaceae, Poales) was the host (Ebsary & Eveleigh, 1983). The nematodes were collected from bottom sediments of the Ottawa River at a depth of 10 cm and 2-3 m from the bank near Deschenes, Quebec, Canada.

DISTRIBUTION

North America: Canada.

BIOCHEMICAL AND MOLECULAR CHARACTERISATION

No available data.

Nematology Monographs & Perspectives, 2010, Vol. 8A, 267-275

Chapter 10

Systematics of the genus *Betulodera*

History

With four other heteroderid species having a very short vulval slit, Mulvey (1972) placed *Heterodera betulae* Hirschmann & Riggs, 1969 in a category designated as group 2 or the *Cacti* group. Because of markedly different morphology, Golden and Raski (1977) had removed this species from the *Cacti* group and placed it on its own, "since it is incompatible with any of the present *Heterodera* groups". Krall and Krall (1978) transferred *H. betulae* to the genus *Cactodera*. Stone (1979) placed it in a separate '*Betulae* group' mainly because of differences in host range compared with species in the *H. cacti* group. Mulvey and Golden (1983) did not assign this species to *Cactodera* or to any of existing groups within the genus *Heterodera*. Subsequently, several authors discussed the peculiarities of *C. betulae*, its uncertain taxonomic position and misclassification, and the clear polyphyletic state of the genus *Cactodera* (Cliff & Baldwin, 1985; Othman *et al.*, 1988; Baldwin & Schouest, 1990; Wouts & Baldwin, 1998; Sturhan, 2000). However, the taxonomic problems were not addressed until Sturhan (2002) erected a new genus, *Betulodera* (from the specific epithet of the type species name = *betulae*, and the Greek *deros* = skin), to accommodate *C. betulae* based on peculiarities of its host range and cyst morphology. Relationships of *Betulodera* with other genera of cyst nematodes as inferred from the analysis of D2-D3 expansion segments of 28S rRNA gene sequences are given in Figure 26.

Genus *Betulodera* Sturhan, 2002

DIAGNOSIS (after Sturhan, 2002)

Cyst: lemon-shaped, pear-shaped or spheroid with insignificant, obtuse, vulval cone. Cyst wall thick, with irregular network-like pattern,

D-layer absent (no punctations in inner, deeper layers of cyst wall), subcrystalline layer heavily developed. Vulva terminal, surrounded by circumfenestration, vulval slit short (<10 μm), underbridge absent, denticles occasionally present, anus without fenestration. *Male:* body twisted, no cloacal tube, spicules with bifid distal tips, phasmid openings punctiform. *J2:* lateral field with three incisures, phasmid openings punctiform, without lens-like structure, labial region with three or four labial annuli and labial disc fused with submedial lips.

TYPE AND ONLY SPECIES

Betulodera betulae (Hirschmann & Riggs, 1969) Sturhan, 2002
= *Heterodera betulae* Hirschmann & Riggs, 1969
= *Cactodera betulae* (Hirschmann & Riggs, 1969) Krall & Krall, 1978

Note

An unknown cyst nematode similar to *B. betulae* and infecting unidentified trees along the south bank of the Merced River was found in Incline, Mariposa County, Sierra National Forest, California, USA. The ITS-rRNA gene sequence of this nematode differed from that of *B. betulae* (Subbotin, unpubl.).

SPECIES DESCRIPTION

***Betulodera betulae* (Hirschmann & Riggs, 1969) Sturhan, 2002**
(Figs 12D; 93, 94)

Common name: birch cyst nematode.

A cyst nematode was isolated from soil samples collected along the bank of the Middle Fork of the White River, east of Fayetteville, Arkansas, USA. Examination of the roots of other plants in the area and preliminary host tests in the glasshouse indicated that river birch, *Betula nigra* L., was the natural host. A detailed description of this nematode under the name *Heterodera betulae* was given by Hirschmann and Riggs (1969). Mulvey (1972) provided additional measurements of cysts, including the cone tops. Stone (1975) added data on head morphology of the J2, Cliff and Baldwin (1985) studied the fine structure of the female cuticle and Othman *et al.* (1988) studied the morphology of the labial region and the posterior end of the J2 and male with SEM.

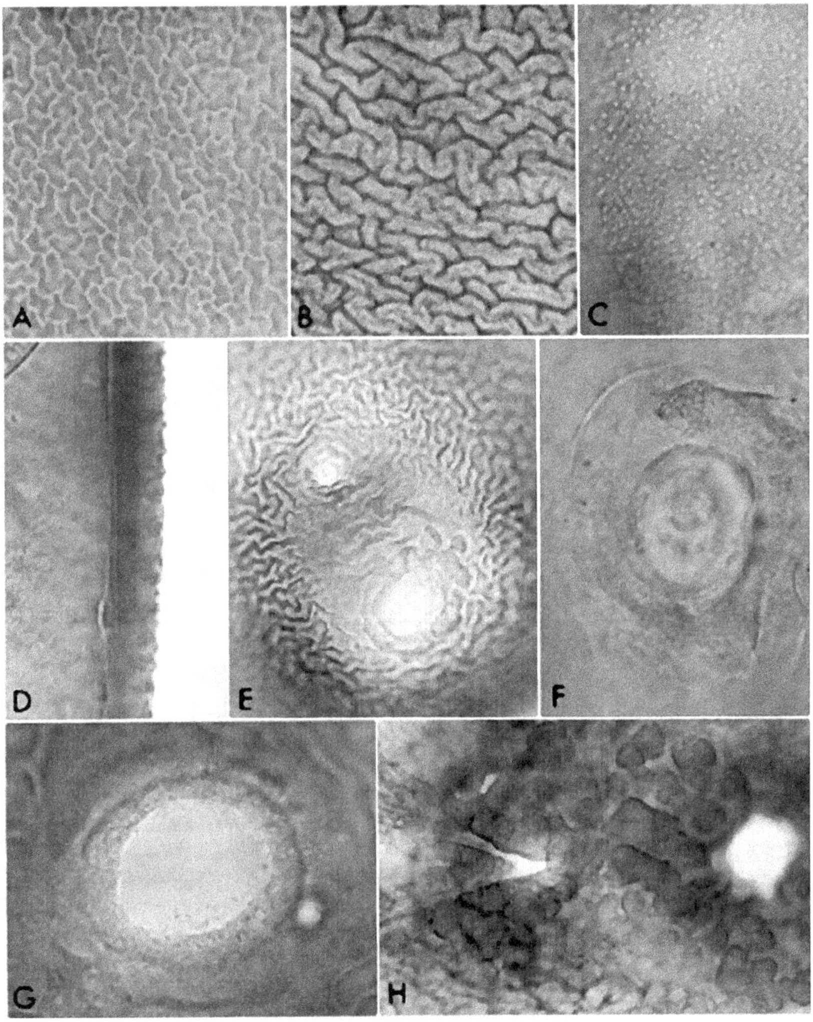

Fig. 93. Betulodera betulae. *Photomicrographs of female and cyst characters. A: Typical network-like cyst wall pattern from middle portion of large cyst; B: Pattern of subcrystalline layer of young white female; C: Pronounced pimples in neck region of cyst; D: Cyst wall in longitudinal section showing rugose surface contour; E: Surface view of cyst cone top with single fenestra (large opening) and anus (small opening); F: Vulval area of white female during early stage of fenestration; G: Fenestra of brown cyst broken through completely; H: Bullae-like structures clustered around fenestra and anus on inside of cyst cone at lower focusing level. After Hirschmann and Riggs (1969).*

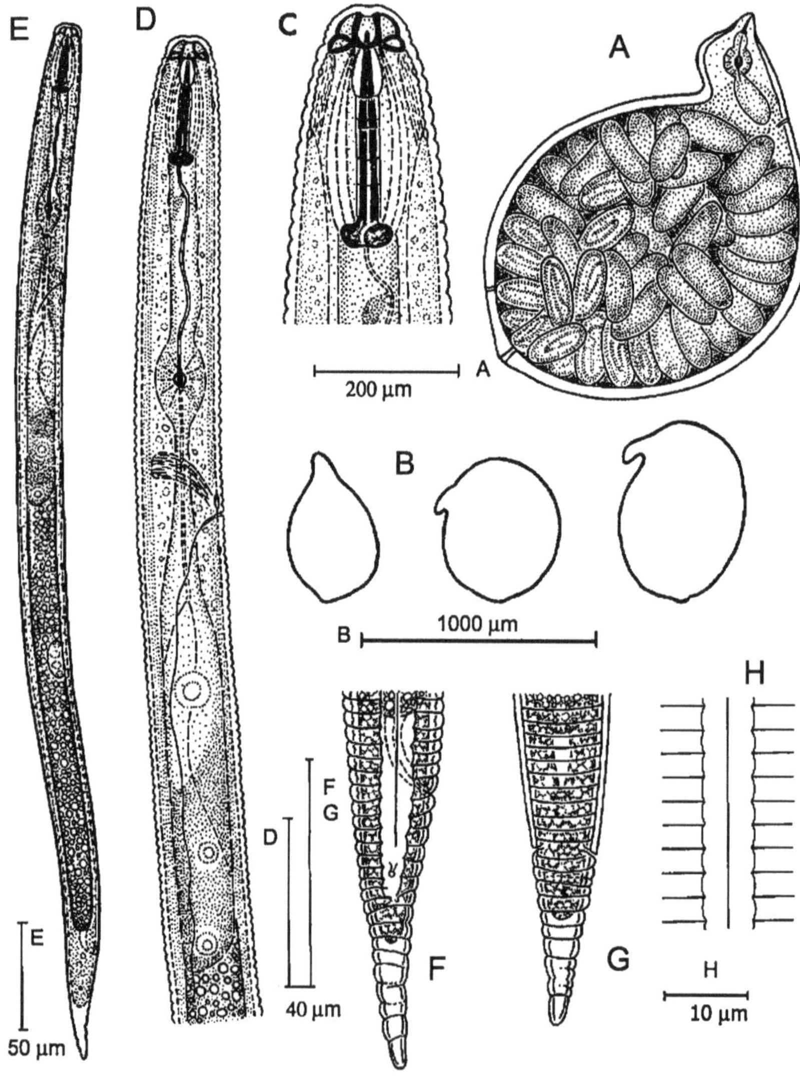

Fig. 94. Betulodera betulae. *Females and cysts. A: Female filled with eggs (lateral); B: Shape and size ranges of cysts. J2; C: Labial region (lateral); D: Pharyngeal region (lateral); E: Full-length outline (lateral); F: Tail (lateral); G: Tail (ventral); H: Lateral field. After Hirschmann and Riggs (1969).*

DESCRIPTION

After Hirschmann and Riggs (1969).

– *Holotype female*: L (excluding neck) = 450 μm; W = 403 μm; L/W ratio = 1.1; stylet = 30.6 μm; DGO = 4.5 μm; median bulb length = 33.2 μm; median bulb diam. = 36.1 μm; anterior end to: end of median bulb = 87.8 μm, excretory pore = 140.4 μm; vulva to anus distance (lateral view) = 78.0 μm; cuticle thickness = 12.7 μm.

– *Paratype females* (n = 20): L (excluding neck): 618.4 (381-945) μm; W = 512.1 (342-720) μm; L/W ratio = 1.2 (1.0-1.5); stylet = 30.2 (28.8-32.6) μm; DGO = 4.80 (4.1-5.3) μm; median bulb length = 36.5 (30.6-45.0) μm; median bulb diam. = 36.4 (30.2-41.0) μm; anterior end to: end of median bulb = 96.3 (74.1-117.0) μm, excretory pore = 166.9 (126.4-253.5) μm; vulva to anus distance (lateral view) = 93.0 (58.5-150.2) μm; cuticle thickness = 14.2 (10.2-19.5) μm. Vulval cones (n = 30); vulval slit = 5.8 (5.1-6.6) μm; width of vulval slit = 2.2 (1.0-5.0) μm; distance from anus to nearest edge of vulva = 80.9 (42.6-153.0) μm.

– *Paratype cysts* (n = 50): L = 763 (435-1160) μm; W = 616.4 (340-920) μm; L/W ratio = 1.2 (1.0-1.8). Vulval cone: long axis of vulval fenestra (A) = 22.0 (11.9-29.8) μm; short axis of vulval fenestra = 18.8 (10.2-27.7) μm; distance from anus to nearest edge of vulval fenestra (B) = 81.0 (44.9-153.0) μm; Granek's ratio B/A = 3.8 (1.8-7.5).

– *Paratype males* (n = 10): L = 1063 (831-1280) μm; W = 30.9 (26.5-36.4) μm; stylet = 29.1 (28.5-29.6) μm; stylet knob height = 3.2 (2.9-3.5) μm; stylet knob width = 6.3 (5.8-6.8) μm; DGO = 4.5 (3.6-5.8) μm; pharyngeal length = 148.5 (136.2-163.8) μm; anterior end to excretory pore = 153.9 (130.1-185.4) μm; tail = 6.2 (5.1-10.2) μm; testis = 538.8 (335-710) μm; spicules = 27.8 (23.3-30.6) μm; gubernaculum = 9.2 (8.2-10.9) μm; sperm = 4.0 (3.6-4.6) μm; a = 34.3 (27.2-38.9); b = 7.2 (5.6-8.7); c = 176.2 (122.5-251.0); T = 50.2 (39.0-59.1).

– *Paratype J2* (n = 108): L = 461.7 ± 18.8 (407.7-508.5) μm; W = 21.8 ± 1.0 (20.4-24.6) μm; a = 21.2 ± 1.0 (18.5-23.3); b = 3.7 ± 0.1 (3.3-4.2); c = 10.6 ± 0.5 (9.0-11.9); stylet = 26.0 ± 0.6 (24.6-27.6) μm; stylet knob height = 2.9 ± 0.3 (2.4-3.7) μm; stylet knob width = 5.9 ± 0.3 (5.3-6.6) μm; DGO = 6.0 ± 0.5 (5.0-7.8) μm; pharyngeal length = 126.6 ± 4.7 (112.7-139.6) μm; anterior end to excretory pore = 107.1 ± 3.9 (95.5-117.2) μm; tail = 43.8 ± 2.5 (39.0-51.0) μm; hyaline region (H) = 20.8 ± 1.6 (14.7-25.5) μm; H/stylet = 0.8 ± 0.1 (0.6-1.0).

– *Paratype eggs* (n = 100): L = 109.0 (97-120) μm; W = 48.4 (45-53) μm; L/W ratio = 2.3 (1.9-2.6).

After Riggs *et al.* (1982) from *Robinia pseudoacacia* L.

– *Cysts* (n = 25): L = 747 μm; W = 549 μm.

– *J2* (n = 4): L = 525 μm; W = 23.0 μm; stylet = 25.6 μm; DGO = 4.9 μm; hyaline region = 24.8 μm.

DESCRIPTION

Female

Body pearly white, varying from slightly lemon-shaped in small individuals to pear-shaped and spherical in large specimens. Protrusion of vulval cone very slight, in some specimens hardly noticeable. Neck frequently slightly subterminal and characteristically curved posteriad. Abrupt bend between neck and body dorsally in vicinity of root, ventral side away from root bearing excretory pore and vulva. Cuticle thickest where neck and body join. Anterior region at level of stylet annulated, neck area with pronounced pimples, neck-to body transition zone with broken longitudinal ridges, cuticle pattern over rest of body slightly zigzag, wavy or network-like depending on size of individual. Subcrystalline layer thick, chalky-looking, crazed deposit, covering body, neck and head completely, with definite pattern in surface view and striated in longitudinal optical section. Labial disc tetraradiate, distinctly sclerotised at all four corners. First labial annulus hexaradiate, indicating presence of six lips. Labial disc and first annulus distinctly offset from rest of neck annuli by constriction. Stylet strong, with well developed posteriorly sloping knobs measuring *ca* 3.5 μm in height. Brownish secretion sometimes adhering to head around oral opening. Dorsal pharyngeal gland orifice slightly posterior to stylet knobs, median bulb large, rounded, with big valve plates, pharyngeal glands contained in single lobe, overlapping ventrally. Excretory pore distinct, located near posterior end of gland lobe or slightly further posteriorly. Germinal zone short, growth zone elongated, oviduct short and narrow, spermatheca well differentiated, hemispherical, uterus long and broad. Both gonads forming pronounced angle at spermatheca-uterus junction. Sperm absent in females examined. Vulva located on very small protrusion, vulval slit short but distinct, surrounded by thin transparent cuticle delimiting boundary of future circumfenestrae. Egg sac present, but without eggs. Anus small but distinct.

Cyst

Size and shape of cyst similar to white female. As in female, neck in many cysts curved posteriad. Cyst light to dark to blackish brown, some cysts appearing almost black throughout. Yellow phase lacking as female ages, white female turning first light brown, then darker. Thick subcrystalline layer also present in young brown cysts, covering whole cyst including neck. Underlying cyst wall semi-transparent, with carefully adjusted light, J2 eggs easily visible through wall. Cyst wall in neck very thick and dark coloured. Wall pattern varying with age and size of cysts, in large cysts appearing as continuous irregular network-like pattern in mid-body with intermittent short broken lines; around vulva and anus exhibiting long, broken zigzag lines; where cyst body merges into neck regular pattern changing to broken longitudinal ridges which, further anteriad in neck region, become arranged in transverse rows of pronounced, wart-like, pimples. Further posteriorly, pattern changes to longitudinal ridges and then a reticulate pattern over most of body. Cyst surface dull, matt, when viewed under low magnification in adjusted light. No punctations observed in inner, deeper layers of cyst wall. Vulval cone very insignificant, obtuse. Vulval protrusion in most cysts darker than rest of cyst. As female ages and turns into brown cyst, thin-walled area surrounding vulval slit breaks down forming a circumfenestra. Vulval fenestra circular to oval-shaped. Occasionally, even in older brown cysts, vulval fenestra not broken through, remaining as thin membrane. Vaginal remnant and underbridge not observed. In several cysts, deeper focus at fenestra and anal openings revealing brown, bullae-like structures clustered in cone.

Male

Body cylindrical, slender, assuming open C shape when killed by gentle heat, tapering more anteriad than posteriad. Posterior part of some males twisted through 90° after killing with gentle heat. Labial region hemispherical, slightly offset. Labial region with 3-5 irregular annuli (not including labial cap) some being discontinuous or anatomosing. Coarse cuticular annulation over rest of body. Lateral field with four equally spaced incisures (inner incisures sometimes slightly closer), outer lines very slightly crenate but in most places straight, a few areolations present, ending as broad band in tail region. Labial framework heavily sclerotised. Cephalids located two and six annuli posterior to labial constriction. Stylet strong with well developed knobs that are anteriorly straight or slightly concave. Dorsal gland orifice just posterior to stylet knobs. Amphidial pores, ducts and pouches distinct in median view, pouches with sensilla extending to mid-level of stylet. Cylindrical procorpus of pharynx followed by oval elongate median bulb with well developed valve plates; gland lobe varying in length, glands individually distinct as in J2, junction with intestine distinct. Hemizonid 1.5-2 annuli long, 0.5-5 annuli anterior to excretory pore. Hemizonion not observed. Testis packed with spermatozoa near cap cell but spermatogonia not observed. Posterior part of *vas deferens* glandular, not filled with sperm. Spicule characteristically straight with slightly bifid end, gubernaculum simple, one-third length of spicule. Tail very short, rounded, coarse annulations extending around tip. Phasmid openings present, subterminal, indistinct.

J2

Body cylindrical, tapering more posteriorly than anteriorly. When killed by gentle heat slightly twisted, true lateral position infrequently encountered. Labial region slightly offset from body, labial annuli indistinct, only three or four very fine annuli anteriad to constriction of labial region observed in lateral or median view. Cuticular annulation over rest of body distinct. Lateral field with three incisures, 4.6 (4.1-5.1) μm wide in region of genital primordium, beginning at level of stylet shaft as two incisures, increasing to three at dorsal gland orifice; outer incisures usually faintly crenate, middle incisure ending slightly anterior to phasmids, outer incisures continuing a short distance posteriad, very slight areolation in anterior and posterior end of lateral field. Labial framework heavily sclerotised, radial blades extending slightly over labial constriction into body. Anterior cephalids at two annuli, posterior cephalids at six annuli posterior to labial constriction. Stylet strong with well developed knobs, slightly concave on anterior surface, shape of knobs variable depending on orientation of J2. Dorsal gland orifice obscure. Amphidial pores, ducts and pouches distinct in median view, pouches with sensilla extending *ca* length of stylet shaft. Pharynx with valvate, oval median bulb and ventrolateral gland overlap, well defined pharyngeal-intestinal junction, intestine located dorsolaterad. Individual pharyngeal glands distinct, dorsal gland clear and light compared to two granular and dark subventral glands. Hemizonid distinct, 1.5 annuli long, located anterior and adjacent to excretory pore, excretory duct visible to posterior end of pharyngeal gland lobe. Hemizonion three-fourths of an annulus long, located three to five annuli posterior to excretory pore. Anus distinct, caudalid, one annulus long, located anterior or adjacent to anus. Tail terminus bluntly rounded. Phasmid opening small but distinct.

TYPE HOST AND LOCALITY

River birch, *Betula nigra* L. (Betulaceae, Fagales), south bank of Middle Fork of White River, 2.8 km upstream from Highway 16 bridge, Washington County, Arkansas, USA.

OTHER HOSTS

In an experimental study of 80 plant species in 20 families, only five species of *Betula* (*Betula nigra* L., *B. lenta* L., *B. pendula* Roth, *B. populifolia* Marsh. and *Betula* sp.), *Cleome spinosa* L. and *Robinia pseudoacacia* L. supported reproduction. Limited reproduction occurred on *Lespedeza striata* (Thunb.) H. & A. and *Alnus glutinosa* (L.) Gaertn., and only occasional females developed on orchard grass, *Dactylis glomerata* L., sweet alyssum, *Lobulana maritime* Desv., and butternut squash, *Cucurbita moschata* Duch. ex Poir (Riggs *et al.*, 1969; Riggs, 1982).

DISTRIBUTION

North America: USA. Recorded from Arkansas, Iowa and Missouri (Golden & Raski, 1977).

BIOCHEMICAL AND MOLECULAR CHARACTERISATION

Ferris *et al.* (2004) was the first to sequence the ITS-rRNA and partly the 18S rRNA gene of *B. betulae* and estimated the phylogenetic position of this species within Heteroderidae.

Nematology Monographs & Perspectives, 2010, Vol. 8A, 277-283

Chapter 11

Systematics of the genus *Paradolichodera*

History

Found in a soil sample collected by D. Sturhan in November, 2000 from the rhizosphere of various plants along a lagoon south of Christchurch, New Zealand, the sample included a few extremely slender tylenchid J2, which were identified as representatives of a new species and genus. In cyst and female characteristics the species most closely resembles *Dolichodera fluvialis* Mulvey & Ebsary, 1980, a species known only from an aquatic habitat in Canada. The slender J2 of the new species from New Zealand werc, however, remarkably different from those of *D. fluvialis* and from all other known heteroderids. Their unique characteristics, including features of the cyst, were considered as justification for proposing a new genus *Paradolichodera* (from Greek *para* = near, and *Dolichodera*) by Sturhan *et al.* (2007).

Genus *Paradolichodera* Sturhan, Wouts & Subbotin, 2007

DIAGNOSIS (after Sturhan *et al.*, 2007)

Mature female and cyst elongate-ovoid, with rounded posterior end. Cuticle transparent, with faint transverse striations on anterior part of body and posteriorly mostly with faint irregular ridges superimposed on distinct punctations. Cuticle turning yellowish to light brown on death, covered by a subcrystalline-like film. Eggs retained in body, egg sac not observed. Labial disc squarish. Stylet well developed. Vulva terminal or subterminal, vulval slit short, circumfenestrate. Anus lacking fenestration. *Male* body not twisted, lateral field with four incisures. Cloacal tube present, spicules rounded at tip. Phasmids lacking. *J2* long, extremely slender for family, lateral fields indistinct. Stylet short (<20 μm). Dorsal gland orifice located more than half stylet length posterior to stylet base, pharyngeal glands long, filling body cavity. Tail long, slender, phasmid openings punctiform.

TYPE AND ONLY SPECIES

Paradolichodera tenuissima Sturhan, Wouts & Subbotin, 2007

SPECIES DESCRIPTION

Paradolichodera tenuissima Sturhan, Wouts & Subbotin, 2007
(Figs 95, 96)

MEASUREMENTS

After Sturhan *et al.* (2007).

– *Holotype cyst*: L (including neck) = 930 μm; L (excluding neck) = 750 μm; W = 360 μm, L/W ratio = 2.1.

– *Paratype cysts* (n = 25): L (including neck) = 860 ± 87 (620-970) μm; L (excluding neck) = 750±95 (470-900) μm; W = 335±63 (205-505) μm; L/W ratio = 2.3 ± 0.3 (1.7-2.7); fenestral diam. = 15-26 μm.

– *Paratype females* (n = 10): L (including neck) = 780 ± 51 (660-880) μm; W = 262±55 (170-340) μm; L/W ratio = 3.1±0.6 (2.4-4.6); stylet = 24.3±0.9 (22.5-25.0) μm; DGO = 10-12 μm; vulval slit = 4-6 μm; vulva to anus distance = 18-36 μm.

– *Paratype males* (n = 10): L = 1370 ± 74 (1220-1455) μm; W = 32.9 ± 0.7 (32-34) μm; a = 41.7 ± 2.3 (38-45); b = 9.0 ± 0.3 (8.5-9.6); b′ = 4.55 ± 0.3 (4.1-5.1); stylet = 25.6 ± 0.6 (25-26.5) μm; DGO = 10.4 ± 0.8 (9.0-12.0) μm; anterior end to: posterior end of pharyngeal glands = 300 ± 21 (270-340) μm, pharyngeal-intestinal valve = 152 ± 9 (131-166) μm, median bulb = 99 ± 4 (92-105) μm, excretory pore = 170±9 (150-185) μm; genital tract length = 750±123 (490-930) μm; spicules = 30 ± 1.1 (28-32) μm.

– *Paratype J2* (n = 25): L = 825±32 (750-880) μm; W = 11.6±0.6 (9.6-12.8) μm; a = 71 ± 4 (61-79); b = 6.9 ± 0.7 (5.8-8.0); b′ = 2.3 ± 0.1 (2.1-2.5); c = 10.2 ± 0.5 (9.1-11.1); c′ = 9.6 ± 0.6 (8.0-10.6); stylet = 18.5 ± 0.5 (17.5-19.5) μm; DGO = 12.9 ± 1.0 (11-15) μm; anterior end to: posterior end of pharyngeal glands = 362 ± 13

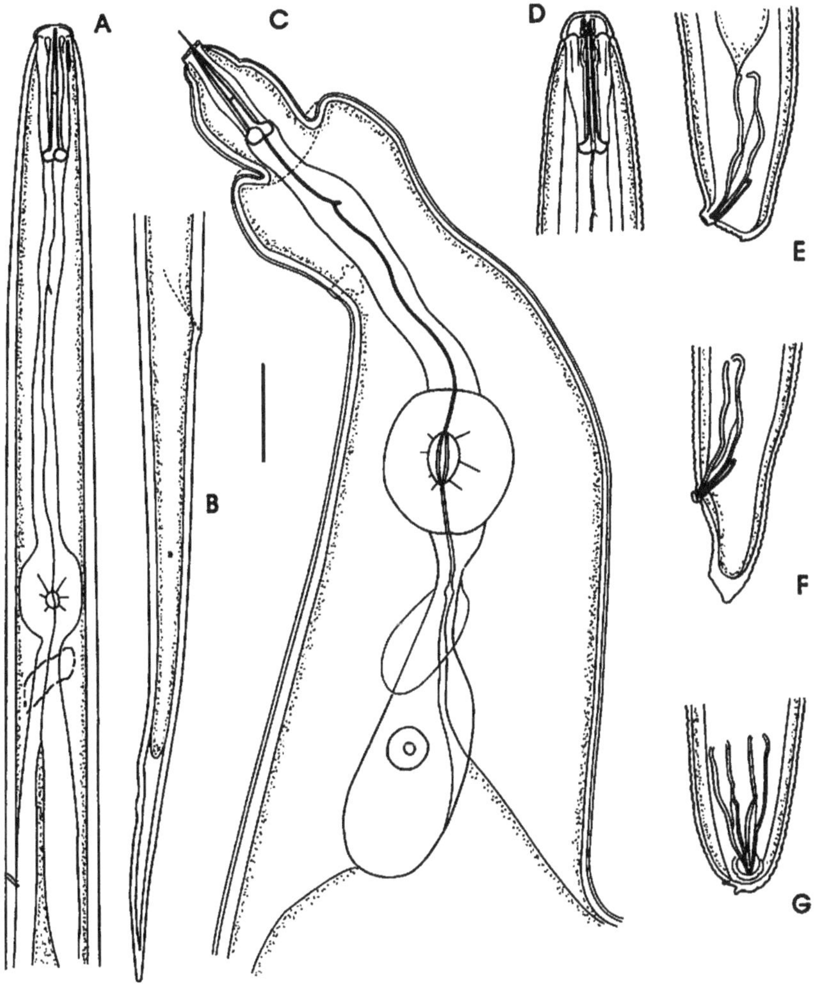

Fig. 95. Paradolichodera tenuissima. *A: Anterior region of J2; B: Tail of J2; C: Pharyngeal region of female; D: Anterior region of male; E, F: Posterior region of male, lateral; G: Posterior end of male, dorsoventral. (Scale bar: A, B = 15 μm; C-G = 20 μm.) After Sturhan* et al. *(2007).*

(330-385) μm, pharyngeal-intestinal valve = 120 ± 14 (98-143) μm, median bulb = 79 ± 3 (70-84) μm; diam. at anus level = 8.5 ± 0.5 (8.0-9.6) μm; anterior end of hyaline region (HBD) = 3.2 ± 0.5 (2.4-4.0) μm; tail = 81.4 ± 4.2 (75-94) μm; hyaline region (H) = 24.1 ± 2.6

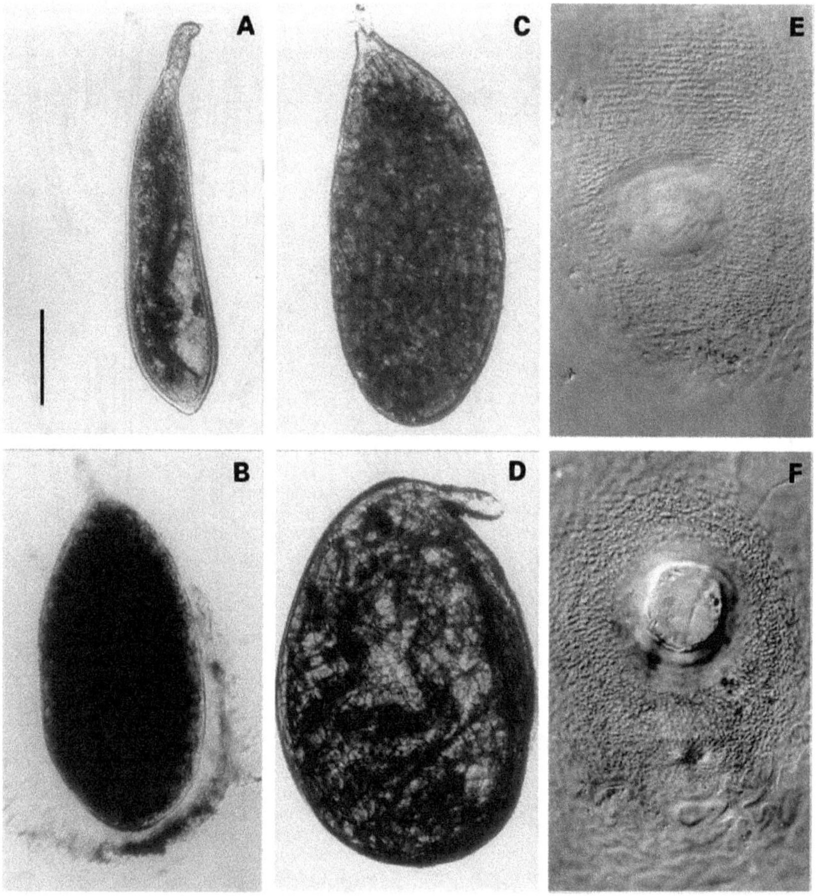

Fig. 96. Paradolichodera tenuissima. *A: Young female; B: Female with dissolved 'subcrystalline' layer; C: Cyst filled with eggs (holotype); D: Older cyst containing eggshells and a few juveniles; E: Female/young cyst posterior end showing vulval slit and developing circumfenestra; F: Cyst posterior end with vulval fenestra and anus (below). (Scale bar: A-D = 200 μm; E, F = 20 μm.) After Sturhan* et al. *(2007).*

(17.6-32) μm; hyaline region (as % of tail) = 30 ± 3.2 (21-38); H/HBD = 7.8 ± 1.4 (5.5-11.7).

– *Paratype eggs* (n = 25): L = 108±4.2 (98-117) μm; W = 38±1.6 (35-41) μm; L/W ratio = 2.9 ± 0.2 (2.5-3.3).

DESCRIPTION

Cyst

Elongate-ovoid, with rounded posterior end, yellowish to light brown, filled with eggs or hatched J2 easily visible through cuticle, neck narrow and mostly straight, offset. Cyst often covered with irregularly developed, heavy 'subcrystalline' layer. No egg sac observed. Vulva and anus terminal or subterminal, not protruding. Vulval area circumfenestrate, bullae and underbridge lacking. Anus small, indistinct, without distinctive surrounding cuticle pattern, located 12-30 μm dorsal to vulval fenestra margin. Cuticle finely striated in anterior part of body, from about mid-body striae gradually transforming into rows of punctations, heaviest and most irregular at posterior end. Posteriorly, pattern may be covered with irregular, faint, longitudinal ridges or reticulate structures. Cuticle 5-12 μm thick at mid-body.

Female

Body elongate to ovoid, posterior end rounded or occasionally weakly conoid. Anterior end tapering, mostly irregularly-shaped. Specimens often becoming creamy or yellowish during development. Layer of unknown origin (= subcrystalline layer?) often covering entire body. Cuticle with transition from striae to punctations as described for cyst. Labial disc squarish, offset, *ca* 5 μm across. Stylet generally slightly curved ventrally, knobs rounded, slightly directed posteriad, stylet base 5-6 μm wide. Median bulb of pharynx globular, 15-29 μm diam. with valve 10-11 μm long. Vulva and anus terminal or subterminal, not protruding.

Male

Body almost straight to C-shaped, posterior part not twisted about longitudinal axis. Cuticle annuli 1.5-2.0 μm wide at mid-body. Lateral field with four faint, equally spaced lines reducing toward anterior and posterior ends of body to irregular central lines or no line at all, outer ridges of lateral field mostly areolated. Deirids, postdeirids and phasmids lacking. Labial region rounded, flattened slightly anteriorly, offset from body, 10-11 μm diam., 5-6 μm high, bearing 6-8 annuli and faint oval labial disc. Labial framework heavily sclerotised. Cephalids located 2-3 annuli and 10-11 annuli posterior to anterior end. Stylet with conus shorter than shaft plus knobs, knobs rounded with flat or slightly posteriorly sloping anterior faces, generally well separated, forming cavity 6 μm wide and 2.5 μm high at stylet base. Median pharyngeal bulb elongate-oval, occupying less than half body diam., valve plates well developed, 6 μm long. Pharyngeal-intestinal valve indistinct. Dorsal pharyngeal gland lobe generally wide, occasionally filling body cavity, subventral gland lobes slender, on ventral side of body cavity. Hemizonid *ca* two cuticle annuli long, *ca* two body diam. from level of centre of median bulb. Excretory pore 6-12 μm posterior to hemizonid, 1.0-1.5 μm wide. Testis well developed, occupying 38-70% of body length. Spicules slightly arcuate with rounded tips, proximal part somewhat offset, gubernaculum almost straight, 9-12 μm long, wide, refractive, cloacal tube present. Tail varying from almost hemispherical to bluntly conoid, 4-10 μm long, often with irregular terminal process.

J2

Body very slender, mostly straight. Cuticle finely annulated; annuli less than 1 μm wide. Lateral field indistinct, visible only with SEM as irregular interruptions of cuticle annulation. Labial region offset, flatly rounded, 5.4-6.0 μm diam. and 2-3 μm high, with few indistinct annuli. Labial region includes fusion of adjacent submedial lips with one another and with labial disc. Labial framework with strong central sclerotisation and rounded at base. Stylet slender, three times labial region diam. long, conus and shaft + knobs of equal length, knobs well separated, rounded, with anterior faces flat or slightly sloping posteriorly. Stylet base 3.4-3.7 μm wide and 1.5-1.7 μm high. Dorsal gland orifice located *ca* two thirds of stylet length posterior to stylet. Cephalids indistinct. Median pharyngeal bulb elongate, filling body cavity, somewhat angular with well developed 4 μm long valves. Excretory pore posterior to level of centre of median bulb at a distance equal to *ca* three corresponding body diam. Hemizonid just anterior to excretory pore, two cuticle annuli long. Deirids absent. Pharyngeal glands long (40-46% of body length), filling body cavity, broadly rounded at posterior end, dorsal gland generally well separated from subventral glands. Genital primordium with 2-6 (mostly four) cells, 12-16 μm long and 5-7 μm diam., 62-70% of body length from anterior end. Tail long, very slender, hyaline region often not demarcated and containing granular inclusions, terminus finely rounded to pointed. Phasmid openings punctiform, located posterior to anus at *ca* one-third to one-half of tail.

Egg

Elongate, uniformly rounded at both ends, with smooth surface, eggshell *ca* 1.4 μm thick, appearing double layered. Body of developed J2 in embryonated eggs folded *ca* seven times.

TYPE HOST AND LOCALITY

Rhizosphere of the rush, *Eleocharis gracilis* R. Br. (Cyperaceae, Poales), growing at the edge of a small lagoon behind the coastal gravel wall. Rakaia Huts, Selwyn, Canterbury, New Zealand.

DISTRIBUTION

Oceania: New Zealand.

RELATIONSHIPS

Paradolichodera tenuissima is distinguished by the characters of the genus, and most distinctly by the extremely slender J2 stage. Cyst are elongate-ovoid with a rounded posterior end, the cuticle is light brown and translucent with heavy punctations and faint striation in the anterior end, the vulva region is circumfenestrate and the anus lacks fenestration. Bullae are lacking. On the basis of these features it is easily distinguished from all other described Heteroderidae species.

BIOCHEMICAL AND MOLECULAR CHARACTERISATION

The D2-D3 expansion segments of 28S rRNA and the ITS1 rRNA of *P. tenuissima* were sequenced and analysed by Subbotin *et al.* (2006) and Sturhan *et al.* (2007), respectively.

Nematology Monographs & Perspectives, 2010, Vol. 8A, 285-290

Appendix I

Techniques for recovery of cysts from soil samples

Extracting cysts from soil or other substrates requires procedures that may be based on sieving, gravity, flotation, centrifugation or combinations thereof. On the basis of these principles, many methods have been developed or adapted in different laboratories and they have been widely used with varying degrees of success for nematology surveys and other research. A widely used and often modified apparatus for cyst extraction, based in the flotation principle, was described by Fenwick (1940). Seinhorst (1964) described an elutriator especially designed for processing moist soil samples. A cyst extraction device, combining elutriation with sieving, was developed by Oostenbrink (1960). Excellent reviews are available on extraction techniques for cyst nematodes from soils and other materials (Shepherd, 1986; Southey, 1986; Turner, 1998; Bellvert *et al.*, 2008; Anon., 2009). There is no single method or device available that can be recommended as being most effective under all circumstances (Cooke *et al.*, 1983; Urek & Sirca, 2003). Herein, we briefly give descriptions of cyst extraction using: *i*) a screening technique with decantation followed by sieving of a soil sample using a water stream; *ii*) the Seinhorst elutriator; and *iii*) the Fenwick can.

SIEVING TECHNIQUE

Sieving is the most basic technique for extraction. Soil is mixed with a large volume of water and heavy particles are allowed to settle for a brief time. The supernatant, containing floating cysts is then poured through several sieves, the first sieve retaining large debris and the second one catching the cysts. The technique as described by Ayoub (1977) is given here with some modifications (Fig. 97). The following equipment and materials are required for a maximum 1 litre (l) of soil: one plastic bucket or steel pan with at least 3 l capacity, 850 μm aperture and 180 μm sieves

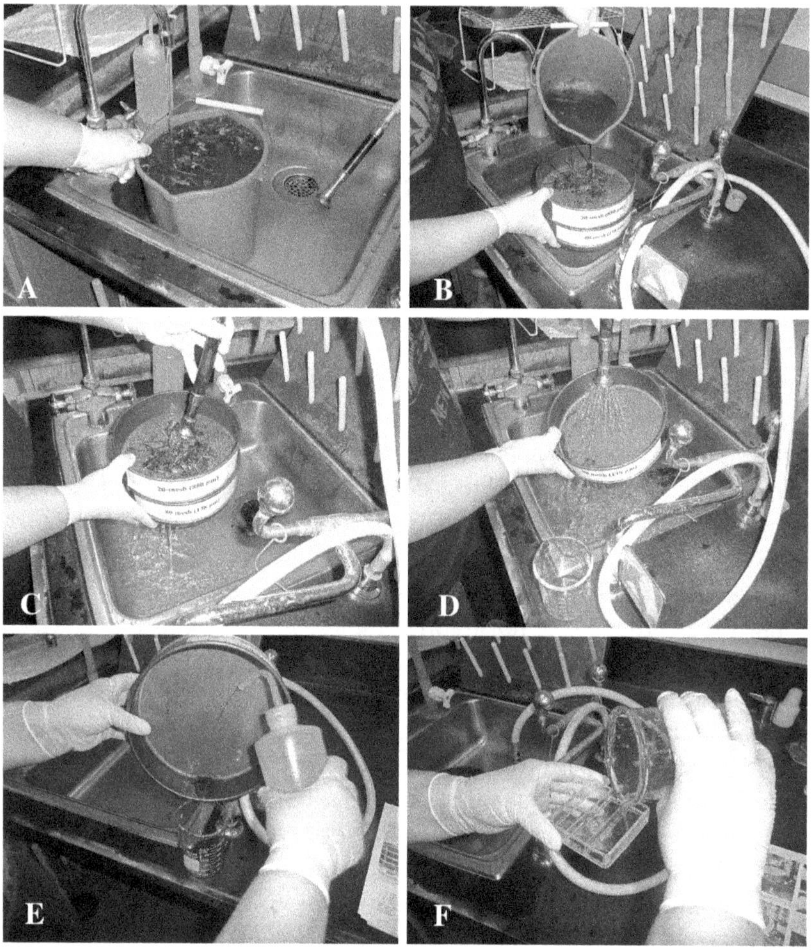

Fig. 97. *Steps of sieving technique process for extraction of cysts from soil samples.*

of 20-25 cm, stereoscopic dissecting microscope, 250 ml beaker, plastic scanning dishes with partitions, dissecting needles and mini tweezers. A 180 μm rather than 250 μm sieve for the second screen is described in some protocols because it has been shown that cysts of some species can pass through a 250 μm sieve (Sharma & Nene, 1986; Hisai *et al.*, 2003). Kim and Riggs (1995) described a wet sieving technique with a list of required items for extracting cysts under field conditions.

Procedure

In order to obtain optimum cyst recovery, the soil must be fully air-dried and subsequently crushed or passed through a 4 mm mesh screen to break down clods and remove stones. Dry cysts will float. However, wet cysts sink in water and this procedure might not be effective for extracting such cysts.

1. Place approximately 1000 cm^3 dry soil in a plastic bucket and create a slurry with a strong stream of water (Fig. 97A).
2. When the solution is uniform, wait 15 s for some of the heavy sediment to settle, lift the bucket and gently pour all floating material through the 850 μm and 180 μm sieves nested together with the 850 μm sieve on top (Fig. 97B).
3. Wash the debris on the 850 μm sieve so that plant material does not clog the mesh and thus prevent cysts from moving through the screen. Discard the coarse material caught on the 850 μm sieve (Fig. 97C).
4. Wash the material retained on the 180 μm sieve with a gentle stream of water from the hose or fogger nozzle. Hold the screen at an angle to concentrate the cysts and debris (Fig. 97D).
5. Backwash material into a 250 ml beaker using a minimal amount of water. Where applicable, it is important to label the beaker with the soil sample number (Fig. 97E).
6. Pour the sample from the beaker into a shallow scanning dish. If there is a large amount of material it can be divided into several dishes (Fig. 97F).
7. If cysts are not readily apparent, search for them using the dissecting microscope. Use the dissecting needle to spread clumps of material and to aid closer inspection of cyst-like objects. Transfer suspected cysts carefully using small tweezers into a sample vial.
8. Upon completion of the procedure, wash all equipment very thoroughly to avoid contamination of subsequent samples.

SEINHORST ELUTRIATOR

This technique was designed by Seinhorst (1964) and is herein given as described by Viaene (2003). The apparatus is made of a funnel in which an upward current keeps the cyst afloat while heavier soil particles sink (Fig. 98). The upward current is kept constant with the aid of a flow regulator (F). Small, light cysts are rinsed into a collar (similar to the Fenwick can) (D) and collected on a 250 μm sieve (M). Heavier

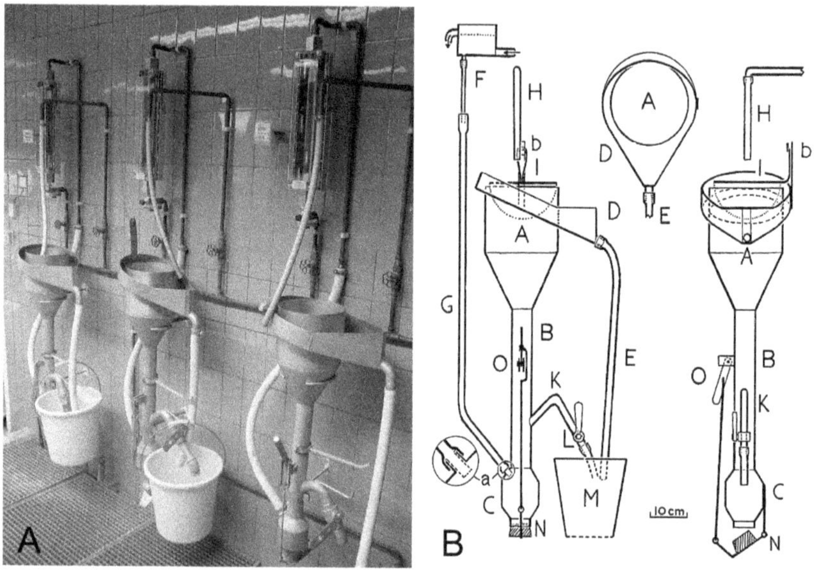

Fig. 98. *A: Photo of Seinhorst elutriator (ILVO, Belgium); B: Diagram of Seinhorst elutriator for separation of* Heterodera *cysts from moist soil. Left and right: as seen from the side. Centre: as seen from above. In circle: inlet enlarged to show placement of sieve for breaking water current (Seinhorst, 1964).*

full cysts remain suspended in the water in the upper half of the funnel (parts A and B) and are released through a side outlet (K and L) onto the same sieve. The sieve (M) is built in a bucket shape because of the large amount of water that leaves the side outlet.

Procedure

This method is suitable for cyst extraction from wet and dried soil samples of up to 1 kg. One person can process a large quantity of samples. The apparatus is difficult and expensive to construct; its operation requires a large quantity of water. Although not described here, Been *et al.* (2007) described an alternative scaled up Seinhorst elutriator, which can process up to 2.5 kg of soil.

1. Place the sieve (M) so that it can capture both the water from the collar and from the side outlet. Open the tap to let water into the apparatus through G, passing the flow meter (F). Adjust the upward current to 3 l per min (F) and maintain this flow rate during the whole process.

2. Open the other tap to let water in the apparatus through H. Adjust the water flow to 3 l per min. Close the main outlet (N) and the side outlet (L) and fill the apparatus with water (G). When the funnel is about half full, rinse the (wet) sample with water through the coarse (2 mm) sieve (I) and into the apparatus, using tap H. Make sure the sample is washed through the sieve before the water starts running in the collar (D). As soon as the water runs in the collar, let the upward current (through G) run for at least 2 min (use timer).
3. Rinse the collar thoroughly (with H), open the side outlet (L) and collect cysts and debris on the sieve. Transfer these from the sieve to the bucket and then concentrate in a smaller sieve or plate and leave to dry.
4. Open both outlets (N and L) and clean the apparatus.

THE FENWICK CAN

The Fenwick can is widely used and is familiar to most nematologists (Fig. 99). The Fenwick can has been adapted by different manufacturers and laboratories and, although it can vary in dimensions, it basically is a metal can that tapers toward the top and has an inclined base. The can has a sloping collar below the rim. Dry soil is placed in a receptor with a <2 mm screen over the can and washed through into the can.

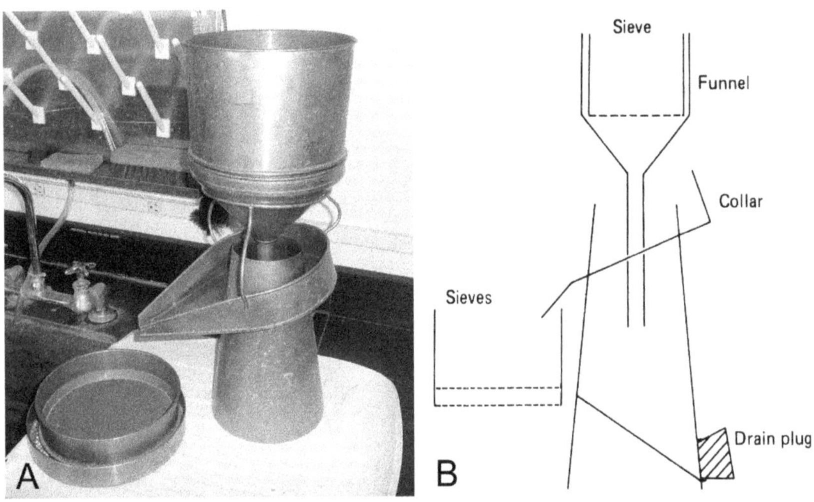

Fig. 99. *A: Photo of Fenwick can (UCD, USA); B: Vertical section diagram of Fenwick can modified by Oostenbrink. After Shepherd (1986).*

Heavy soil particles fall to the bottom of the can, whereas dried cysts and light soil debris float to the surface where they are siphoned over the rim to collecting sieves. These sieves include an 840 μm on top followed below by a 250 μm sieve. The cysts are collected on the 250 μm sieve. Most of the cysts from the soil sample are collected at this stage. Then, the funnel above the can is removed and the soil at the base of the can is elutriated by means of water flowing rapidly through metal tube for 1 min. The tube is inserted deep into the can to stir the sediment and so release any trapped cysts (Fenwick, 1940; Shepherd, 1986; Turner, 1998; USDA, 2008; Anon., 2009). Caswell *et al.* (1985) described a Fenwick technique and separation of cysts by using an ethanol-glycerin mixture.

Nematology Monographs & Perspectives, 2010, Vol. 8A, 291-294

Appendix II

Protocol for preparation of the vulval cones of cyst nematodes

Morphological identification of cyst nematodes requires observation of diagnostic structures, often including the vulval-anal terminal area. To observe all these structures clearly, the posterior region should be dissected from the cyst and mounting on a slide (Figs 100, 101). Viewing diagnostic structures generally requires that the vulval-anal terminal area be mounted with the posterior end up toward the cover glass, and maintaining this position requires special mounting materials. The excised posterior region is then processed to one of several embedding and mounting materials, such as Euparal, glycerin agar, glycerin-gelatine or Canadian balsam. The terminal regions of the round cyst species (*Globodera*, *Punctodera*, *Dolichodera* and *Paradolichodera*), which are almost flat, can be placed in a drop of glycerin. Several methods are outlined including, for example, Correia and Abrantes (1997), who describe mounting *Heterodera* cysts in glycerin-agar. Herein we give a short description of protocols for slide preparation, including embedding vulval cones of lemon-shaped cysts in glycerin-gelatine. This method has an advantage because it is allows relative ease in adjusting orientation. Glycerin-gelatine is prepared with gelatin, glycerol and phenol and is very similar in composition to Kaiser's glycerin jelly. The product must be warmed to 55-60°C so that it is liquefied during mounting. At approximately 37°C the product returns to a semi-solid state and thus secures the cone in position. Although glycerin-gelatine can be easily prepared in any laboratory it can also be ordered from Sigma-Aldrich. Phenol is highly toxic to humans and it is recommended that this, and glycerin-agar, be handled under ventilated conditions, such as a fume hood.

Preparation of glycerin-gelatine (after Kirjanova & Krall, 1968 with modification)

- Gelatin 3.5 g
- Distilled water 21 ml

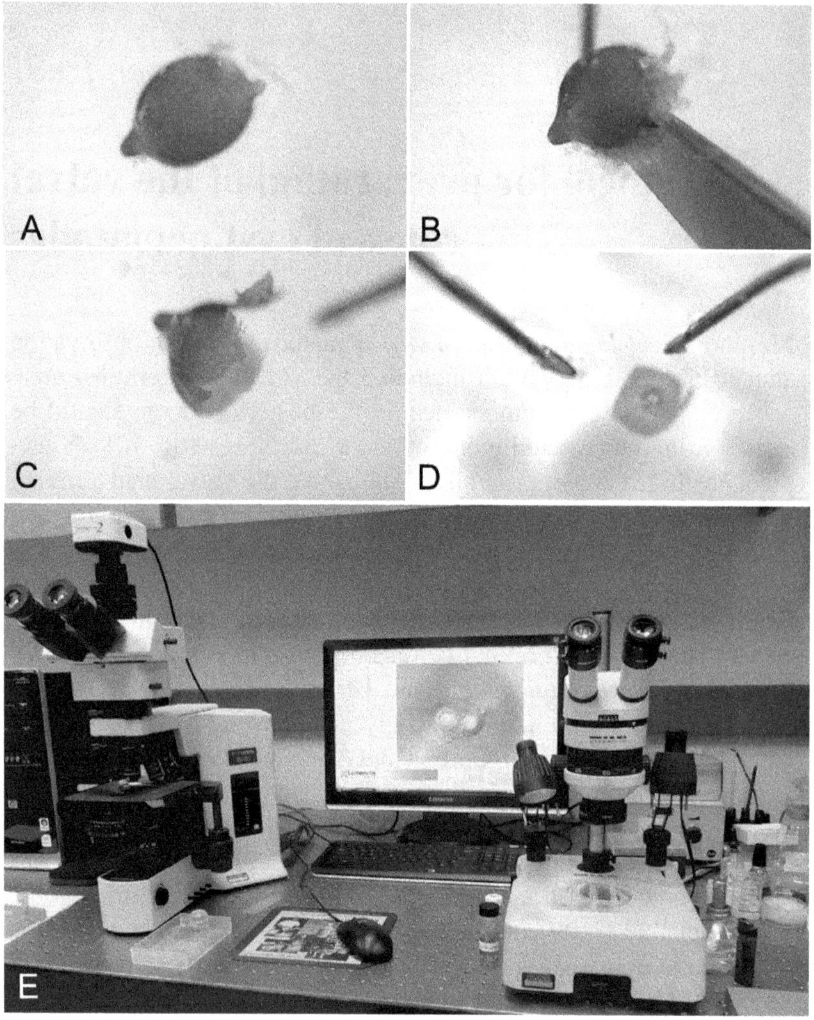

Fig. 100. *Techniques for preparing cyst cones for light microscope observation. A-D: Steps in cutting cone area of cyst; E: General view of nematological laboratory with light microscopes.*

- Glycerin 25 ml
- Phenol 0.3 g

– Measure 21 ml distilled water into a glass flask, add 3.5 g gelatin and leave for a few min.

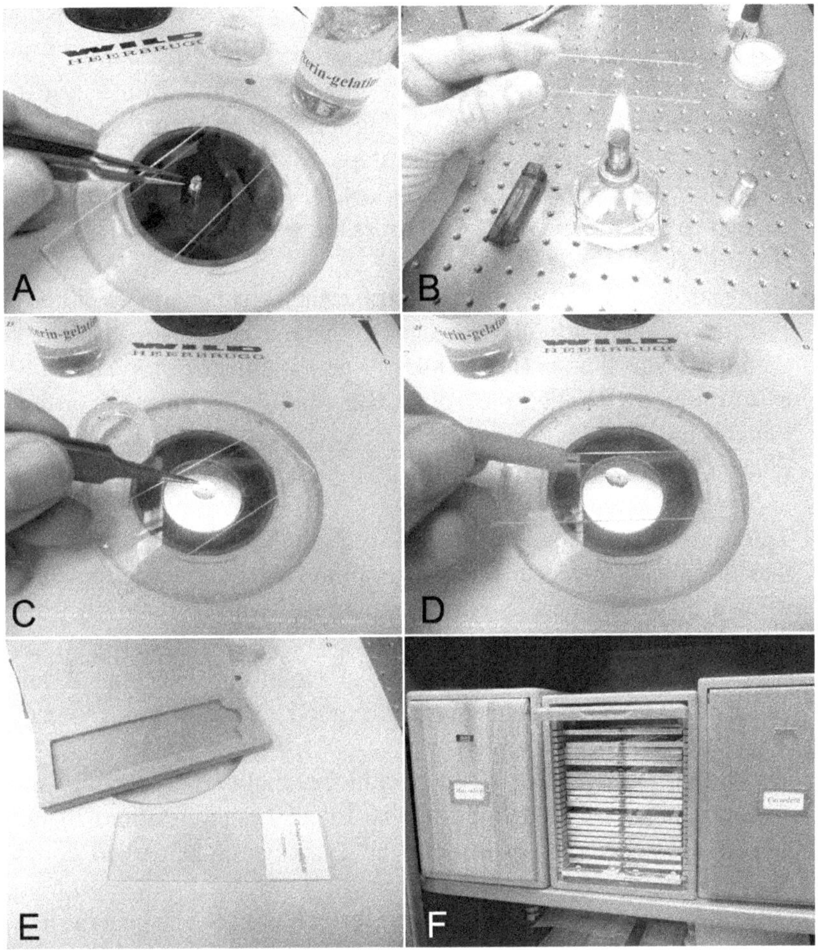

Fig. 101. *Techniques for preparing cyst cones for light microscope observation. A-E: Steps in mounting of vulval cone in glycerin-gelatine; F: Slide storage boxes.*

– Add 25 ml glycerin using a pipette.

– Heat the mixture, using a water bath and stirring carefully, until it is completely dissolved.

– Add 0.3 g phenol.

– Store at room temperature in a brown bottle.

Procedure

Dry cysts should be soaked in water for 15 min up to several hours.

1. Using small tweezers, place a cyst in a drop of water on a plastic Petri plate on the stage of a dissecting stereomicroscope (Fig. 100A).
2. Using a sharp dissecting needle, gently make a hole in the cyst to relieve pressure; this makes damaging fenestral or other vulval cone structures less likely in subsequent steps.
3. Using a scalpel with a sharp blade, cut off the vulval cone, trim the cut end so that it is between 5-10 times the width of the fenestral area (Fig. 100B, C).
4. Using very fine forceps, clean the surface and any adhering soft body contents, taking particular care not to damage diagnostic structures within the vulval cone (Fig. 100D). Where pigmentation is so heavy as to obscure contents, the vulval cone can be bleached for a few minutes in H_2O_2.
5. Place a piece of glycerin-gelatine on a glass slide and gently heat it to 55-60°C (Fig. 101A, B).
6. Transfer the vulval cone using a dissecting needle or mini tweezers into the melted glycerin-gelatine drop and then correct its position using a dissecting needle. Any air bubbles should be removed from the vulval cone (Fig. 101C, D).
7. After the glycerin-gelatine solidifies in 15-20 min, place a 12 mm cover glass on top, taking care to make good contact near the apex of the vulval cone (Fig. 101E).
8. The posterior region of cysts should be analysed under the microscope in transmitted light at ×40 to ×80 magnification. Image analysis systems can assist with the examination of vulval cones (Fig. 100F).
9. Glass slides carrying the mounted posterior region can be kept in slide storage boxes (Fig. 101F).

Nematology Monographs & Perspectives, 2010, Vol. 8A, 295-297

Appendix III

Protocol for identification of *Globodera rostochiensis* and *G. pallida* using IEF of proteins with Pharmacia PhastSystem®

GENERAL INFORMATION

The Pharmacia PhastSystem® (Uppsala, Sweden) consists of a Separation-Control unit for system control and electrophoresis, and a Development Unit for gel staining (Fig. 32A). Multi-step programming capability allows gels to be pre run, loaded and run automatically. Controlled parameters include temperature, voltage, current, power, and volt-hours. Electrophoresis is performed on a thermostatic plate capable of maintaining temperatures accurately from 0 to +70°C.

The method identification of cysts of *G. rostochiensis* and *G. pallida* by protein separation with isoelectric focusing (IEF) using cysts was originally described by Fleming and Marks (1983) and modified by Karssen *et al.* (1995). Cysts filled with eggs and juveniles are the protein source. One or several cysts are required for this identification. The protocol described here was developed by Karssen *et al.* (1995).

Sample preparation

1. Cysts are presoaked in 1% glycerin for at least 12 h at room temperature (21°C).
2. PhastGel sample-well stamp is placed on ice and used for protein extraction.
3. Small sample wells are each filled with one presoaked cyst and 0.7 μl of 1% glycerin.
4. Using a small glass rod the cysts are crushed and macerated for 15 s per well to release the egg contents.
5. The Pharmacia broad pI calibration kit (mixtures of stable, salt-free, highly purified proteins) with pI 3.5-9.3 is used as reference and for pI measuring. Marker (0.7 μl) is added in one or two middle wells.

6. Well contents are loaded on the 12/0.3 sample applicator (0.3 μl sample/well) and inserted in the middle position of the apparatus applicator arm.

IEF run

IEF is run with PhastGel IEF 3-9 or 5-8 with pH ranges of 3-9 and 5-8, respectively. The following standard programs are used:

i) PhastGel IEF 3-9

Sample appl. down at	1.2	0 Vh			
Sample appl. up at	1.3	0 Vh			
Sep 1.1	2000 V	2.5 mA	3.5 W	15°C	75 Vh
Sep 1.2	200 V	2.5 mA	3.5 W	15°C	15 Vh
Sep 1.3	2000 V	2.5 mA	3.5 W	15°C	410 Vh

ii) PhastGel IEF 5-8

Sample appl. down at	2.2	0 Vh			
Sample appl. up at	2.3	0 Vh			
Sep 2.1	2000 V	2.0 mA	3.5 W	15°C	75 Vh
Sep 2.2	200 V	2.0 mA	3.5 W	15°C	15 Vh
Sep 2.3	2000 V	5.0 mA	3.5 W	15°C	510 Vh

For more specific information about IEF running conditions see the Pharmacia manual. The total electrophoresis time is approximately 30 min.

Staining

An electrophoresed PastGel is stained with the PhastGel Silver Kit. The automated developer is used to stain the gels. The procedures are performed according to Pharmacia PhastSystem protocols. This improved staining method is described in the silver kit instruction manual. The background reducer step (Table 1 of the manual, program step No. 14) is changed from 1.5 min to 1.3 min, while 5 ml acetic acid is added instead of Tris-HC1.

RESULT INTERPRETATION

Figure 33 shows the overall picture and the essential differences between the two *Globodera* species. *Globodera rostochiensis* shows a stable major band at p*I* 5.7 (C) and a weaker band at p*I* 8.7 (D), which is absent in the *G. pallida* pattern. For *G. pallida* the stable major band occurs at p*I* 5.9 (A) and a unique weaker band at p*I* 6.9 (B). The differences between these bands for the same samples with a narrower pH interval (PhastGel 5-8) are shown in Figure 33A and B. Single cysts of *G. tabacum* showed band at p*I* 5.8, and *G. artemisiae* showed two major bands between p*I* 5.85 and p*I* 5.8 (Sumiya *et al.*, 2002).

Nematology Monographs & Perspectives, 2010, Vol. 8A, 299-301

Appendix IV

Protocols for identification of cyst nematode species with PCR-ITS-RFLP using AB28 and TW81 primers

General information

These protocols are designed to amplify the internal transcribed spacer (ITS1-5.8S-ITS2) regions of the rRNA gene and characterise the obtained amplified product using a series of restriction enzymes. Identification of nematodes using PCR-RFLP requires standard molecular biological equipments: DNA thermal cycler, set of calibrated pipettes, vortex, centrifuge, horizontal gel electrophoresis box with power supply, freezer, microwave oven, UV light box with camera and water bath. Juveniles, eggs, females, cysts filled with juveniles and eggs are the DNA source. A single cyst or one or several juveniles and eggs can be used for identification. The present protocols are described in several publications (Subbotin *et al.*, 2000a; Amiri *et al.*, 2002; Tanha Maafi *et al.*, 2003) and Work Instructions (Nakhla, 2008a, b), and are given here with some modifications.

Sample preparation and DNA extraction

1. Transfer one cyst or several juveniles into an Eppendorf tube containing 18 μl or more 1X PCR buffer and crush their contents using a microhomogeniser or ultrasonicator. Nakhla (2008a) recommended using a Mini BeadBeater-1 (BioSpec products) for destroying juveniles. Sample should be homogenised with one glass bead in a tube for 30 s using a Mini BeadBeater-1 (time setting 3 and speed setting 25). Alternatively, crush cyst and cut juveniles or eggs using scalpel under binocular microscope and transfer nematode fragments into a tube using a pipette.

2. Add 2 μl of proteolytic enzyme proteinase K (600 μg ml^{-1}) into this tube. Incubate the tube at 65°C for 1 h and 95°C for 15 min.
3. Centrifuge the tube for 1 min at 13000 rpm.
4. Use supernatant with DNA immediately, or store overnight at 4°C or at −20 to −80°C for longer period.

PCR

Using Qiagen Taq *PCR Core Kit*

Prepare PCR master mixture (23 μl per tube) using Qiagen *Taq* PCR Core Kit as described by Subbotin *et al.* (2000a) and Tanha Maafi *et al.* (2003):

Double distilled water – 14.6 μl
Q-Solution – 5 μl
10× PCR buffer – 2.5 μl
dNTP-Mix (10 mM each) – 0.5 μl
Forward primer TW81 (1.5 μM) – 0.15 μl
Reverse primer AB28 (1.5 μM) – 0.15 μl
Taq DNA Polymerase (5 units μl^{-1}) – 0.1 μl

1. Distribute 23 μl PCR master mixture using a pipette with filter in 0.2 μl thin-wall PCR tubes.
2. Add 2 μl or more of DNA suspension in a tube.
3. A negative control (a tube without DNA) should be included in every assay.
4. Run PCR with following temperature profile:
   4 min at 94°C
   35 cycles of 1 min at 94°C, 1.5 min at 55°C, 2 min at 72°C
   10 min at 72°C

Using Invitrogen Platinum Taq *Polymerase*

Prepare PCR master mixture (45 μl per tube) using with Invitrogen Platinum *Taq* Polymerase as described by Nakhla (2008b):

Molecular grade water – 35.2 μl
10× Platinum *Taq* buffer – 5.0 μl
$MgCl_2$ (50 mM) – 1.5 μl
dNTP mixture (10 mM each) – 1 μl
Forward primer TW81 (10 μM) – 1 μl
Reverse primer AB28 (10 μM) – 1 μl

Platinum *Taq* DNA Polymerase – 0.3 μl

1. Distribute 45 μl PCR master mixture using a pipette with filter in 0.2 μl thin-wall PCR tubes.
2. Add 5 μl DNA suspension in a tube.
3. A negative control (a tube without DNA) should be included in every assay.
4. Run PCR with following temperature profile:
   2 min at 95°C
   35 cycles of 30 min at 95°C, 30 min at 55°C, 1 min 30 s at 72°C
   5 min at 72°C
5. Run PCR product on a 0.8% buffered agarose gel and visualise under UV light.
6. The length of obtained amplified product should be fallen in the range of 970-1100 bp.
7. Use PCR product for digestions and store remaining PCR product at −20°C until next use.

RFLP

1. Prepare restriction mixtures (concentration per 15 μl reaction) for several enzymes according to supplier's conditions.
2. Distribute restriction mixtures using pipette in 0.2 μl PCR tubes.
3. Add 2 μl or more (depending on the PCR concentration) PCR product in each tube.
4. Incubate a tube in water bath from 5 min to several h and at temperature for digestion according to the supplier's recommendations.
5. Run digested DNA mixture in a 1.5% agarose gel, visualise and photograph resulting fragments under UV light.

RESULT INTERPRETATION

Figure 39 in Volume 8A, and Figures 1-6 in Volume 8B show the RFLP diagnostic profiles for several species of cyst nematode. Lengths of RFLP fragments generated by several restriction enzymes for cyst nematodes are given in Tables 15 and 25 in Volume 8A and Table 4 in Volume 8B.

References

ADAM, W. (1932). Note sur *Heterodera schachtii* Schm. parasite des cactus. *Bulletin du Musée Royal d'Histoire Naturelle de Belgique* 8, 1-10.

ADAMS, B.J. (1998). Species concepts and the evolutionary paradigm in modern nematology. *Journal of Nematology* 30, 1-21.

AL HAZMI, A.S., COOK, R. & IBRAHIM, A.A.M. (2001). Pathotype caracterisation of the cereal cyst nematode, *Heterodera avenae*, in Saudi Arabia. *Nematology* 3, 379-382.

AMBROGIONI, L. (1969). Due casi di infestazioni miste da nematodi dei gen. *Heterodera* e *Meloidogyne*. *Redia* 51, 159-168.

AMBROGIONI, L. & D'ERRICO, F.P. (1998). Studies on the biology of *Globodera tabacum* in southern Italy. *Nematologia Mediterranea* 26, 117-121.

AMIRI, S., SUBBOTIN, S.A. & MOENS, M. (2001). An efficient method for identification of the *Heterodera schachtii sensu stricto* group using PCR with specific primers. *Nematologia Mediterranea* 29, 241-246.

AMIRI, S., SUBBOTIN, S.A. & MOENS, M. (2002). Identification of the beet cyst nematode *Heterodera schachtii* by PCR. *European Journal of Plant Pathology* 108, 497-506.

ANDERSEN, S. (1959). Resistance of barley to various populations of the cereal root eelworm (*Heterodera avenae*). *Nematologica* 4, 91-98.

ANDERSSON, S. & ANDERSEN, K. (1982). Suggestions for determination and terminology of pathotypes and genes for resistance in cyst-forming nematodes, especially *Heterodera avenae*. *EPPO Bulletin* 12, 379-386.

ANDRÁSSY, I. (1976). *Evolution as a basis for the systematization of nematodes*. London, UK, Pitman Publishing, 288 pp.

ANDRÉS, M.F., ROMERO, M.D., MONTES, M.J. & DELIBES, A. (2001). Genetic relationships and isozyme variability in the *Heterodera avenae* complex determined by isoelectrofocusing. *Plant Pathology* 50, 270-279.

ANGIOSPERM PHYLOGENY GROUP (2003). An update of the Angiosperm Phylogeny Group classification for the orders and families of flowering plants: APG II. *Botanical Journal of the Linnean Society* 141, 399-436.

ANON. (2004). *Globodera tabacum. Distribution maps of plant diseases.* Issue: October (1st edition). Map 929. Wallingford, UK, CABI Publishing.

ANON. (2009). *Globodera rostochiensis* and *Globodera pallida*. *EPPO Bulletin* 39, 354-368.

AWAN, F.A. & HOMINICK, W.H. (1982). Observation on tanning of the potato cyst nematode, *Globodera rostochiensis*. *Parasitology* 85, 61-71.

AYOUB, S.M. (1977). *Plant nematology, an agricultural training aid.* Department of Food and Agriculture, Division of Plant Industry, State of California, 157 pp.

BACIC, J., STARE, B.G., SIRCA, S. & UREK, G. (2008). Analyses of *Globodera rostochiensis* and *G. pallida* populations from Serbia by morphometrics and real-time PCR. *Russian Journal of Nematology* 16, 63-65.

BAKER, A.D. (1957). Notes on some nematodes in Canada, 1956. *Canadian Insect Pest Review* 34, 120-122.

BAKKER, J., FOLKERTSMA, R.T., ROUPPE VAN DER VOORT, J.N., DE BOER, J.M. & GOMMERS, F.J. (1993). Changing concepts and molecular approaches in the management of virulence genes in potato cyst nematodes. *Annual Review of Phytopathology* 31, 169-190.

BALDWIN, J.G. (1982). Fine structure of the esophagus of males of *Sarisodera hydrophila* (Heteroderoidea). *Journal of Nematology* 14, 279-291.

BALDWIN, J.G. (1983). Fine structure of body wall cuticle of females of *Meloidodera charis*, *Atalodera lonicerae* and *Sarisodera hydrophila* (Heteroderidae). *Journal of Nematology* 15, 370-381.

BALDWIN, J.G. (1985). Fine structure of the phasmid of second-stage juveniles of *Heterodera schachtii* (Tylenchida: Nematoda). *Canadian Journal of Zoology* 63, 534-542.

BALDWIN, J.G. (1992). Evolution of cyst and non-cyst-forming Heteroderinae. *Annual Review of Phytopathology* 30, 271-290.

BALDWIN, J.G. & BELL, A.H. (1985). *Cactodera eremica* n. sp., *Afenestrata africana* (Luc *et al.*, 1973) n. gen., n. comb., and an emended diagnosis of *Sarisodera* Wouts & Sher, 1971 (Heteroderidae). *Journal of Nematology* 17, 187-201.

BALDWIN, J.G. & EDDLEMAN, C.D. (1992). Fine structure of the body wall cuticle of females of *Bellodera utahensis* and *Ekphymatodera thomasoni* (Sarisoderini, Heteroderinae). *Canadian Journal of Zoology* 70, 268-273.

BALDWIN, J.G. & HIRSCHMANN, H. (1973). Fine structure of cephalic sense organs in *Meloidogyne incognita* males. *Journal of Nematology* 5, 285-302.

BALDWIN, J.G. & HIRSCHMANN, H. (1975a). Body wall fine structure of the anterior region of *Meloidogyne incognita* and *Heterodera glycines* males. *Journal of Nematology* 7, 175-193.

BALDWIN, J.G. & HIRSCHMANN, H. (1975b). Fine structure of cephalic sense organs in *Heterodera glycines* males. *Journal of Nematology* 7, 40-53.

BALDWIN, J.G. & HIRSCHMANN, H. (1976). Comparative fine structure of the stomatal region of males of *Meloidogyne incognita* and *Heterodera glycines*. *Journal of Nematology* 8, 1-17.

BALDWIN, J.G. & MUNDO-OCAMPO, M. (1991). Heteroderinae, cyst and non-cyst-forming nematodes. In: Nickle, W.R. (Ed.). *Manual of agricultural nematology*. New York, NY, USA, Marcel Dekker Inc., pp. 275-362.

BALDWIN, J.G. & SCHOUEST JR, L.P. (1990). Comparative detailed morphology of the Heteroderinae Filipjev & Schuurmans Stekhoven, 1941, *sensu* Luc *et al.* (1988): Phylogenetic systematics and revised classification. *Systematic Parasitology* 15, 81-106.

BALDWIN, J.G., HIRSCHMANN, H. & TRIANTAPHYLLOU, A.C. (1977). Comparative fine structure of the esophagus of males of *Heterodera glycines* and *Meloidogyne incognita*. *Nematologica* 23, 239-252.

BALDWIN, J.G., MUNDO-OCAMPO, M. & MCCLURE, M.A. (1997). *Cactodera salina* n. sp. from the Estuary Plant, *Salicornia bigelovii*, in Sonora, Mexico. *Journal of Nematology* 29, 465-473.

BALDWIN, J.G., RAGSDALE, E.J. & BUMBARGER, D. (2004). Revised hypotheses for phylogenetic homology of the stomatostylet. *Nematology* 6, 23-632.

BATES, J.A., TAYLOR, E.J., GANS, P.T. & THOMAS, J.E. (2002). Determination of relative proportions of *Globodera* species in mixed populations of potato cyst nematodes using PCR product melting peak analysis. *Molecular Plant Pathology* 3, 153-161.

BEEN, T.H., VAN BEKKUM, P.J., VAN BEERS, T.G. & BENIERS, A. (2007). A scaled-up Seinhorst elutriator for extraction of cyst nematodes from soil. *Nematology* 9, 431-435.

BEHRENS, E. (1972). [Morphological investigation of *Heterodera* spp. on *Solanum tuberosum*.] *Probleme der Phytonematologie Vortrage anlasslich der 11. Tagung über Probleme der Phytonematologie im Institut für Pflanzenzuchtung Gross Lusewitz der Akademie der Landwirtschaftswissenschaften der Deutschen Demokratischen Republik am 2 Juni 1972*, pp. 76-85.

BEHRENS, E. (1975). *Globodera* Skarbilovic, 1959, eine selbstandige Gattung in der Unterfamilie Heteroderinae Skarbilovic, 1947 (Nematoda: Heteroderidae). *Vortragstagung zu Aktuellen Problemen der Phytonematologie am 29.5.1975 in Rostock. Manuskriptdruck der Vortrage. Biologische Gesellschaft der DDR, Sektion Phytopathologie und Universitat, Rostock*, pp. 12-26.

BEKAL, S., GAUTHIER, J.P. & RIVOAL, R. (1997). Genetic diversity among a complex of cereal cyst nematodes inferred from RFLP analysis of the ribosomal internal transcribed spacer region. *Genome* 40, 479-486.

BÉLAIR, G. & MILLER, S. (2006). First report of *Globodera tabacum* infecting tobacco plants in Quebec, Canada. *Plant Disease* 90, 527.

BELLVERT, J., CROMBIE, K. & HORGAN, F.G. (2008). Effect of sample size on cyst recovery using flotation methods: Recommendations for soil processing during EU monitoring of potato cyst nematodes (*Globodera* spp.). *EPPO Bulletin* 38, 205-210.

BENDEZU, I.F., EVANS, K., BURROWS, P.R., DE POMERAI, D. & CANTO-SAENZ, M. (1998). Inter and intra-specific variability of the potato cyst nematodes *Globodera pallida* and *G. rostochiensis* from Europe and S. America using RAPD-PCR. *Nematologica* 44, 49-61.

BERGE, J.B., DALMASSO, A., PERSON, F., RIVOAL, R. & THOMAS, D. (1981). Isoestérases chez le nématode *Heterodera avenae*: 1-Polymorphisme chez différentes races françaises. *Revue de Nématologie* 4, 99-105.

BERNARD, E., DONALD, P.A., HANDOO, Z.A., HEINZ, R. & POWERS, T. (2007). Characterization of a new species of cyst nematode parasitizing corn. *Journal of Nematology* 39, 74.

BERT, W., LELIAERT, F., VIERSTRAETE, A.R., VANFLETEREN, J.R. & BORGONIE, G. (2008). Molecular phylogeny of the Tylenchina and evolution of the female gonoduct (Nematoda: Rhabditida). *Molecular Phylogenetics and Evolution* 48, 728-744.

BISHNOI, S.P., SHASHI SINGH, SATISH MEHTA & BAJAJ, H.L. (2004). Isozyme patterns of *Heterodera avenae* and *H. filipjevi* populations of India. *Indian Journal of Nematology* 34, 33-36.

BLAXTER, M.L., DE LEY, P., GAREY, J.R., LIU, L.X., SCHELDEMAN, P., VIERSTAETE, A., VANFLETEREN, J.R., MACKEY, L.Y., DORRIS, M., FRISSE, L.M. ET AL. (1998). A molecular evolutionary framework for the phylum Nematoda. *Nature* 392, 71-75.

BLOK, V.C. & PHILLIPS, M.S. (1995). The use of repeat sequence primers for investigating genetic diversity between populations of potato cyst nematodes with differing virulence. *Fundamental and Applied Nematology* 18, 575-582.

BLOK, V.C., PHILLIPS, M.S. & HARROWER, B.E. (1997). Comparison of British populations of potato cyst nematodes with populations from continental Europe and South America using RAPDs. *Genome* 40, 286-293.

BLOK, V.C., MALLOCH, G., HARROWER, B., PHILLIPS, M.S. & VRAIN, T.C. (1998). Intraspecific variation in ribosomal DNA in populations of the potato cyst nematode *Globodera pallida*. *Journal of Nematology* 30, 262-274.

BORGONIE, G., CLAEYS, M., DE WAELE, D. & COOMANS, A. (1995). Ultrastructure of the intestine of the bacteriophagous nematoes *Caenorhabditis elegans*, *Panagrolaimus superbus* and *Acrobeloides maximus* (Nematoda: Rhabditida). *Fundamental and Applied Nematology* 20, 173-186.

BOSSIS, M. & MUGNIÉRY, D. (1993). Specific status of six *Globodera* parasites of solanaceous plants studied by means of two-dimensional gel electrophoresis with a comparison of gel patterns by a computed system. *Fundamental and Applied Nematology* 16, 47-56.

BOSSIS, M. & RIVOAL, R. (1989). Esterase-polymorphism in *Heterodera avenae* Woll – interfield and intrafield population variation. *Nematologica* 35, 331-339.

BOUWMAN, L.A. & ROSS, H. (1972). Differentiation between *Heterodera rostochiensis* and an undescribed allied species by female colour, morphometrics and pathogenicity. *Nematologica* 18, 265-269.

BRODIE, B.B. (1984). Nematode parasites of potatoes. In: Nickle, W.R. (Ed.). *Plant and insect nematodes*. New York, NY, USA, Marcel Dekker Inc., pp. 167-212.

BROWN, E.B. (1969). Assesment of the damage caused to potatoes by potato cyst eelworm, *Heterodera rostochiensis* Woll. *Annals of Applied Biology* 63, 493-502.

BROWN, G., CALLOW, R.K., GREEN, C.D., JONES, F.G.W., RAYNER, J.H., SHEPHERD, A.M. & WILLIAMS, T.D. (1971). The structure, composition and origin of the subcrystalline layer in some species of the genus *Heterodera*. *Nematologica* 17, 591-599.

BRZESKI, M.W. (1998). *Nematodes of Tylenchida in Poland and temperate Europe*. Warsaw, Poland, Muzeum i Instytut Zoologii Polska Akademia Nauk, 397 pp.

BRZESKI, M.W., LAUGHLIN, C.W., VARGAS JR, J.M., MCCALLU, D.K. & HANNA, M.K. (1971). Occurrence of cyst nematodes (*Heterodera* spp.) in Michigan. *Plant Disease Reporter* 55, 399.

BULMAN, S.R. & MARSHALL, J.W. (1997). Differentiation of Australasian potato cyst nematode (PCN) populations using the polymerase chain reaction (PCR). *New Zealand Journal of Crop and Horticultural Science* 25, 123-129.

BUMBARGER, D.J., CRUM, J., ELLISMAN, M.H. & BALDWIN, J.G. (2006). Three dimensional reconstruction of the nose epidermal cells in the microbial feeding nematode, *Acrobeles complexus* (Nematoda: Rhabditida). *Journal of Morphology* 267, 1257-1272.

BUMBARGER, D.J., CRUM, J., ELLISMAN, M.H. & BALDWIN, J.G. (2007). Three-dimensional fine structural reconstruction of the nose sensory structures of *Acrobeles complexus* compared to *Caenorhabdtis elegans* (Nematoda: Rhabditida). *Journal of Morphology* 268, 649-663.

BUMBARGER, D.J., WIJERATNE, S., CARTER, C., CRUM, J., ELLISMAN, M.H. & BALDWIN, J.G. (2009). Three dimensional reconstruction of the amphid sensilla in the microbial feeding nematode, *Acrobeles complexus* (Nematoda: Rhabditida). *Journal of Comparative Neurology* 512, 271-181.

BURROWS, P.R. & STONE, A.R. (1985). Heterodera glycines. *CIH Descriptions of plant-parasitic nematodes, Set 4, No 46*. St Albans, UK, Commonwealth Institute of Helminthology, 3 pp.

CAB INTERNATIONAL (2005a). Globodera rostochiensis. Crop Protection Compendium, Global Module, 5th edition. Wallingford, UK, CAB International.

CAB INTERNATIONAL (2005b). *Globodera pallida*. Crop Protection Compendium, Global Module, 5th edition. Wallingford, UK, CAB International.

CAMPOS-VELA, A. (1967). *Taxonomy, life cycle and host range of* Heterodera mexicana *n. sp. (Nematoda: Heteroderidae)*. Ph.D. Thesis, University of Wisconsin, 70 pp.

CANTO, M. & DE SCURRAH, M.M. (1977). Races of the potato cyst nematode in the Andean region and a new system of classification. *Nematologica* 23, 340-349.

CARES, J.E. & BALDWIN, J.G. (1994a). Comparative fine structure of sperm in *Verutus volvingentis* and *Meloidodera floridensis* (Heteroderinae, Nematoda). *Canadian Journal of Zoology* 72, 1481-1491.

CARES, J.E. & BALDWIN, J.G. (1994b). Fine structure of sperm of *Ekphymatodera thomasoni* (Heteroderinae, Nemata). *Journal of Nematology* 26, 375-383.

CARES, J.E. & BALDWIN, J.G. (1995). Comparative fine structure of sperm in *Heterodera schachtii* and *Punctodera chalcoensis*, with phylogenetic implications for Heteroderinae (Nematoda). *Canadian Journal of Zoology* 73, 309-320.

CARTA, L.K. & BALDWIN, J.G. (1990). Phylogenetic implications of phasmid absence in males of three genera in Heteroderinae. *Journal of Nematology* 22, 386-394.

CASTILLO, P., DOLORES ROMERO, M. & GOMEZ BARCINA, A. (1989). New record of *Punctodera punctata* from Spain. *Nematologia Mediterranea* 17, 67-70.

CASWELL, E.P., THOMASON, I.J. & MCKINNEY, H.E. (1985). Extraction of cysts and eggs of *Heterodera schachtii* from soil with an assessment and extraction efficiency. *Journal of Nematology* 17, 337-340.

CHAVES, E. (1987). Cyst nematodes (Heteroderidae) from Argentina. *Nematologica* 33, 22-33.

CHEN, P.S., ZHENG, J.W. & MA, C.Z. (1994). [A new record of the genus *Globodera* (*G. artemisiae*) from Beijing, P. R. China.] *Scientia Agricultura Sinica* 27, 89.

CHITWOOD, B.G. (1949a). Root-knot nematodes. I. A revision of the genus *Meloidogyne* Goeldi, 1887. *Proceedings of the Helminthological Society of Washington* 16, 90-104.

CHITWOOD, B.G. (1949b). Cyst forming *Heterodera* encountered in soil sampling. *Plant Disease Reporter* 33, 130-131.

CHITWOOD, B.G. (1950). An outline classification of the nematodes. In: Chitwood, B.G. & Chitwood, M.B. (Eds). *An introduction to nematology. I. Anatomy.* Baltimore, MD, USA, Monumental Printing Co., pp. 12-27.

CHITWOOD, B.G., HANNON, C.J. & ESSER, R.P. (1956). A new nematode genus *Meloidodera* linking the genera *Heterodera* and *Meloidogyne*. *Phytopathology* 46, 364-366.

CHIZHOV, V.N. & IDEKH, S.B. (1980). [The cyst nematode, *Punctodera punctata* (Thorne, 1928) Mulvey *et* Stone, 1976 in the Moscow region.] *Byulleten' Vsesoyuznogo Instituta Gel'mintologii im. K.I. Skryabina* 26, 96-98.

CHIZHOV, V.N., CHUMAKOVA, O.A., SUBBOTIN, S.A. & BALDWIN, J.G. (2006). Morphological and molecular characterization of foliar nematodes of the genus *Aphelenchoides*: *A. fragariae* and *A. ritzemabosi* (Nematoda: Aphelenchoididae) from the Main Botanical Garden of the Russian Academy of Sciences, Moscow. *Russian Journal of Nematology* 14, 179-184.

CHIZHOV, V.N., UDALOVA, ZH.V. & NASONOVA, L.V. (2008). [*Globodera arenaria* n. sp. and *Cactodera radicale* n. sp. (Nematoda: Tylenchida) from rizosphere of meadows in Mid-Volga region.] *Russian Parasitological Journal* 2, 109-116.

CHOI, Y.E. & KIM, S.E. (2001). Two unrecorded species, *Globodera hypolysi* and *Sarisodera hydrophila* (Nematoda: Heteroderidae) from Korea. *Korean Journal of Applied Entomology* 40, 5-13.

CID DEL PRADO, V.I. & MIRANDA, B.L. (2008). A second cyst-forming nematode parasite of barley (*Hordeum vulgare* L. var. Esmeralda) from Mexico. *Nematropica* 38, 105-114.

CID DEL PRADO, V.I. & ROWE, J.A. (2000). *Cactodera evansi* sp. n. and *Meloidodera astonei* sp. n. (Tylenchida: Heteroderidae) from Mexico. *International Journal of Nematology* 10, 159-168.

CLARK, S.A., SHEPHERD, A.M. & KEMPTON, A. (1973). Spicule structure in some *Heterodera* spp. *Nematologica* 19, 242-247.

CLIFF, G.M. & BALDWIN, J.G. (1985). Fine structure of body wall cuticle of females of eight genera of Heteroderidae. *Journal of Nematology* 17, 286-296.

COBB, G.S. & TAYLOR, A.L. (1953). *Heterodera leptonepia* n. sp., a cyst-forming nematode found in soil with stored potatoes. *Proceedings of the Helminthological Society of Washington* 20, 13-15.

COOK, R. & NOEL, G.R. (2002). Cyst Nematodes: *Globodera* and *Heterodera* species. In: Starr, J.L., Cook, R. & Bridge, J. (Eds). *Plant resistance to parasitic nematodes.* Wallingford, UK, CABI Publishing, pp. 71-105.

COOK, R. & RIVOAL, R. (1998). Genetics of resistance and parasitism. In: Sharma, S.B. (Ed.). *The cyst nematodes.* Dordrecht, The Netherlands, Kluwer Academic Publishers, pp. 322-352.

COOKE, D.A., MATHIAS, P.L., CHWARSZCZYNSKA, D.M. & COPPOCK, L.J. (1983). Comparison of methods of assessing field populations of *Heterodera schachtii*. *Plant Pathology* 32, 339-343.

COOMANS, A. (1979). General principles of systematics with particular reference to speciation; 1-19. In: Lamberti, F. & Taylor, C.E. (Eds). *Root-knot nematodes (*Meloidogyne *species). Systematics, biology and control.* London, UK, Academic Press, pp. 1-19.

COOMANS, A. (1983). General principles for the phylogenetic systematics of nematodes. In: Stone, A.R., Platt, H.M. & Khalil, L.F. (Eds). *Concepts in nematode systematics.* New York, NY, USA, Academic Press, pp. 1-10.

COOPER, B.A. (1955). A preliminary key to British species of *Heterodera* for use in soil examination. In: Kevan, D.K.McE. (Ed.). *Soil zoology.* London, UK, Butterworths, pp. 269-280.

CORDERO, C.D.A. (1989). *Compartive morphology and development of the cone of* Heterodera schachtii *and* Cactodera cacti. Ph.D. Thesis, University of California, Riverside.

CORDERO, C.D.A. & BALDWIN, J.G. (1990). The effect of age on body wall cuticle morphology of *Heterodera schachtii* Schmidt females. *Journal of Nematology* 22, 356-361.

CORDERO, C.D.A. & BALDWIN, J.G. (1991). Fine structure of the posterior cone of *Heterodera schachtii* Schmidt (Heteroderinae) with emphasis on musculature and fenestration. *Journal of Nematology* 23, 110-121.

CORDERO, C.D.A., BALDWIN, J.G. & MUNDO-OCAMPO, M. (1991). Fine structure of the posterior cone of females of *Cactodera cacti* Filip'ev and Schuurmans Stekhoven (Nemata: Heteroderinae). *Revue de Nématologie* 14, 455-465.

CORREIA, F.J.S. & ABRANTES, I.M. DE O. (1997). An improved technique for mounting *Heterodera* cysts for light microscopy. *Nematologica* 43, 507-509.

CORREIA, F.J.S. & ABRANTES, I.M. DE O. (2005). Characterization of *Heterodera zeae* populations from Portugal. *Journal of Nematology* 37, 328-335.

CURRAN, J., DRIVER, F., BALLARD, J.W.O. & MILNER, R.J. (1994). Phylogeny of *Metarhizium*: analysis of ribosomal DNA sequence data. *Mycological Research* 98, 547-555.

DAVIS, E.L., HUSSEY, R.S. & BAUM, T.J. (2004). Getting to the roots of parasitism by nematodes. *Trends in Parasitology* 20, 134-141.

DE LEY, P. & BERT, W. (2002). Video capture and editing as a tool for the storage, distribution and illustration of morphological characters of nematodes. *Journal of Nematology* 34, 296-302.

DICKSON, D.W., HUISINGH, D. & SASSER, J.N. (1971). Dehydrogenases, acid and alkaline phosphatases, and esterases for chemotaxonomy of selected *Meloidogyne*, *Ditylenchus*, *Heterodera* and *Aphelenchus* spp. *Journal of Nematology* 3, 1-16.

DOBOSZ, R. (2007). [Life cycle of *Globodera artemisiae* (Eroshenko et Kazachenko, 1972) on mugwort (*Artemisia vulgaris* L.).] *Progress in Plant Protection* 47, 198-199.

DOBOSZ, R., OBREPALSKA-STEPLOWSKA, A. & KORNOBIS, S. (2006). *Globodera atermisiae* (Eroshenko *et* Kazachenko, 1972) (Nematoda: Heteroderidae) from Poland. *Journal of Plant Protection Research* 46, 403-407.

DOLINSKI, C. & BALDWIN, J.G. (2003). Fine structure of the stoma of *Bunonema* sp. *Teratorhabditis palmarum* (Nematoda) and its phylogenetic significance. *Journal of Nematology* 35, 244-251.

DOLINSKI, C., BALDWIN, J.G. & THOMAS, W.K. (2001). Comparative survey of early embryogenesis of Secernentea (Nematoda) with phylogenetic implications. *Canadian Journal of Zoology* 79, 82-94.

DONEGHAN, J. & MCKENZIE, M. (1987). Distinguishing *Globodera achilleae* from the potato cyst nematodes *G. rostochiensis* and *G. pallida*, using isoelectric focusing. In: *Abstracts, Meeting of the Nematology Group, Association of Applied Biologists, Imperial College, London*. Wellesbourne, UK, Association of Applied Biologists, 18. [Abstr.]

EBSARY, B.A. & EVELEIGH, E.S. (1983). *Meloidogyne aquatilis* n. sp. (Nematoda: Meloidogynidae) from *Spartina pectinata* with a key to the Canadian species of *Meloidogyne. Journal of Nematology* 15, 349-353.

ELLENBY, C. (1946). Nature of the cyst wall of the potato-root eelworm, *Heterodera rostochiensis* Wollenweber, and its permeability to water. *Nature* 157, 302-303.

ELLENBY, C. (1954). Tuber forming species and varieties of the genus *Solanum* tested for resistance to the potato root eelworm *Heterodera rostochiensis* Wollenweber. *Euphytica* 3, 195-202.

ELSTON, D.A., PHILLIPS, M.S. & TRUDGILL, D.L. (1991). The relationship between initial population density of potato cyst nematode *Globodera pallida* and the yield of partially resistant potatoes. *Revue de Nématologie* 14, 231-219.

ENDO, B.Y. (1980). Ultrastucture of the anterior neurosensory organs of the larvae of the soybean cyst nematode, *Heterodera glycines. Journal of Ultrastructural Research* 72, 349-366.

ENDO, B.Y. (1983). Ultrastructure of the stomatal region of the juvenile stage of the soybean cyst nematode, *Heterodera glycines. Proceedings of the Helminthological Society of Washington* 50, 43-61.

ENDO, B.Y. (1984). Ultrastructure of the esophagus of larvae of the soybean cyst nematode, *Heterodera glycines. Proceedings of the Helminthological Society of Washington* 51, 1-24.

ENDO, B.Y. (1985). Ultrastructure of the head region of molting second-stage juveniles of *Heterodera glycines* with emphasis on stylet formation. *Journal of Nematology* 17, 112-123.

ENDO, B.Y. (1988). Ultrastructure of the intestine of second and third juvenile stages of the soybean cyst nematode, *Heterodera glycines. Proceedings of the Helminthological Society of Washington* 55, 117-131.

ENDO, B.Y. (1991). Ultrastructure of initial responses of susceptible and resistant soybean roots to infection by *Heterodera glycines. Revue de Nématologie* 14, 73-94.

ENDO, B.Y. (1998). *Atlas on ultrastructure of infective juveniles of the soybean cyst nematode*, Heterodera glycines. US Department of Agriculture, Agriculture Handbook No. 711, 224 pp.

ENDO, B.Y. & WYSS, U. (1992). Ultrastructure of cuticular exudations of *Heterodera schachtii*, as related to cuticle structure. *Protoplasma* 166, 67-77.

ENDO, B.Y., ZUNKE, U. & WERGIN, W.P. (1997). Ultrastructure of the lesion nematode, *Pratylenchus penetrans* (Nemata: Pratylenchidae). *Journal of the Helminthological Society of Washington* 64, 59-95.

EPPO QUARANTINE PEST. Data Sheets on Quarantine Pests. *Globodera rostochiensis* and *Globodera pallida*, under Contract 90/399003. http://www.eppo.org/QUARANTINE/nematodes/Globodera_pallida/HETDSP_ds.pdf

EPPS, J.M. & DUCLOS, L.A. (1970). Races of soybean cyst nematode in Missouri and Tennessee. *Plant Disease Reporter* 54, 319-320.

EROSHENKO, A.S. & KAZACHENKO, I.P. (1972). [*Heterodera artemisiae* sp. n. (Nematoda: Heteroderidae) – a new cyst forming nematode species from Primorsk Territory.] *Parazitologiya* 6, 166-170.

EROSHENKO, A.S. & KAZACHENKO, I.P. (1983). [Description of male of *Globodera artemisiae* (Nematoda, Heteroderidae) from a type locality.] *Parazitologiya* 15, 415-417.

EROSHENKO, A.S., SUBBOTIN, S.A. & KAZACHENKO, I.P. (2001). *Heterodera vallicola* sp. n. (Tylenchida: Heteroderidae) from elm trees, *Ulmus japonica* (Rehd.) Sarg. in the Primorsky territory, the Russian Far East, with rDNA identification of closely related species. *Russian Journal of Nematology* 9, 9-17.

ESBENSHADE, P.R. & TRIANTAPHYLLOU, A.C. (1988). Genetic analysis of esterase polymorphism in the soybean cyst nematode, *Heterodera glycines*. *Journal of Nematology* 20, 486-492.

ESPARRAGO, G., BLANCO, I. & PULIDO, I. (2002). Distribution of *Globodera tabacum* in the La Vera region of Spain. *Nematology* 4, 265-266.

ESSER, R.P. (1992). Cactus cyst nematode, *Cactodera cacti*. Nematology Circular, Florida Department Agriculture and Consumer Service, Division of Plant Industry 197, 3 pp.

EVANS, K. (1969). Changes in a *Heterodera rostochiensis* population through the growing season. *Annals of Applied Biology* 64, 31-41.

EVANS, K. (1970). Longevity of males and fertilisation of females of *Heterodera rostochiensis*. *Nematologica* 16, 369-374.

EVANS, K. & FRANCO, J. (1977). Morphological variation in some populations of potato cyst-nematodes from Europe and South America. *Nematologica* 21, 417-430.

EVANS, K. & STONE, A.R. (1977). A review of the distribution and biology of the potato cyst-nematodes *Globodera rostochiensis* and *G. pallida*. *International Journal of Pest Management* 23, 178-189.

EVANS, K., FRANCO, J. & DE SCURRAH, M.M. (1975). Distribution of species of potato cyst nematodes in South America. *Nematologica* 21, 365-369.

FENWICK, D.W. (1940). Methods for the recovery and counting of cysts of *Heterodera schachtii* from soil. *Journal of Helminthology* 18, 155-172.
FERRIS, V.R. (1979). Cladistic approaches in the study of soil and plant parasitic nematodes. *American Zoologist* 19, 1195-1215.
FERRIS, V.R. (1985). Evolution and biogeography of cyst-forming nematodes. *EPPO Bulletin* 15, 123-129.
FERRIS, V.R. (1998). Evolution, phylogeny and systematics. In: Sharma, S. (Ed.). *The cyst nematodes*. Dordrecht, The Netherlands, Kluwer Academic Publishers, pp. 57-82.
FERRIS, V.R., FERRIS, J.M. & FAGHIHI, J. (1993). Variation in spacer ribosomal DNA in some cyst-forming species of plant parasitic nematodes. *Fundamental and Applied Nematology* 16, 177-184.
FERRIS, V.R., FERRIS, J.M., FAGHIHI, J. & IREHOLM, A. (1994). Comparisons of isolates of *Heterodera avenae* using 2-D PAGE protein patterns and ribosomal DNA. *Journal of Nematology* 26, 144-151.
FERRIS, V.R., KRALL, E., FAGHIHI, J. & FERRIS, J.M. (1999a). Phylogenetic relationships of *Globodera millefolii*, *G. artemisiae*, and *Cactodera salina* based on ITS region of ribosomal DNA. *Journal of Nematology* 31, 498-507.
FERRIS, V.R., SUBBOTIN, S.A., IREHOLM, A., SPIEGEL, Y., FAGHIHI, J. & FERRIS, J.M. (1999b). Ribosomal DNA sequence analysis of *Heterodera filipjevi* and *H. latipons* isolates from Russia and comparisons with other nematode isolates. *Russian Journal of Nematology* 7, 121-125.
FERRIS, V.R., SABO, A., BALDWIN, J.G., MUNDO-OCAMPO, M., INSERRA, R.N. & SHARMA, S. (2004). Phylogenetic relationships among selected Heteroderoidea based on 18S and ITS ribosomal DNA. *Journal of Nematology* 36, 202-206.
FILIPJEV, I.N. & SCHUURMANS STEKHOVEN, J.H. (1941). *A manual of agricultural helminthology*. Leiden, The Netherlands, E.J. Brill, 878 pp.
FLEMING, C.C. & DOLAN, S. (1986). The identification of cyst nematodes using agarose isoelectric focusing and silver staining. *Record of Agricultural Research (Belfast)* 34, 47-52.
FLEMING, C.C. & MARKS, R.J. (1982). A method for the quantitative estimation of *Globodera rostochiensis* and *Globodera pallida* in mixed species samples. *Record of Agricultural Research* 30, 67-70.
FLEMING, C.C. & MARKS, R.J. (1983). The identification of the potato cyst nematodes *Globodera rostochiensis* and *G. pallida* by isoelectric focusing of proteins on polyacrylamide gels. *Annals of Applied Biology* 103, 227-281.
FLEMING, C.C. & POWERS, T.O. (1998). Potato cyst nematode diagnostics: morphology, differential hosts and biochemical techniques. In: Marks, R.J. & Brodie, B.B. (Eds). *Potato cyst nematodes, biology, distribution and control*. Wallingford, UK, CABI Publishing, pp. 91-114.

FLEMING, C.C., RAO, J., MORELAND, B., CRAIG, D. & TURNER, S.J. (2000). Diagnostics of cyst nematodes and tephritid fruit flies using mitochondrial and ribosomal DNA. *EPPO Bulletin* 30, 585-590.

FOTEDAR, D.N. & HANDOO, Z.A. (1978). A revised scheme of classification to order Tylenchida Thorne, 1949 (Nematoda). *Journal of Science, University of Kashmir* 3 (1975), 55-82.

FOX, P.C. & ATKINSON, H.J. (1984). Isoelectric focusing of general protein and specific enzymes from pathotypes of *Globodera rostochiensis* and *Globodera pallida*. *Parasitology* 66, 131-136.

FOX, P.C. & ATKINSON, H.J. (1988). Non-specific esterase variation in field populations of the potato cyst nematode *Globodera rostochiensis* and *G. pallida*. *Nematologica* 34, 156-163.

FRANCO, J. (1977). *Studies on the taxonomy and biology of potato cyst nematodes* Globodera *spp.* Ph.D. Dissertation, University of London, London, UK.

FRANCO, J., OROS, R., MAIN, G. & OPTUNO, N. (1998). Potato cyst nematodes (*Globodera* species) in South America. In: Marks, R.J. & Brodie, B.B. (Eds). *Potato cyst nematodes, biology, distribution and control.* Wallingford, UK, CABI Publishing, pp. 239-269.

FRANCO, N.F., CID DEL PRADO VERA, I. & LAMOTHE-ARGUMEDO, R. (2000). *Globodera bravoae* sp. n. (Tylenchida: Heteroderidae) from Mexico. *International Journal of Nematology* 10, 169-176.

FRANKLIN, M.T. (1939). On the structure of the cyst wall of *Heterodera schachtii* (Schmidt). *Journal of Helminthology* 17, 127-134.

FRANKLIN, M.T. (1951). *The cyst-forming species of* Heterodera. Farnham Royal, UK, Commonwealth Agricultural Bureaux, 147 pp.

FUCHS, O. (1911). Beitrage zur Biologie der Rubennematoden. *Zeitschrift für das landwirtschaftliche Versuchswesen in Deutschösterreich* 14, 923-949.

FULLAONDO, A., BARRENA, E., VIRIBAY, M., BARRENA, I., SALAZAR, A. & RITTER, E. (1999). Identification of potato cyst nematode species *Globodera rostochiensis* and *G. pallida* by PCR using specific primer combinations. *Nematology* 1, 157-163.

GÄBLER, C., STURHAN, D., SUBBOTIN, S.A. & RUMPENHORST, H.J. (2000). *Heterodera pratensis* sp. n., a new cyst nematode of the *H. avenae* complex (Nematoda: Heteroderidae). *Russian Journal of Nematology* 8, 115-126.

GANGULY, A.K., DASGUPTA, D.R. & RAJASEKHAR, S.P. (1990). ß-esterase variation in three common species of *Heterodera*. *Indian Journal of Nematology* 20, 113-114.

GIAMMARA, B.L. & HANKER, J.S. (1986). Epoxy-slide embedment of cytochemically stained tissues and cultured cells for light and electron microscopy. *Stain Technology* 61, 51-58.

GOFFART, H. (1936). *Heterodera schachtii* Schmidt an gemeiner Hanfnessel (*Galeopsis tetrahit* L.) und an Kakteen. *Zeitschrift für Parasitenkunde* 8, 528-532.

GOLDEN, A.M. (1971). Classification of the genera and higher categories of the order Tylenchida (Nematoda). In: Zuckerman, B.M., Mai, W.F. & Rohde, R.A. (Eds). *Plant parasitic nematodes. Vol. 1. Morphology, anatomy, taxonomy and ecology*. New York, USA, Academic Press, pp. 191-232.

GOLDEN, A.M. (1986). Morphology and identification of cyst nematodes. In: Lamberti, F. & Taylor, C.E. (Eds). *Cyst nematodes*. New York, USA, Plenum Press, pp. 23-79.

GOLDEN, A.M. & ELLINGTON, D.M.S. (1972). Redescription of *Heterodera rostochiensis* (Nematoda: Heteroderidae) with a key and notes on closely related species. *Proceedings of the Helminthological Society of Washington* 39, 64-77.

GOLDEN, A.M. & KLINDIĆ, O. (1973). *Heterodera achilleae* n. sp. (Nematoda, Heteroderidae) from yarrow in Yugoslavia. *Journal of Nematology* 5, 196-201.

GOLDEN, A.M. & RASKI, D.J. (1977). *Heterodera thornei* n. sp. (Nematoda: Heteroderidae) and a review of related species. *Journal of Nematology* 9, 93-112.

GOLDEN, A.M., EPPS, J.M., RIGGS, R.D., DUCLOS, L.A., FOX, J.A. & BERNARD, R.L. (1970). Terminology and identity of infraspecific forms of the soybean cyst nematode (*Heterodera glycines*). *Plant Disease Reporter* 54, 544-546.

GONZALES, J.A., PHILLIPS, M.S. & HACKETT, C.A. (1992). Morphological variability of potato cyst nematodes (*Globodera* spp.) in the Canary Islands. *Annals of Applied Biology* 120, 315-328.

GRANEK, I. (1955). Additional morphological differences between the cysts of *Heterodera rostochiensis* and *Heterodera tabacum*. *Plant Disease Reporter* 39, 716-718.

GRANEY, L.S.O. & BIRD, G.W. (1990). Descriptions and comparative morphology of *Cactodera milleri* n. sp. (Nematoda: Heteroderidae) and *Cactodera cacti* with a review and key to the genus *Cactodera*. *Journal of Nematology* 22, 457-480.

GRECO, N. (1988). *Potato cyst nematodes:* Globodera rostochiensis *and* G. pallida. Nematology Circular, Florida Department Agriculture and Consumer Service, Division of Plant Industry 149, 4 pp.

GRECO, N., INSERRA, R.N., BRANDONISIO, A., TIRRO, A. & DE MARINIS, G. (1988). Life-cycle of *Globodera rostochiensis* on potato in Italy. *Nematologia Mediterranea* 16, 69-70.

GREEN, C.D. (1971). The morphology of the terminal area of the round cyst nematodes, S.G. *Heterodera rostochiensis* and allied species. *Nematologica* 17, 34-46.

GREEN, C.D. (1975). The vulval cone and associated structures of some cyst nematodes (genus *Heterodera*). *Nematologica* 21, 134-144.

GREEN, C.D., GREET, D.N. & JONES, F.G.W. (1970). The influence of multiple mating on the reproduction and genetics on *Heterodera rostochiensis* and *H. schachtii*. *Nematologica* 16, 309-326.

GREET, D.N. (1972). Electrophoresis and morphometrics of round-cyst nematodes. *Annals of Applied Biology* 71, 283-286.

GRENIER, E., BOSSIS, M., FOUVILLE, D., RENAULT, L. & MUGNIÉRY, D. (2001). Molecular approaches to the taxonomic position of Peruvian potato cyst nematodes and gene pool similarities in indigenous and imported populations of *Globodera*. *Heredity* 86, 277-290.

GRENIER, E., FOURNET, S., PETIT, E., ANTHOINE, G. (2010). A cyst nematode 'species factory' called the Andes. *Nematology* 12, 163-169.

GROSSHEIM, A.A. (1945). [On question of graphical image of the system of flowering plants.] *Sovetskaya Botanika* 13, 3-27.

GUILE, C.T. (1966). Cyst chromogenesis in potato cyst-eelworm pathotypes. *Plant Pathology* 15, 125-128.

GUILE, C.T. (1970). Further observations on cyst colour changes in potato cyst eelworm pathotypes. *Plant Pathology* 19, 1-6.

HEINZ, R., MITCHUM, M., NATHAN, M. & SWEETS, L. (2007). Missouri corn nematode survey. *Integrated Pest Crop Management, Plant Protection Programs, University of Missouri* 18, 1-2.

HENNIG, W. (1966). *Phylogenetic systematics*. Urbana, Illinois, USA, University of Illinois Press, 263 pp.

HERNANDEZ ARAGON, L. (1965). *Comportamiento de tres variedades de maiz al ataque de* Heterodera punctata *Thorne*. M.Sc. Thesis, Escuela Nacional de Agricultura, Chapingo, Mexico.

HESLING, J.J. (1959). The identification of *Heterodera* cysts. *London, UK, Ministry of Agriculture, Fisheries and Food Technical Bulletin* 7, 64-70.

HESLING, J.J. (1973). The estimation of Granek's ratio in round-cyst. *Heterodera. Nematologica* 19, 119-120.

HESLING, J.J. (1978). Cyst nematodes: morphology and identification of *Heterodera, Globodera* and *Punctodera*. In: Southey, J.F. (Ed.). *Plant nematology*. London, UK, Her Majesty's Stationery Office, pp. 125-155.

HESLING, J.J. & ELLIS, P.R. (1974). The fenestralia of *Heterodera rostochiensis* and *H. pallida*. *Nematologica* 20, 43-51.

HIRSCHMANN, H. (1956). Comparative morphological studies on the soybean cyst nematode, *Heterodera glycines* and the clover cyst nematode, *H. trifolii*

(Nematoda: Heteroderidae). *Proceeding of the Helminthological Society of Washington* 23, 140-151.

HIRSCHMANN, H. & RIGGS, R.D. (1969). *Heterodera betulae* n. sp. (Heteroderidae), a cyst-forming nematode from river birch. *Journal of Nematology* 1, 169-179.

HIRSCHMANN, H. & TRIANTAPHYLLOU, A.C. (1979). Morphological comparison of members of the *Heterodera trifolii* species complex. *Nematologica* 25, 458-481.

HISAI, J., HIRATA, K. & SUZUKI, K. (2003). [The relationship between cyst shape and pore size of sieves for Fenwick method.] *Research Bulletin of the Plant Protection Service, Japan* 39, 61-63.

HLAOUA, W., HORRIGUE-RAOUANI, N., FOUVILLE, D. & MUGNIÉRY, D. (2008). Morphological and molecular characterization of potato cyst nematode populations from Tunisia and survey of their probable geographical origin. *Biotechnology* 7, 651-659.

HOLGADO, R., ROWE, J., ANDERSSON, S. & MAGNUSSON, C. (2004). Electrophoresis and biotest studies on some populations of cereal cyst nematode, *Heterodera* spp. (Tylenchida: Heteroderidae). *Nematology* 6, 857-865.

HOOPER, D.J. (1978). Structure and classification of nematodes. In: Southey, J.F. (Ed.). *Plant nematology*. London, UK, Her Majesty's Stationery Office, pp. 3-45.

HORNE, C.W. (1965). *The taxonomic status, morphology and biology of a cyst nematode (Nematoda: Heteroderidae) found attacking* Poa annua *L.* Ph.D. Thesis, Texas A&M University, College Station, Texas, USA.

HORNE, C.W. & THAMES JR, W.H. (1966). Notes on occurrence and distribution of *Heterodera punctata. Plant Disease Reporter* 50, 869-871.

HUSAIN, S.I. (1976). Phylogeny and inter-relationship of the superfamily Heteroderoidea (Skarbilovich, 1947) Golden, 1971. *Geobios* 3, 9-12.

IBRAHIM, S.K. & PERRY, R.N. (1993). Use of esterase patterns of females and galled roots for the identification of species of *Meloidogyne. Fundamental and Applied Nematology* 16, 187-191.

IBRAHIM, S.K. & ROWE, J.A. (1995). Use of isoelectric focusing and polyacrylamide gel electrophoresis of nonspecific esterase phenotypes for the identification of cyst nematodes *Heterodera* species. *Fundamental and Applied Nematology* 18, 189-196.

IBRAHIM, S.K., MINNIS, S.T., BARKER, A.D.P., RUSSELL, M.D., HAYDOCK, P.P.J., EVANS, K., GROVE, I.G., WOODS, S.R. & WILCOX, A. (2001). Evaluation of PCR, IEF and ELISA techniques for the detection and identification of potato cyst nematodes from field soil samples in England and Wales. *Pest Management Science* 57, 1068-1074.

INADA, H., ITO, H., SATTERLEE, J., SENGUPTA, P., MATSUMOTO, K. & MORI, I. (2006). Identification of guanylyl cylases that function in thermosensory neurons of *Caenorhabditis elegans*. *Genetics* 172, 2239-2252.

JATALA, P. & GAVILANO, L. (1991). Electrophoretic studies of protein patterns of the *Globodera pallida* and *G. rostochiensis* eggs. *Nematropica* 21, 127.

JOHNSON, C.S., WAY, J. & BARKER, K.R. (2005). Nematode parasites of tobacco. In: Luc, M., Sikora, R.A. & Bridge, J. (Eds). *Plant parasitic nematodes in subtropical and tropical agriculture*, 2nd edition. Wallingford, UK, CABI Publishing, pp. 675-708.

JONES, F.G.W. (1950). Observations on the beet eelworm and other cyst forming species of *Heterodera*. *Annals of Applied Biology* 37, 407-440.

JONES, F.G.W. (1952). Differentiation of species of *Heterodera*. *Proceedings of the International Symposium, Harpenden, UK, September 3-14, 1951.* Rome, Italy, FAO, pp. 3-14.

JONES, F.G.W. (1981). Host cell responses to endoparasitic nematode attack: structure and function of giant cells and syncytia. *Annals of Applied Biology* 97, 353-372.

JONES, F.G.W., CARPENTER, J.M., PARROTT, D.M., STONE, A.R. & TRUDGILL, D.L. (1970). Potato cyst nematode: one species or two? *Nature* 227, 83-84.

JONES, M.G.K. & DROPKIN, V.H. (1975). Cellular alternations induced in soybean roots by three endoparasitic nematodes. *Physiological Plant Pathology* 5, 119-124.

JONES, P.W., TYLKA, G.L. & PERRY, R.N. (1998). Hatching. In: Perry, R.N. & Wright, D.J. (Eds). *The physiology and biochemistry of free-living and plant-parasitic nematodes*. Wallingford, UK, CABI Publishing, pp. 181-212.

KARNOVSKY, M.J. (1965). A formaldehyde-glutardehyde fixative of high osmolarity for use in electron microscopy. *Journal of Cell Biology* 27, 137-138A.

KARSSEN, G., VAN HOENSELAAR, T., VERKERK-BAKKER, B. & JANSSEN, R. (1995). Species identification of cyst and root-knot nematodes from potato by electrophoresis of individual females. *Electrophoresis* 16, 105-109.

KAZACHENKO, I.P. (1993). [*Cyst-forming nematodes of Far East and their control.*] Vladivostok, USSR, Dalnauka, 77 pp.

KIM, D.G. & RIGGS, R.D. (1995). Portable cyst extractor: detecting cyst nematodes in the field. *Journal of Nematology* 27, 125-126.

KIRJANOVA, E.S. (1963). [Collection and diagnosis of root nematodes of the family Heteroderidae (Skarbilovich, 1947) Thorne 1949.] *Sbornik rabot po nematodam selskoknozjaistvenykh rastenii* 3, 6-32.

KIRJANOVA, E.S. (1969). [On the structure of the subcrystalline layer of the nematode genus *Heterodera* (Nematoda: Heteroderidae) with a description of two new species.] *Parazitologiya* 3, 81-91.

KIRJANOVA, E.S. & BORISENKO, A.V. (1975). [A cyst forming nematode – *Heterodera mali* sp. n. – a new species a parasite of apple trees in the Kazakhstan.] *Parazitologiya* 9, 335-338.

KIRJANOVA, E.S. & KRALL, E. (1963). [*Heterodera estonica* n. sp. (Nematodes: Heteroderidae), Estonian cyst forming nematode.] *Izvestija Akademii Nauk Estonskoj SSR, Biologicheskaja Seria* 12, 219-223.

KIRJANOVA, E.S. & KRALL, E. (1965). [The milfoil cyst nematode – *Heterodera millefolii* n. sp. (Nematoda: Heteroderidae).] *Izvestija Akademii Nauk Estonskoi SSR, Biologicheskaja Seria* 14, 325-328.

KIRJANOVA, E.S. & KRALL, E.L. (1968). [*Plant parasitic nematodes and their control.*] Vol. I. Leningrad, USSR, Nauka, 443 pp.

KIRJANOVA, E.S. & KRALL, E.L. (1971). [*Plant parasitic nematodes and their control.*] Vol. II. Leningrad, USSR, Nauka, 522 pp.

KIRJANOVA, E.S. & TERENTEVA, T.G. (1961). [The cactus nematode, *Heterodera cacti* (Adam, 1932) Filipjev et Schuurmans-Stekhoven, 1941 (Nematoda: Heteroderidae) in glasshouses of Leningrad.] *Voprosy Fitogel'mintologii*, GELAN, Moscow, pp. 74-83.

KIRJANOVA, E.S., BALBAEVA, Z.A., KYANSHALIEVA, E.N., SAGITOV, A.O. & PETKO, G.I. (1976). [Cyst nematodes of the genus *Heterodera* in Kazakhstan.] *Izvestiya Akademii Nauk Kazakhskoi SSR (Biol.)* 4, 14-17.

KNOETZE, R., MALAN, A.P. & MOUTON, C. (2006). Differentiation of South African potato cyst nematodes (PCN) by analysis of the rDNA internal transcribed spacer region. *African Plant Protection* 12, 103-110.

KORT, J., ROSS, H., RUMPENHORST, H.J. & STONE, A.R. (1977). An international scheme for identifying and classifying pathotypes of potato cyst-nematodes *Globodera rostochiensis* and *G. pallida*. *Nematologica* 23, 333-339.

KOVALEVA, E.S., SUBBOTIN, S.A., MASLER, E.P. & CHITWOOD, D.J. (2005). Molecular characterization of the actin gene from cyst nematodes in comparison to those from other nematodes. *Comparative Parasitology* 72, 39-49.

KRALL, E. (1977). Compendium of cyst nematodes in the USSR. *Nematologica* 23, 311-332.

KRALL, E. (1990). Different approaches to and recent developments in the systematics and co-evolution of the family Heteroderidae (Nematoda: Tylenchida) with host plants. *Proceedings of the Estonian Academy of Sciences, Biology* 39, 259-270.

KRALL, E.L. & KRALL, H.A. (1970). [On the question of evolution of relations of parasitic phytohelminths of the family Heteroderidae with their host plants.] *Sbornik Nauchnykh Trydov Estonskoi Selskokhozjaistevoi Akademii* 70, 152-154.

KRALL, E.L. & KRALL, H.A. (1973). [On the principles of studying phylogenetic relationships in phytohelminths of the genus *Heterodera* Schmidt, 1871.] In: *Materialy Pribaltiiskoi Nauchnoi Koordinatsionnoi Konferentsii po Voprosam Parazitologii, Vulnyus*, pp. 85-86.

KRALL, E.L. & KRALL, H.A. (1978). [Revision of the plant nematodes of the family Heteroderidae on the basis of trophic specialization of these parasites and their co-evolution with their host plants.] In: *Fitogel'mintologicheskie Issledovaniya.* Moscow, USSR, Nauka, pp. 39-56.

KÜHN, J. (1881). *Die Ergebnisse der Versuche zur Ermittelung der Rübenmüdigkeit und zur Erforschung der Natur der Nematoden.* Schönfeld's Verlagsbuchhandlung Dresden 1881. Berichte aus dem physiologischen Laboratorium und der Versuchsanstalt des landwirthschaftlichen Instituts der Universität Halle H. 3, 153.

LAMONDIA, J.A. (1990). Effect of oxamyl on *Globodera tabacum* population dynamics and shade tobacco growth and yield. *Journal of Nematology* 22, 654-657.

LAMONDIA, J.A. (1995). Shade tobacco yield loss and *Globodera tabacum tabacum* population changes in relation to initial nematode density. *Journal of Nematology* 27, 114-119.

LANGESLAG, M., MUGNIÉRY, D. & FAYET, G. (1982). Développement embryonaire de *Globodera rostochiensis* et *G. pallida* en fonction de la température, en conditions contrôlées et naturelles. *Revue de Nématologie* 5, 103-109.

LAWSON, E.C., CARTER, G.C. & LEWIS, S.A. (1984). Application of isoelectric focusing to taxonomic identification *of Meloidogyne* spp. *Journal of Nematology* 16, 91-96.

LEE, Y.B., PARK, W.M. & PAIK, S.B. (1988). A comparative study on the protein and some isoenzyme patterns between races of the soybean cyst nematode (*Heterodera glycines* Ichinohe). *Korean Journal of Plant Pathology* 4, 49-53.

LIEBSCHER, G. (1892). Beobachtungen über das Aufreten eines Nematoden an Erbsen. *Journal für Landwirtschaft* 40, 357-368.

LILLEY, C.J., ATKINSON, H.J. & URWIN, P.E. (2005). Molecular aspects of cyst nematodes. *Molecular Plant Pathology* 6, 577-588.

LOWNSBERY, B.F. & LOWNSBERY, J.W. (1954). *Heterodera tabacum* new species, a parasite of Solanaceaous plants in Connecticut. *Proceedings of the Helminthological Society of Washington* 21, 42-47.

LUC, M. (1986a). Cyst nematodes in equatorial and hot tropical regions. In: Lamberti, F. & Taylor, C.E. (Eds). *Cyst nematodes.* New York & London, Plenum Press, pp. 355-372.

LUC, M. (1986b). Heterodera sacchari. *CIH Descriptions of plant-parasitic nematodes, Set 3, No 48*. St Albans, UK, Commonwealth Institute of Helminthology, 3 pp.

LUC, M. & MERNY, G. (1963). *Heterodera sacchari* n. sp. (Nematoda: Tylenchoidea) parasite de la canne à sucre au Congo-Brazzaville. *Nematologica* 9, 31-37.

LUC, M., TAYLOR, D.P. & CADET, P. (1978). Description of a new tropical Heteroderidae, *Hylonema ivorense* n. gen., n. sp., and a new outlook on the family Heteroderidae (Nematoda: Tylenchida). *Revue de Nématologie* 1, 73-86.

LUC, M., WEISCHER, B., STONE, A.R. & BALDWIN, J.G. (1986). On the definition of heteroderid cysts. *Revue de Nématologie* 9, 418-421.

LUC, M., MAGGENTI, A.R. & FORTUNER, R. (1988). A reappraisal of Tylenchina (Nemata). 9. The family Heteroderidae Filipjev & Schuurmans Stekhoven, 1941. *Revue de Nématologie* 11, 159-176.

MADANI, M., VOVLAS, N., CASTILLO, P., SUBBOTIN, S.A. & MOENS, M. (2004). Molecular characterization of cyst nematode species (*Heterodera* spp.) from the Mediterranean basin using RFLPs and sequences of ITS-rDNA. *Journal of Phytopathology* 152, 229-234.

MADANI, M., SUBBOTIN, S.A. & MOENS, M. (2005). Quantitative detection of the potato cyst nematode, *Globodera pallida*, and the beet cyst nematode, *Heterodera schachtii*, using Real-Time PCR with SYBR green I dye. *Molecular and Cellular Probes* 19, 81-86.

MADANI, M., WARD, L.J. & DE BOER, S.H. (2008). Multiplex real-time polymerase chain reaction for identifying potato cyst nematodes, *Globodera pallida* and *Globodera rostochiensis*, and the tobacco cyst nematode, *Globodera tabacum*. *Canadian Journal of Plant Pathology* 30, 554-564.

MADANI, M., SUBBOTIN, S.A., WARD, L.J. & DE BOER, S.H. (2010). Molecular characterization of Canadian populations of potato cyst nematodes, *Globodera rostochiensis* and *G. pallida* using ribosomal nuclear RNA and cytochrome *b* genes. *Canadian Journal of Plant Pathology* 32, 142-153.

MAI, W.F. (1952). Susceptibility of *Lycopersicon* species to the golden nematode. *Phytopathology* 42, 461.

MAI, W.F. & LOWNSBERY, B.F. (1948). Studies on the host range of the golden nematode of potatoes, *Heterodera rostochiensis* Wollenweber. *The American Potato Journal* 25, 290-294.

MANDURIC, S. & ANDERSSON, S. (2004). The identity of a Swedish *Globodera* (Nematoda: Heteroderidae) population, following comparisons with known populations of *G. artemisiae* (Eroshenko & Kazachenko, 1972) Behrens, 1975. *Russian Journal of Nematology* 12, 39-44.

MANDURIC, S., OLSSON, E., ENGLUND, J.-E. & ANDERSSON, S. (2004). Separation of *Globodera rostochiensis* and *G. pallida* (Tylenchida: Heteroderidae) using morphology and morphometrics. *Nematology* 6, 171-181.

MARCHÉ, L., VALETTE, S., GRENIER, E. & MUGNIÉRY, D. (2001). Intraspecies DNA polymorphism in the tobacco cyst-nematode complex (*Globodera tabacum*) using AFLP. *Genome* 44, 941-946.

MARKS, R.J. & FLEMING, C.C. (1985). The use of isoelectric focusing as a tool in the identification and management of potato cyst nematode populations. *EPPO Bulletin* 15, 289-298.

MAYR, E. (1942). *Systematics and the origin of species*. Columbia New York, NY, USA, University Press, 334 pp.

MEHER, H.C., KAUSHAL, K.K., KHAN, E. & NAVED, S.H. (1998). Use of esterase phenotypes of females for precise diagnosis of four *Heterodera* species. *Indian Journal of Nematology* 28, 81-84.

MEHER, H.C., SHARMA, S.B. & SINGH, G. (2004). A rapid technique for esterase and malate dehydrogenase analysis from cyst and root-knot nematodes. *Annals of Plant Protection Sciences* 12, 380-383.

MILLER, L.I. (1983). Diversity of selected taxa of *Globodera* and *Heterodera* and their interspecific and intergeneric hybrids. In: Stone, A.R., Platt, H.M. & Khalil, L.F. (Eds). *Concepts in nematode systematics. The Systematics Association Special Volume No. 22.* London, UK, Academic Press, pp. 207-220.

MILLER, L.I. (1986). Economic importance of cyst nematodes in North America. In: Lamberti, F. & Taylor, C.E. (Eds). *Cyst nematodes.* New York, NY, USA, Plenum Press, pp. 373-385.

MILLER, L.I. & GRAY, B.J. (1968). Horsenettle cyst nematode, *Heterodera virginiae* n. sp., a parasite of solanaceous plants. *Nematologica* 14, 535-543.

MILLER, L.I. & GRAY, B.J. (1972). *Heterodera solanacearum* n. sp., a parasite of solanaceaous plants. *Nematologica* 18, 404-413.

MILLER, L.I., HARRISON, M.B. & SCHINDLER, A.F. (1962). Horsenettle and Ostborne's cyst nematodes – two undescribed nematodes occurring in Virginia. *Phytopathology* 52, 743.

MOKABLI, A., VALLETTE, S., GAUTHIER, J.P. & RIVOAL, R. (2002). Variation in virulence of cereal cyst nematode populations from North Africa and Asia. *Nematology* 4, 521-525.

MOLINARI, S., EVANS, K., ROWE, J. & RUSSELL, M. (1996). Identification of *Heterodera* cysts by SOD isozyme electrophoresis profiles. *Annals of Applied Biology* 129, 361-368.

MORI, I. & OHSHIMA, Y. (1995). Neural regulation of thermotaxis in *Caenorhabditis elegans*. *Nature* 376, 344-348.

MORRIS, R.F. (1971). Distribution and biology of the golden nematode *Heterodera rostochiensis* in Newfoundland. *Nematologica* 17, 370-376.

MOTA, M. & EISENBACK, J. (1993a). Morphology of second-stage juveniles and males of *Globodera tabacum tabacum*, *G. t. virginiae*, and *G. t. solanacearum* (Nemata: Heteroderinae). *Journal of Nematology* 25, 27-33.

MOTA, M. & EISENBACK, J. (1993b). Morphology of females and cyst of *Globodera tabacum tabacum*, *G. t. virginiae*, and *G. t. solanacearum* (Nemata: Heteroderinae). *Journal of Nematology* 25, 136-147.

MOTA, M. & EISENBACK, J. (1993c). Morphometrics of *Globodera tabacum tabacum*, *G. t. virginiae*, and *G. t. solanacearum* (Nemata: Heteroderinae). *Journal of Nematology* 25, 148-160.

MUGNIÉRY, D. (1978). Vitesse de développement, en fonction de le température, de *Globodera rostochiensis* et *G. pallida* (Nematoda: Heteroderidae). *Revue de Nématologie* 1, 3-12.

MUGNIÉRY, D. & PHILLIPS, M. (2007). The nematode parasites of potato. In: Vreugdenhil, D. (Ed.). *Potato biology and biotechnology: advances and perspectives*. The Netherlands, Elsevier, pp. 569-594.

MUGNIÉRY, D., PHILLIPS, M.S., RUMPENHORST, H.J., STONE, A.R., TREUR, A. & TRUDGILL, D.L. (1989). Assessment of partial resistance of potato to, and pathotype and virulence differences in potato cyst nematodes. *EPPO Bulletin* 19, 7-25.

MUGNIÉRY, D., BOSSIS, M. & PIERRE, J.-S. (1992). Hybridations entre *Globodera rostochiensis* (Wollenweber), *G. pallida* (Stone), *G. virginiae* (Miller and Gray), *G. solanacearum* (Miller and Gray), et *G. "mexicana"* (Campos-Vela). Description et devenir des hybrides. *Fundamental and Applied Nematology* 15, 375-382.

MULDER, A., VAN DER WAL, A.F. & VELEMA, R.A.J. (1997). Relationship between potato cyst nematodes and their principal host. II. Evaluation of mathematical descriptors relating potato yield and nematode density. *Potato Research* 40, 327-334.

MULHOLLAND, V., CARDE, L., O'DONNELL, K.L., FLEMING, C.C. & POWERS, T.O. (1996). Use of the polymerase chain reaction to discriminate potato cyst nematode at the species level. In: Marshall, C. (Ed.). *Proceedings of diagnostics in crop production symposium*. Tarnham, UK, British Crop Production Council, pp. 247-252.

MULK, M.M. (1977). Heterodera cacti. *CIH Descriptions of plant-parasitic nematodes, Set 7, No 96*. St Albans, UK, Commonwealth Institute of Helminthology, 4 pp.

MULVEY, R.H. (1959). Giant eggs of the clover cyst nematode, *Heterodera trifolii* Goffart, 1932. *Nature* 184, 1662-1663.

MULVEY, R.H. (1960). Giant larvae of the clover cyst nematode, *Heterodera trifolii* (Nematoda: Heteroderidae). *Nematologica* 5, 53-55.

MULVEY, R.H. (1972). Identification of *Heterodera* cysts by terminal and corn top structure. *Canadian Journal of Zoology* 50, 1277-1292.

MULVEY, R.H. (1973). Morphology of the terminal areas of the white females and cysts of the genus *Heterodera* (s.g. *Globodera*). *Journal of Nematology* 5, 303-311.

MULVEY, R.H. (1974). Cone-top morphology of the white females and cysts of the genus *Heterodera* (subgenus *Heterodera*), a cyst-forming nematode. *Canadian Journal of Zoology* 52, 77-81.

MULVEY, R.H. & EBSARY, B.A. (1980). *Dolichodera fluvialis* n. gen., n. sp. (Nematoda: Heteroderidae) from Québec, Canada. *Canadian Journal of Zoology* 58, 1697-1702.

MULVEY, R.H. & GOLDEN, A.M. (1983). An illustrated key to the cyst-forming genera and species of Heteroderidae in the western hemisphere with species morphometrics and distribution. *Journal of Nematology* 15, 1-59.

MULVEY, R.H. & STONE, A.R. (1976). Description of *Punctodera matadorensis* n. gen., n. sp. (Nematoda: Heteroderidae) from Saskatchewan, with lists of species and generic diagnosis of *Globodera* (n. rank), *Heterodera* and *Sarisodera*. *Canadian Journal of Zoology* 54, 772-785.

MUNDO-OCAMPO, M. & BALDWIN, J.G. (1983a). Host response to *Meloidodera* spp. (Heteroderidae). *Journal of Nematology* 15, 544-554.

MUNDO-OCAMPO, M. & BALDWIN, J.G. (1983b). Host parasite relationship of *Atalodera* spp. (Heteroderidae). *Journal of Nematology* 15, 234-243.

MUNDO-OCAMPO, M. & BALDWIN, J.G. (1983c). Host response to *Sarisodera hydrophila* Wouts and Sher, 1971. *Journal of Nematology* 15, 259-268.

MUNDO-OCAMPO, M. & BALDWIN, J.G. (1984). Comparison of host response of *Cryphodera utahensis* with other Heteroderidae and a discussion of phylogeny. *Proceedings of the Helminthological Society of Washington* 51, 25-31.

MUNDO-OCAMPO, M. & BALDWIN, J.G. (1992). Comparison of host response of *Ekphymatodera thomasoni* with other Heteroderinae. *Fundamental and Applied Nematology* 15, 63-70.

MUNDO-OCAMPO, M., TROCOLLI, A.S., SUBBOTIN, A., DEL CID, J., BALDWIN, J.G. & INSERRA, R.N. (2008). Synonomy of *Afenestrata* with *Heterodera* supported by phylogenetics with molecular and morphological characterization of *H. koreana* comb. n. and *H. orientalis* comb. n. (Tylenchida: Heteroderidae). *Nematology* 10, 611-632.

NAKHLA, M. (2008a). *Rapid* Globodera *spp. sample preparation and nucleic acid extraction*. Work Instruction, USDA, APHIS, PPQ, CPHST, Document Control Number WI-B-T-2-5.

NAKHLA, M. (2008b). *General nematode PCR (ITS1 & ITS2) amplification for identification of* Globodera *spp.* Work Instruction, USDA, APHIS, PPQ, CPHST, Document Control Number WI-B-T-1-15.

NARBAEV, Z.N. (1973). [Detection of the cactus nematode in Tashkent.] *Materialy Nauchnykh Konferentsii Vsesoyuznogo Obshchestva Gel'mintologov 1971-1972* 25, 164-167.

NARBAEV, Z.N. & MIRSALIMOWA, M.S. (1989). [Parasitic nematodes of greenhouse ornamental plants.] *Uzbekistan Biologiya Zhurnal* 5, 46-49.

NIBLACK, T.L., ARELLI, P.R., NOEL, G.R., OPPERMAN, C.H., ORF, J.H., SCHMITT, D.P., SHANNON, J.G. & TYLKA, G.L. (2002). A revised classification scheme for genetically diverse populations of *Heterodera glycines*. *Journal of Nematology* 34, 279-288.

NIBLACK, T.L., LAMBERT, K.N. & TYLKA, G.L. (2006). A model plant pathogen from the kingdom Animalia: *Heterodera glycines*, the soybean cyst nematode. *Annual Review of Phytopathology* 44, 283-303.

NOBBS, J.M., IBRAHIM, S.K. & ROWE, J. (1992). A morphological and biochemical comparison of the four cyst nematode species, *Heterodera elachista*, *H. oryzicola*, *H. oryzae* and *H. sacchari* (Nematoda: Heteroderidae) known to attack rice (*Oryza sativa*). *Fundamental and Applied Nematology* 15, 551-562.

NOEL, G.R. & LIU, Z.L. (1998). Esterase allozymes of soybean cyst nematode, *Heterodera glycines*, from China, Japan, and the United States. *Journal of Nematology* 30, 468-476.

NORGREN, R.L. & GOLDON, A.M. (1986). First report of *Cactodera estonica* in the United States. *Plant Disease* 70, 1159.

NOWACZYK, K., DOBOSZ, R., KORNOBIS, S. & OBREPALSKA-STEPLOWSKA, A. (2008). TaqMan REAL-Time PCR-based approach for differentiation between *Globodera rostochiensis* (golden nematode) and *Globodera artemisiae* species. *Parasitology Research* 103, 577-581.

NUNN, G.B. (1992). *Nematode molecular evolution*. Ph.D. Dissertation, University of Nottingham, Nottingham, UK.

O'BRIEN, P.C. & FISHER, J.M. (1979). Reactions of cereals to populations of *Heterodera avenae* in South Australia. *Nematologica* 25, 261-267.

OGAWA, Y., OHSHIMA, Y. & ICHINOHE, M. (1983). *Globodera hypolysi* n. sp., a mugwort cyst nematode from Japan. *Japanese Journal of Nematology* 12, 41-46.

OHMS, J.P. & HEINICKE, D.H.K. (1983). Pathotypes of the potato cyst-nematode. 1. Rapid identification of *Globodera rostochiensis* and *Globodera pallida* by isoelectric focusing. *Zeitschrift für Pflanzenkrankheiten und Pflanzenschutz* 90, 258-264.

ONIONS, T.G. (1953). Giant larvae of the potato eelworm *Heterodera rostochiensis* Wollenweber. *Nature, London* 172, 249-250.

OOSTENBRINK, M. (1960). Estimating nematode populations by some selected methods. In: Saser, J.N. & Jenkins, W.R. (Eds). *Nematology*. Chapel Hill, NC, USA, University of North Carolina Press, pp. 85-102.

OOSTENBRINK, M. & DEN OUDEN, H. (1954). De structuur van de kegeltop als taxonomisch kenmerk bij *Heterodera*-soorten met citroenvormige cysten. *Tijdschrift over Plantenziekten* 60, 146-151.

ORUI, Y. (1997). [Discrimination of *Globodera rostochiensis* and four *Heterodera* species (Nematoda: Heteroderidae) by PCR-RFLP analysis.] *Japanese Journal of Nematology* 27, 67-75.

OTHMAN, A.A. (1985). *Comparison of characters with SEM, for systematics and phylogeny of Heteroderidae (Nematoda: Tylenchida).* Ph.D. Dissertation, Department of Plant Pathology, University of California, Riverside, USA, 183 pp.

OTHMAN, A.A. & BALDWIN, J.G. (1985). Comparative morphology of *Meloidodera* spp. and *Verutus* sp. (Heteroderidae) with scanning electron microscopy. *Journal of Nematology* 17, 297-309.

OTHMAN, A.A. & BALDWIN, J.G. (1986). Comparative morphology of *Sarisodera hydrophilia*, *Rhizonema sequoiae*, and *Afenestrata africana* (Heteroderidae) with scanning electron microscopy. *Proceedings of the Helminthological Society of Washington* 53, 69-79.

OTHMAN, A.A., BALDWIN, J.G. & BELL, A.H. (1986). Comparative morphology of *Atalodera* spp. and *Thecavermiculatus* spp. (Heteroderidae) with scanning electron microscopy. *Journal of Nematology* 18, 275-287.

OTHMAN, A.A., BALDWIN, J.G. & MUNDO-OCAMPO, M. (1988). Comparative morphology of *Globodera*, *Cactodera*, and *Punctodera* spp. (Heteroderidae) with scanning electron microscopy. *Revue de Nématologie* 11, 53-63.

OU, S., PENG, D., LIU, X., LI, Y. & MOENS, M. (2008). Identification of *Heterodera glycines* using PCR with sequence characterised amplified region (SCAR) primers. *Nematology* 10, 397-403.

ØYDVIN, J. (1973). The usefulness of some larval dimensions to distinguish *Heterodera rostochiensis* and *H. pallida*. *Nematologica* 19, 435-442.

PAN, C.S., LIN, J. & XUE, R. (1997). [Description of *Cactodera cacti* and their observation by scanning electron microscope.] *Acta Parasitologica et Medica Entomologica Sinica* 4, 214-217.

PANEQUE, M. & SAMPEDRO, J. (1986). Reporte de nematodo cistogeno *Cactodera cacti* en la Provincia de Holguin, Cuba. *Ciencia y Técnica en la Agricultura, Protección de Plantas* 9, 123-129.

PARAMONOV, A.A. (1967). [A critical review of the suborder Tylenchina (Filip'jev, 1934) (Nematoda: Secernentea).] *Trudy Gel'mintologicheskoi Laboratorii Akademii Nauk SSSR* 18, 78-101.

PARROT, D.M. & MILLER, L.I. (1977). NC-95 flue-cured tobacco as a host of *Globodera pallida*. *Virginia Journal of Science* 28, 54.

PENG, D.L. & VOVLAS, N. (1994). Occurrence of the cyst-forming nematode *Cactodera thornei* in China. *Nematologia Mediterranea* 22, 75-78.

PERRY, R.N. (1997). Plant signals in nematode hatching and attraction. In: Fenoll, C., Grundler, F.M.W. & Ohl, S.A. (Eds). *Cellular and molecular aspects of plant-nematode interactions*. Dordrecht, The Netherlands, Kluwer Academic Publishers, pp. 38-50.

PERRY, R.N. (2002). Hatching. In: Lee, D.L. (Ed.). *Biology of nematodes*. London, UK, Taylor & Francis, pp. 147-169.

PHILIS, J. (1980). Life history of the potato cyst-nematode *Globodera rostochiensis* in Cyprus. *Nematologica* 26, 295-301.

PHILLIPS, M.S. & TRUDGILL, D.L. (1983). Variations in the ability of *Globodera pallida* to produce females on potato clones bred from *Solanum vernei* or *S. tuberosum* ssp. *andigena* CPC2802. *Nematologica* 29, 217-226.

PICARD, D., SEMPERE, T. & PLANTARD, O. (2007). A northward colonisation of the Andes by the potato cyst nematode during geological times suggests multiple host-shifts from wild to cultivated potatoes. *Molecular Phylogenetics and Evolution* 42, 308-316.

PICARD, D., SEMPERE, T. & PLANTARD, O. (2008). Direction and timing of uplift propagation in the Peruvian Andes deduced from molecular phylogenetics of highland biotaxa. *Earth and Planetary Science Letters* 271, 326-336.

PLANTARD, O., PICARD, D., VALETTE, S., SCURRAH, M., GRENIER, E. & MUGNIÉRY, D. (2008). Origin and genetic diversity of western European populations of the potato cyst nematode *Globodera pallida* inferred from mitochondrial sequences and microsatellite loci. *Molecular Ecology* 17, 2208-2218.

POGHOSSIAN, E.E. & KARAPETYAN, D.A. (1975). [Description of males of *Heterodera artemisiae* (Nematoda, Heteroderidae) from the Armenian SSR.] *Doklady Akademii Nauk Armjanskoi SSR* 61, 306-311.

POWERS, T.O. & FLEMING, C.C. (1998). Biochemical and molecular characterization. In: Perry, R.N. & Wright, D.J. (Eds). *The physiology and biochemistry of free-living and plant-parasitic nematodes*. Wallingford, UK, CABI Publishing, pp. 355-380.

PYLYPENKO, L.A., PHILLIPS, M.S. & BLOK, V.C. (2008). Characterisation of two Ukrainian populations of *Globodera pallida* in terms of their virulence and mtDNA, and the biological assessment of a new resistant cultivar Vales Everest. *Nematology* 10, 585-590.

QUADER, M., NAMBIAR, L. & CUNNINGTON, J. (2008). Conventional and real-time PCR-based species identification and diversity of potato cyst nematodes (*Globodera* spp.) from Victoria, Australia. *Nematology* 10, 471-478.

QUIÑONES, J.A. & SOSA MOSS, C. (1977). [Morphometric characterization of three Mexican populations of the genus *Globodera*.] *Nematropica* 7, 13-14.

RADICE, A.D., MYERS, R.F. & HALISKY, P.M. (1983). The grass cyst nematode in New Jersey. *Journal of Nematology* 15, 4-8.

RADICE, A.D., MYERS, R.F. & HALISKY, P.M. (1985). Studies on the host range, biology, and pathogenicity of *Punctodera punctata* infecting turfgrasses. *Journal of Nematology* 17, 162-165.

RADICE, A.D., RIGGS, R.D. & HUANG, F.H. (1988). Detection of intraspecific diversity of *Heterodera glycines* using isozyme phenotypes. *Journal of Nematology* 20, 29-39.

RADZIWINOWICZ, J. (1977). [Host plants of Polish cyst-forming nematodes.] *Lista roslin zywicielskich krajowych gatunkow matwikow*, 14 pp.

RAGSDALE, E.J., CRUM, J., ELLISMAN, M.H. & BALDWIN, J.G. (2008). Phylogenetic implications of three dimensional reconstruction of the stomatostylet and anterior epidermis in *Aphelenchus avenae. Journal of Morphology* 269, 1181-1196.

RAGSDALE, E.J., NGO, P.T., CRUM, J., ELLISMAN, M.H. & BALDWIN, J.G. (2009). Comparative, three-dimensional anterior sensory reconstruction of *Aphelenchus avenae* (Nematoda: Tylenchomorpha). *Journal of Comparative Neurology* 517, 616-632.

RASKI, D.J. (1950). The life history and morphology of the sugar-beet nematode, *Heterodera schachtii* Schmidt. *Phytopathology* 40, 135-152.

REID, A. & PICKUP, J. (2005). Molecular characterization of a morphologically unusual potato cyst nematode. *EPPO Bulletin* 35, 69-70.

REIS, L.G.L. (1997). Some morphological characters of a remarkable *Globodera* species (Nematoda: Heteroderidae) occurring in Portugal. *Acta Parasitologica Portuguesa* 4, 126.

REYNOLDS, E.S. (1963). The use of lead citrate at high pH as an electron opaque stain for electron microscopy. *Journal Cell Biology* 17, 208-212.

RIGGS, R.D. (1982). Cyst nematodes in the southern U.S.A. In: Riggs, R.D. (Ed.). *Nematology in the southern region of the United States*. Southern Cooperative Series Bulletin 276. Fayetteville, AR, University of Arkansas, pp. 77-95.

RIGGS, R.D. & SCHMITT, D.P. (1988). Complete characterization of the race scheme for *Heterodera glycines. Journal of Nematology* 20, 392-395.

RIGGS, R.D., HIRSCHMANN, H. & HAMBLEN, M.L. (1969). Life cycle, host range, and reproduction of *Heterodera betulae. Journal of Nematology* 4, 180-183.

RIGGS, R.D., HAMBLEN, M.L. & RAKES, L. (1981). Infra-species variation in reactions to hosts in *Heterodera glycines* populations. *Journal of Nematology* 13, 171-179.

RIGGS, R.D., RAKES, L. & HAMBLEN, M.L. (1982). Morphometric and serologic comparisons of a number of populations of cyst nematodes. *Journal of Nematology* 14, 188-198.

RIVOAL, R., VALETTE, S., BEKAL, S., GAUTHIER, J.-P. & YAHYAOUI, A. (2003). Genetic and phenotypic diversity in the graminaceous cyst nematode complex, inferred from PCR-RFLP of ribosomal DNA and morphometric analysis. *European Journal of Plant Pathology* 109, 227-241.

ROBBINS, R.T. (1978). A new Ataloderinae (Nematoda: Heteroderidae), *Thecavermiculatus gracililancea* n. gen., n. sp. *Journal of Nematology* 10, 250-254.

ROBERTS, P.A. & STONE, A.R. (1981). Host ranges of *Globodera* species within *Solanum* subgenus *Leptostemonum. Nematologica* 27, 172-189.

ROBINSON, A.J., STONE, A.R., HOOPER, D.J. & ROWE, J.A. (1996). A redescription of *Heterodera arenaria* Cooper 1955, a cyst nematode from marram grass. *Fundamental and Applied Nematology* 19, 109-117.

ROMERO, M.D., ANDRÉS, M.F., LÓPEZ-BRAÑA, I. & DELIBES, A. (1996). A pathogenic and biochemical comparison of two Spanish populations of the cereal cyst nematode. *Nematologia Mediterranea* 24, 235-244.

RUMPENHORST, H.J. (1985). Vergleichende elektrophoretische Untersuchungen von Proteinen einiger Zystennematoden von Getreide und Gräsern. *Mitteilungen aus der Biologischen Bundesaustalt fur Land- und Fortmirtschaft Berlin-Dahlem* 226, 64-74.

RUMPENHORST, H.J. & AYUB, M. (2001). Gibt es einc dritte Art beim Kartoffelnematoden? *Die 29 Tagung des Arbeitskreises Nematologie, Monheim, 14-15, März, 2001.*

SABO, A., VOVLAS, N. & FERRIS, V.R. (2001). Phylogenetic relationships based on ribosomal DNA data for four species of cyst nematodes from Italy and one from Syria. *Journal of Nematology* 33, 183-190.

SABO, A., REIS, L.G.L., KRALL, E., MUNDO-OCAMPO, M. & FERRIS, V.R. (2002). Phylogenetic relationships of a distinct species of *Globodera* from Portugal and two *Punctodera* species. *Journal of Nematology* 34, 263-266.

SANTOS, M.S.N. DE A., EVANS, K., ABREU, C.A., MARTINS, F.F. & ABRANTES, I.M. DE O. (1989). Os nemátodes de quisto da batateira em Portugal. *Ciência Biológica Ecology and Systematics* 9, 1-26.

SANTOS, M.S.N. DE A., EVANS, K., ABREU, C.A., MARTINS, F.F. & ABRANTES, I.M. DE O. (1995). A review of potato cyst nematodes in Portugal. *Nematologia Mediterranea* 23, 35-42.

SCHACHT, H. (1859). Über einige Feinde der Rübenfelder. *Zeitschrift des Vereines für die Rübenzucker-Industrie im Zollverein* 9, 175-179.

SCHMIDT, A. (1871). Über den Rübennematoden. *Zeitschrift der Vereinte Rübenzuckerindustrie Zollverein* 21, 1-19.

SCHROEDER, N.E., GERHARDT, D.J., PHIBBS, A. & MACGUIDWIN, A.E. (2008). First report of *Cactodera milleri* in Wisconsin. *Plant Disease* 92, 656.

SCHUSTER, M.L. & BREZINA, L. (1979). [Association of soil-borne pathogens with soybean Platte Valley Yellows. 1. *Heterodera acnidae* n. sp. (Heteroderidae: Nematoda) a parasite of *Acnida altissima*.] *Fitopatologia Brasileira* 4, 379-389.

SCOGNAMIGLIO, A., CAPUTI, F. & CIANCIO, G. (1985). Piante succulente, nuovi ospiti di *Cactodera cacti*. *Osservatorio per le Malattie delle Piante per la Campania, Napoli. Circolare* 15 (n.s.), 4-21.

SEIDEL, M. (1973). *Zur Biologie von* Heterodera longicaudata *Seidel, 1972 und* Heterodera punctata *Thorne, 1928 sowie zu deren Verbreitung und wirtschaftlichen Bedeutung im Bezirk Rostock (DDR)*. Dissertation, Universität Rostock.

SEINHORST, J.W. (1964). Methods for the extraction of *Heterodera* cysts from not previously dried soil samples. *Nematologica* 10, 87-94.

SEINHORST, J.W. (1965). The relation between nematode density and damage to plants. *Nematologica* 11, 137-154.

SEINHORST, J.W. (1982). The relationship in field experiments between population density of *Globodera rostochiensis* before planting potatoes and yield of potato tubers. *Nematologica* 28, 277-284.

SHAHINA, F. & MAQBOOL, M.A. (1989). Embryonic and postembryonic development stages of *Heterodera zeae* Koshy *et al.*, 1971 on *Zea mays*. *Pakistan Journal of Nematology* 7, 3-16.

SHAHINA, F. & MAQBOOL, M.A. (1995). *Cyst nematodes of Pakistan (Heteroderidae)*. Karachi, Pakistan, Karachi University Press, 155 pp.

SHARMA, S.B. & NENE, Y.L. (1986). Extraction of *Heterodera cajani* cysts using modified sieving technique. *Indian Journal of Nematology* 16, 123-124.

SHARMA, S.B. & SWARUP, G. (1984). *Cyst forming nematodes of India*. New Delhi, India, Cosmos Publications, 152 pp.

SHARMA, S.B., KAUSHAL, K.K., SINGH, M., PANDE, S., POKHAREL, R. & UPRETI, R.P. (2001). Description of a new species *Cactodera johanseni* (Nematoda: Heteroderidae) from Nepal. *International Journal of Nematology* 11, 39-42.

SHASHI SINGH, SATISH KUMAR, BAJAJ, H.K. & GANGULY, A.K. (1998). Esterase and malate dehydrogenase patterns of races of *Heterodera cajani* and *H. zeae*. *Indian Journal of Nematology* 28, 105-109.

SHEPHERD, A.M. (1986). Extraction and estimation of cyst nematodes. In: Southey J.F. (Ed.). *Laboratory methods for works with plant and soil nematodes*. London, UK, Her Majesty's Stationery Office, pp. 31-46.

SHEPHERD, A.M., CLARK, S.A. & DART, P.J. (1972). Cuticle structure in the genus *Heterodera*. *Nematologica* 18, 1-17.

SHEPHERD, A.M., CLARK, S.A. & KEMPTON, A. (1974). Spermatogenesis and sperm ultrastructure in some cyst nematodes, *Heterodera* spp. *Nematologica* 19, 551-560.

SHMALKO, V.F. (1959). [Kaktusovaya nematoda – *Heterodera cacti* Filipjev *et* Schuurmans-Stekhoven, 1941.] *Trudi Gelmintologov Laboratorii Akademiyi Nauk Soyuza Sovetskikh Sotsialisticheskikh Respublik* 9, 289-290.

SIDDIQI, M.R. (1986). *Tylenchida parasites of plants and insects.* Farnham Royal, UK, Commonwealth Agricultural Bureaux, 645 pp.

SIDDIQI, M.R. (2000). *Tylenchida parasites of plants and insects*, 2nd edition. Wallingford, UK, CABI Publishing, 833 pp.

SIRCA, S. & UREK, G. (2004). Morphometrical and ribosomal DNA sequence analysis of *Globodera rostochiensis* and *Globodera achilleae* from Slovenia. *Russian Journal of Nematology* 12, 161-168.

SIRCA, S., UREK, G. & MEGLIC, V. (2003). Molecular and biochemical methods used for the identification of *Globodera* species in Slovenia. *Plant Protection Science* 39, 151-153.

SKANTAR, A.M. & CARTA, L.K. (2005). Phylogenetic evaluation of nucleotide and protein sequences from the heat shock protein 90 gene of selected nematodes. *Journal of Nematology* 36, 136-145.

SKANTAR, A.M., HANDOO, Z.A., CARTA, L.K. & CHITWOOD, D.J. (2007). Morphological and molecular identification of *Globodera pallida* associated with potato in Idaho. *Journal of Nematology* 39, 133-144.

SKARBILOVICH, T.S. (1947). [Revision of the systematics of the nematode family Anguillulidae Baylis and Daubney, 1926.] *Doklady Akademii Nauk SSSR* 57, 307-309.

SKARBILOVICH, T.S. (1959). On the structure of the systematics of nematodes of the order Tylenchida Thorne, 1949. *Acta Parasitologica Polonica* 7, 117-132.

SMILEY, R., SHEEDY, J., PINKERTON, J., EASLEY, S., THOMPSON, A. & YAN, G. (2007). Cereal cyst nematode: distribution, yield reduction and crop management strategies. *Oregon Agricultural Experimental Station Special Report* 1074, 15-29.

SOLOVJEVA, G.I. & VASILJEVA, A.P. (1973). [*Heterodera punctata* a parasite of wild grasses in South Karelia.] *Parazitologiya* 7, 521-525.

SOSA-MOSS, C. (1987). Cyst nematodes in Mexico, Central and South America. *Nematologia Mediterranea* 15, 1-12.

SOSA-MOSS, C. & GONZALES, P.C. (1973). [Response of fertilized and non-fertilized corn to four different population levels of the Mexican race of *Heterodera punctata* (Nematoda: Heteroderidae).] *Nematropica* 3, 13-14.

SOUTHEY, J.F. (Ed.) (1986). *Laboratory methods for work with plant and soil nematodes.* London, UK, Ministry of Agriculture, Fisheries and Food, Reference Book 402. Her Majesty's Stationery Office, 202 pp.

SPEARS, J.F. (1956). Occurrence of the grass cyst nematode, *Heterodera punctata*, and *Heterodera cacti* group cysts in North Dakota and Minnesota. *Plant Disease Reporter* 40, 583-584.

SPEARS, J.F. (1968). The golden nematode, *Heterodera rostochiensis*, found in western New York. *Plant Disease Reporter* 52, 184.

STEINER, G. (1949). Plant nematodes the grower should know. *Proceedings of the Soil Science Society of Florida* 4, 82-117.

STELTER, H. (1984). *Die Arten der Unterfamilie Heteroderinae (Nematoda: Heteroderidae) und ihre Verbreitung*. Berlin, Akademie der Landwirtschaftswissenschaften der DDR, 56 pp.

STELTER, H. (1987). Die Wirtseignung von *Solanum*-species für drei *Globodera*-Arten. *Nematologica* 33, 310-315.

STELTER, H. & ENGEL, K.-H. (1973). Wirtsplanzen von *Heterodera rostochiensis* Woll., Rasse A, und *Heterodera pallida* Stone, Rasse E, aus der Familie der Solanaceae. *Archiv für Phytopathologie und Pflanzenschutz* 11, 233-244.

STONE, A.R. (1971). Effect of processing on measurements *of Heterodera rostochiensis* larvae. *Nematologica* 17, 167-171.

STONE, A.R. (1973a). *Heterodera pallida* n. sp. (Nematoda: Heteroderidae), a second species of potato cyst nematode. *Nematologica* 18 (1972), 591-606.

STONE, A.R. (1973b). Heterodera rostochiensis. *CIH Descriptions of plant-parasitic nematodes, Set 2, No 16*. St Albans, UK, Commonwealth Institute of Helminthology, 3 pp.

STONE, A.R. (1973c). Heterodera pallida. *CIH Descriptions of plant-parasitic nematodes, Set 2, No 17*. St Albans, UK, Commonwealth Institute of Helminthology, 2 pp.

STONE, A.R. (1975). Head morphology of second-stage juveniles of some Heteroderidae (Nematoda: Tylenchoidea). *Nematologica* 21, 81-88.

STONE, A.R. (1977). Recent developments and some problems in the taxonomy of cyst nematodes, with a classification of the Heteroderoidea. *Nematologica* 23, 273-288.

STONE, A.R. (1979). Co-evolution of nematodes and plants. *Symbolae Botanicae Upsaliensis* 22, 46-61.

STONE, A.R. (1983). Three approaches to the status of a species complex, with a revision of some species of *Globodera*. In: Stone, A.R., Platt, H.M. & Khalil, L.F. (Eds). *Concepts in nematode systematics*. London, UK, Academic Press, pp. 221-223.

STONE, A.R. (1985). Co-evolution of potato cyst nematodes and their hosts: implications for pathotypes and resistance. *EPPO Bulletin* 15, 131-137.

STONE, A.R. (1986). Taxonomy and phylogeny of cyst nematodes. In: Lamberti, F. & Taylor, C.E. (Eds). *Cyst nematodes*. London, UK, Plenum Press, pp. 1-21.

STONE, A.R., SOSA MOSS, C. & MULVEY, R.H. (1976). *Punctodera chalcoensis* n. sp. (Nematoda: Heteroderidae) a cyst nematode from Mexico parasitising *Zea mays*. *Nematologica* 22, 381-389.

STOYANOV, D. (1972). [*Heterodera amaranthi* n. sp. (Tylenchida: Heteroderidae); a cyst-forming nematode in Cuba.] *Poeyana, Instituto de Zoologia, Cuba* 97, 12 pp.

STOYANOV, D. (1973). [*Heterodera amaranthi*. Life cycle, hosts and distribution.] *Poeyana, Instituto de Zoologia, Cuba* 111, 26 pp.

STOYANOV, D. (1977). [Results of the study of the occurrence of cyst nematodes in Bulgaria. 4. Some species of the genus *Heterodera* Schmidt found in Bulgaria.] *Cheterideset godini nyauchnoizsledovatelska i prilozhna deinost na instituta za zashchita na rasteniyata*, pp. 99-112.

STURHAN, D. (1980). The occurrence of nematode cysts of the genus *Globodera* on yarrow in Germany. *Nachrichtenblatt der Deutschen Pflanzenschutzdienstes, Braunschweig* 32, 38-39.

STURHAN, D. (1983). The use of the subspecies and the superspecies categories in nematode taxonomy. In: Stone, A.R., Platt, H.M. & Khalil, L.F. (Eds). *Concepts in nematode systematics. Systematics Association Special Volume 22*. London & New York, Academic Press, pp. 41-53.

STURHAN, D. (1984). Nachweis des Schajgarbenzystennematoden *Globodera millefolii* (Kirjanova & Krall) für die Bundesrepublik Deutschland. *Nachrichtenblatt des Deutschen Pflanzenschutzdienstes* 36, 81-82.

STURHAN, D. (1985). Species, subspecies, race and pathotype problems in nematodes. *EPPO Bulletin* 15, 139-144.

STURHAN, D. (1988). [First record of the mugwort cyst nematode *Globodera artemisiae*.] *Gesunde Pflanzen* 40, 205-206.

STURHAN, D. (2000). Wirts-Spezifität bei Zystennematoden und anderen Heteroderiden. *Mitteilungen aus der Biologischen Bundesanstalt für Land- und Forstwirtschaft* 376, 299.

STURHAN, D. (2002). Notes on the genus *Cactodera* Krall & Krall, 1978 and proposal of *Betulodera betulae* gen. nov., comb. nov. (Nematoda: Heteroderidae). *Nematology* 4, 875-882.

STURHAN, D. (2006). Zystenbildende Nematoden und verwandte Heteroderiden in Deutschland. *Mitteilungen aus der Biologischen Bundesanstalt für Land- und Forstwirtschaft Berlin-Dahlem* 404, 18-30.

STURHAN, D. (2010). Notes on morphological characteristics of 25 cyst nematodes and related Heteroderidae. *Russian Journal of Nematology*, in press.

STURHAN, D. & LISKOVA, M. (2004). Cyst nematodes in the Slovak Republic. *Helminthologia* 41, 217-219.

STURHAN, D. & RUMPENHORST, H.J. (1996). Untersuchungen über den *Heterodera avenae* – Komplex. *Mitteilungen aus der Biologischen Bundesanstalt für Land- und Forstwirtschaft Berlin-Dahlem* 317, 75-91.

STURHAN, D., WOUTS, W.M. & SUBBOTIN, S.A. (2007). An unusual cyst nematode from New Zealand, *Paradolichodera tenuissima* gen. n., sp. n. (Tylenchida, Heteroderidae). *Nematology* 9, 561-571.

SUBBOTIN, S.A., RUMPENHORST, H.J. & STURHAN, D. (1996). Morphological and electrophoretic studies on populations of the *Heterodera avenae* complex from the former USSR. *Russian Journal of Nematology* 4, 29-38.

SUBBOTIN, S.A., STURHAN, D., WAEYENBERGE, L. & MOENS, M. (1997). *Heterodera riparia* sp. n. (Tylenchida: Heteroderidae) from common nettle, *Urtica dioica* L., and rDNA-RFLP separation of species from the *H. humuli* group. *Russian Journal of Nematology* 5, 143-157.

SUBBOTIN, S.A., WAEYENBERGE, L., MOLOKANOVA, I.A. & MOENS, M. (1999a). Identification of species from the *Heterodera avenae* group by morphomometrics and ribosomal DNA RFLPs. *Nematology* 1, 195-207.

SUBBOTIN, S.A., HALFORD, P.D. & PERRY, R.N. (1999b). Identification of populations of potato cyst nematodes from Russia using protein electrophoresis, rDNA-RFLPs and RAPDs. *Russian Journal of Nematology* 7, 57-63.

SUBBOTIN, S.A., WAEYENBERGE, L. & MOENS, M. (2000a). Identification of cyst-forming nematodes of the genus *Heterodera* (Nematoda: Heteroderidae) based on the ribosomal DNA-RFLPs. *Nematology* 2, 153-164.

SUBBOTIN, S.A., HALFORD, P.D., WARRY, A. & PERRY, R.N. (2000b). Variations in ribosomal DNA sequences and phylogeny of *Globodera* parasitising Solanaceae. *Nematology* 2, 591-604.

SUBBOTIN, S.A., VIERSTRAETE, A., DE LEY, P., ROWE, J., WAEYENBERGE, L., MOENS, M. & VANFLETEREN, J.R. (2001a). Phylogenetic relationships within the cyst-forming nematodes (Nematoda, Heteroderidae) based on analysis of sequences from the ITS regions of ribosomal DNA. *Molecular Phylogenetics and Evolution* 21, 1-16.

SUBBOTIN, S.A., PENG, D. & MOENS, M. (2001b). A rapid method for the identification of the soybean cyst nematode *Heterodera glycines* using duplex PCR. *Nematology* 3, 365-370.

SUBBOTIN, S.A., STURHAN, D., RUMPENHORST, H.J. & MOENS, M. (2002). Description of Australian cereal cyst nematode *Heterodera australis* sp. n. (Tylenchida: Heteroderidae). *Russian Journal of Nematology* 10, 139-148.

SUBBOTIN, S.A., STURHAN, D., RUMPENHORST, H.J. & MOENS, M. (2003). Molecular and morphological characterisation of the *Heterodera avenae* complex species (Tylenchida: Heteroderidae). *Nematology* 5, 515-538.

SUBBOTIN, S.A., STURHAN, D., CHIZHOV, V.N., VOVLAS, N. & BALDWIN, J.G. (2006). Phylogenetic analysis of Tylenchida Thorne, 1949 as inferred from D2 and D3 expansion fragments of the 28S rRNA gene sequences. *Nematology* 8, 455-474.

SULLIVAN, M.J., INSERRA, R.N., FRANCO, J., MORENO-LEHEUDE, I. & GRECO, N. (2007). Potato cyst nematodes: plant host status and their regulatory impact. *Nematropica* 37, 193-201.

SUMIYA, T., HIRATA, K. & YAEGASHI, T. (2002). Identification of three *Globodera* species in Japan by using isoelectric focusing method. *Research Bulletin of the Plant Protection Service, Japan* 38, 49-51.

SUZUKI, K., HIRATA, K. & HISAI, J. (2003). [Some solanaceous host plants of the tobacco cyst nematode, *Globodera tabacum* (Tylenchida: Heteroderidae).] *Research Bulletin of the Plant Protection Service, Japan* 39, 57-59.

SYKES, G.B. & WEBLEY, D.P. (1979). *Globodera achilleae* (Golden & Klindić) Mulvey & Stone: first records for Wales and England. *Plant Pathology* 28, 54.

SYRACUSE, A.J., JOHNSON, C.S., EISENBACK, J.D., NESSLER, C.L. & SMITH, E.P. (2004). Intraspecific variability within *Globodera tabacum solanacearum* using random amplified polymorphic DNA. *Journal of Nematology* 36, 433-439.

SZALANSKI, A.L., SUI, D.D., HARRIS, T.S. & POWERS, T.O. (1997). Identification of cyst nematodes of agronomic and regulatory concern with PCR-RFLP of ITS1. *Journal of Nematology* 29, 255-267.

TANHA MAAFI, Z., SUBBOTIN, S.A. & MOENS, M. (2003). Molecular identification of cyst-forming nematodes (Heteroderidae) from Iran and a phylogeny based on ITS-rDNA sequences. *Nematology* 5, 99-111.

TANHA MAAFI, Z., STURHAN, D., KHEIRI, A., GERAERT, E., SUBBOTIN, S.A. & MOENS, M. (2004). Morphology of some cyst-forming nematodes from Iran. *Russian Journal of Nematology* 11, 59-78.

TARJAN, A.C. & SASSER, J.N. (1953). Observations on *Heterodera weissi* Steiner, 1949 (Heteroderidae, Nematoda). *Proceedings of the Helminthological Society of Washington* 20, 62-64.

THIÉRY, M. & MUGNIÉRY, D. (1996). Interspecific rDNA restriction fragment length polymorphism in *Globodera* species, parasites of Solanaceaous plants. *Fundamental and Applied Nematology* 19, 472-479.

THIÉRY, M. & MUGNIÉRY, D. (2000). Microsatellite loci in the phytoparasitic nematode *Globodera*. *Genome* 43, 160-165.

THIÉRY, M., MUGNIÉRY, D., FOUVILLE, D. & SCHOTS, A. (1996a). Natural hybridization between *Globodera rostochiensis* and *G. pallida*. *Fundamental and Applied Nematology* 19, 437-442.

THIÉRY, M., FOUVILLE, D. & MUGNIÉRY, D. (1997a). Intra- and interspecific variability in *Globodera*, parasites of Solanaceous plants, revealed by Random Amplified Polymorphic DNA (RAPD) and correlation with biological features. *Fundamental and Applied Nematology* 20, 495-504.

THIÉRY, M., MUGNIÉRY, D., BOSSIS, M. & SOSA-MOSS, C. (1997b). Résultats de croisements entre *Globodera pallida* Stone et *G. "mexicana"* Campos-Vela: héritabilité du développement sur pomme de terre et noyion d'espèce. *Fundamental and Applied Nematology* 20, 551-556.

THORNE, G. (1928). *Heterodera punctata* n. sp., a nematode parasite on wheat roots from Saskatchewan. *Scientific Agriculture* 8, 707-710.

THORNE, G. (1943). *Cacopaurus pestis*, nov. gen., nov. spec. (Nematoda: Criconematinae), a destructive parasite of the walnut, *Juglans regia* Linn. *Proceedings of the Helminthological Society of Washington* 10, 78-83.

THORNE, G. (1949). On the classification of Tylenchida, new order (Nematoda: Phasmidia). *Proceedings of the Helminthological Society of Washington* 16, 37-73.

THORNE, G. (1961). *Principles of nematology.* New York, NY, USA, McGraw-Hill, 553 pp.

THORNE, G. & MALEK, R.B. (1968). Nematodes of the northern Great Plains. Part I. Tylenchida (Nemata: Secernentea). *Technical Bulletin of the South Dakota Agricultural Experiment Station* 31, 1-111.

TOVAR SOTO, A., CID DEL PRADO VERA, I., NICOL, J.M., EVANS, K., SANDOVAL ISLAS, J.S. & MARTINEZ GARZA, A. (2003). *Cactodera galinsogae* n. sp. (Tylenchida: Heteroderinae) on barley (*Hordeum vulgare* L.) in the high valleys of Mexico. *Nematropica* 33, 41-54.

TOVAR SOTO, A., PRADO VERA, I.C., SANDOVAL ISLAS, J.S., MARTINEZ GARZA, A., NICOL, J.M. & EVANS, K. (2006). Los nematodos formadores de quistes en Mexico. *Revista Mexicana de Fitopatologia* 24, 145-151.

TOVAR SOTO, A., CID DEL PRADO VERA, I., GUTIERREZ AGUILAR, M., GARCIA ZUNIGA, J. & EVANS, K. (2008). Postembryonic development of *Cactodera galinsogae* (Tylenchida: Heteroderinae) in barley (*Hordeum vulgare* L.) and the role of weeds as hosts. *Nematropica* 38, 145-153.

TOYOTA, K., SHIRAKASHI, T., SATO, E., WADA, S. & MIN, Y.Y. (2008). Development of a real-time PCR method for the potato-cyst nematode *Globodera rostochiensis* and the root-knot nematode *Meloidogyne incognita*. *Soil Science and Plant Nutrition* 54, 72-76.

TRIANTAPHYLLOU, A.C. & HIRSCHMANN, H. (1962). Oogenesis and mode of reproduction in the soybean cyst nematode, *Heterodera glycines*. *Nematologica* 7, 235-241.

TRIANTAPHYLLOU, A.C. & HIRSCHMANN, H. (1980). Cytogenetics and morphology in relation to evolution and speciation of plant-parasitic nematodes. *Annual Review of Phytopathology* 18, 333-359.

TRINDADE, Z.M.C. DE A., CONCEIÇÃO, I.L.P.M. DA, SANTOS, M.S.N. DE A. & ABRANTES, I.M. DE O. (1992). Isoelectric focusing of soluble proteins of *Heterodera* populations. *Nematologica* 38, 439.

TRUDGILL, D.L. (1985). Potato cyst nematodes: critical review of the current pathotyping scheme. *EPPO Bulletin* 15, 273-279.

TRUDGILL, D.L. & CARPENTER, J.M. (1971). Disk electrophoresis of proteins of *Heterodera* species and pathotypes of *Heterodera rostochiensis*. *Annals of Applied Biology* 69, 35-41.

TRUDGILL, D.L., PARROTT, D.M. & STONE, A.R. (1970). Morphometrics of males and larvae of ten *Heterodera rostochiensis* populations and the influence of resistant hosts. *Nematologica* 16, 410-416.

TURNER, S.J. (1998). Sample preparation, soil extraction and laboratory facilities for the detection of potato cyst nematodes. In: Marks, R.J. & Brodie, B.B. (Eds). *Potato cyst nematodes: biology, distribution and control.* Wallingford, UK, CABI Publishing, pp. 75-90.

TURNER, S.J. & ROWE, J.A. (2006). Cyst nematodes. In: Perry, R.N. & Moens, M. (Eds). *Plant nematology*. Wallingford, UK, CABI Publishing, pp. 91-122.

TZORTZAKAKIS, E.A., COCEICAO, I.L.P.M. DA, ABRANTES, I.M. DE O. & SANTOS, M.S.N. DE A. (2004). Characterisation and identification of potato cyst nematode populations from Crete, Greece, by isoelectric focusing of proteins. *Nematology* 6, 153-154.

UEHARA, T., KUSHIDA, A., ITOU, K., NARABU, T. & MOMOTA, Y. (2005). Discrimination of three cyst forming nematodes of the genus *Globodera* (Nematode: Heteroderidae) from Japan based on PCR-RFLP of ribosomal DNA. *Applied Entomology and Zoology* 40, 537-543.

UEHARA, T., KUSHIDA, A., BLOK, V., PHILLIPS, M., NEILSON, R., NARABU, T. & MOMOTA, Y. (2006). [Discrimination of *Globodera pallida* and three *Globodera* species based on PCR-RFLP method.] *Japanese Journal of Nematology* 36, 33-37.

UMARAO & VANGAPANDU, S. (2008). Molecular characterization of Indian populations of *Heterodera filipjevi* in tomato using PCR-RFLP of rDNA. *International Journal of Nematology* 18, 118-122.

UMARAO, GRACE, T., KAUSHAL, K.K. & VANGAPANDU, S. (2008). Molecular characterization of Indian populations of maize cyst nematode, *Heterodera zeae* using PCR-RFLP of ribosomal DNA. *Indian Journal of Nematology* 38, 81-85.

UREK, G. & SIRCA, S. (2003). Efficiency of the system of nematode cyst extraction from soil samples used in Slovenia. *Acta Phytopathologica et Entomologica Hungarica* 38, 209-215.

USDA (2008). Golden Nematode Program Manual. 08/2008-01. Available at http://www. aphis.usda.gov/import_export/plants/manuals/domestic/downloads/gnpm.pdf

VAZQUEZ, J.T. (1976). [*Infestations of parasitic nematodes as a factor limiting the production of maize in the Mexican altiplano.*] Veracruz, Mexico, DGEM, Xalapa, 79 pp.

VENKATESAN, P., MEHER, H.C., SRIVASTAVA, A.N. & SINGH, G. (2004). Isozyme, RAPD and microsatellite markers for the quick diagnosis of four *Heterodera* species. *Annals of Plant Protection Sciences* 12, 99-105.

VIAENE, N. (2003). *General techniques used in nematology*. Lecture Notes, PINC, Ghent University, 45 pp.

VRAIN, T.C., WAKARCHUK, D.A., LEVESQUE, A.C. & HAMILTON, R.I. (1992). Intraspecific rDNA restriction fragment length polymorphism in the *Xiphinema americanum* group. *Fundamental and Applied Nematology* 15, 563-573.

WALSH, J.A. & SHEPHERD, A.M. (1983). A further observation on sperm structure in a *Heterodera* sp. *Revue de Nématologie* 6, 148-150.

WEBLEY, D. (1970). A morphometric study of the three pathotypes of the potato cyst eelworm (*Heterodera rostochiensis*) recognized in Great Britain. *Nematologica* 16, 107-112.

WEBLEY, D.P. & LEWIS, S. (1977). Punctodera punctata. *CIH Descriptions of plant-parasitic nematodes, Set 7, No 102*. St Albans, UK, Commonwealth Institute of Helminthology, 3 pp.

WILSON, E.M. (1968). Vulval bodies in certain species of *Heterodera. Nature* 217, 879.

WOLLENWEBER, H. (1923). Krankheiten und Beschädingung der Kartoffel. *Arbeiten Forschungs Institut für Kartoffel. Berlin* 7, 1-56.

WOUTS, W.M. (1973). A revision of the family Heteroderidae (Nematoda: Tylenchoidea). I. The family Heteroderidae and its families. *Nematologica* 18, 218-235.

WOUTS, W.M. (1984). *Globodera zelandica* n. sp. (Nematoda: Heteroderidae) from New Zealand, with a key to the species of the genus *Globodera. New Zealand Journal of Zoology* 11, 129-135.

WOUTS, W.M. (1985). Phylogenetic classification of the family Heteroderidae (Nematoda: Tylenchida). *Systematic Parasitology* 7, 295-328.

WOUTS, W.M. & BALDWIN, J.G. (1998). Taxonomy and identification. In: Sharma, S.B. (Ed.). *The cyst nematodes*. Dordrecht, Boston & London, Kluwer Academic Publishers, pp. 83-122.

WOUTS, W.M. & SHER, S.A. (1971). The genera of the subfamily Heteroderinae (Nematoda: Tylenchoidea) with a description of two new genera. *Journal of Nematology* 3, 129-144.

WOUTS, W.M. & WEISCHER, B. (1977). Eine Klassifizierung von fünfzehn in Westeuropa häufigen Arten der Heteroderinae auf Grund von Larvenmerkmalen. *Nematologica* 23, 289-310.

WOUTS, W.M., WEISCHER, B. & TRIGGS, C.M. (1986). On the identity of European populations of *Punctodera punctata* (Nematoda: Heteroderidae). *Nematologica* 32, 79-88.

WOUTS, W.M., RUMPENHORST, H.J. & STURHAN, D. (2001). *Heterodera betae* sp. n., the yellow beet cyst nematode (Nematoda: Heteroderidae). *Russian Journal of Nematology* 9, 33-42.

WRIGHT, D.J. & PERRY, R.N. (2006). Reproduction, physiology and biochemistry. In: Perry, R.N. & Moens, M. (Eds). *Plant nematology*. Wallingford, UK, CABI Publishing, pp. 187-209.

WYSS, U. & GRUNDLER, F.M.W. (1992). Feeding behaviour of sedentary plant parasitic nematodes. *Netherlands Journal of Plant Pathology* 98, 165-172.

WYSS, U., STENDER, C. & LEHMANN, H. (1984). Ultrastructure of feeding sites of the cyst nematode *Heterodera schachtii* Schmidt in roots of susceptible and resistant *Raphanus sativus* L. var. *oleiformis* Pers. cultivars. *Physiological Plant Pathology* 25, 21-37.

YAMADA, E. & TAKAKURA, S. (2003). [Discovery of *Globodera hypolysi* from mugwort, *Artemisia montana*, in the east of Hokkaido.] *Japanese Journal of Nematology* 33, 14-22.

YUAN, H.X., SUN, J.W., YANG, W.X., XING, X.P., WANG, Z.Y., RILEY, I.T. & LI, H.L. (2010). New pathotypes of *Heterodera avenae* (cereal cyst nematode) from winter wheat in Zhengzhou, Henan, China. *Australasian Plant Pathology* 39, 107-111.

ZAHEER, K., FLEMING, C. & TURNER, S.J. (1996). Genetic variation in potato cyst nematodes in Northern Ireland. *Biochemical Systematics and Ecology* 24, 509-519.

ZHANG, Y. & BALDWIN, J.G. (1999). Ultrastructure of the esophagus of *Diplenteron* sp. (Diplogasterida) to test hypotheses of homology with Rhabditida and Tylenchida. *Journal of Nematology* 31, 1-19.

ZHENG, J., SUBBOTIN, S.A., WAEYENBERGE, L. & MOENS, M. (2000). Molecular characterization of Chinese *Heterodera glycines* and *H. avenae* populations based on RFLPs and sequences of rDNA-ITS regions. *Russian Journal of Nematology* 8, 109-113.

ZIMMERMANN, H. (1927). Versuche über die Kartofelnematode (*Heterodera schachtii*, forma *solani*). *Arbeiten der Forschungsinstitut für Kartoffelanbau*, Berlin 8, 151-154.

ZUNKE, U. (1986). Zur Bildung der subkriustallinen Schich bei *Heterodera schachtii* unter aseptischen Bedingungen. *Nematologica* 3, 117-120.

Nematology Monographs & Perspectives, 2010, Vol. 8A, 341-345

Index of nematode Latin names

Index of plant-host Latin names

Index of country names

Index of plant common names

www.ingramcontent.com/pod-product-compliance
Lightning Source LLC
LaVergne TN
LVHW081515300926
849161LV00006B/1192
* 9 7 8 9 0 0 4 1 6 2 2 5 9 *